Monographs on Endocrinology

Volume 30

H.L.J.Makin R.Newton (Eds.)

High-Performance Liquid Chromatography in Endocrinology

With 83 Figures and 21 Tables

Springer-Verlag
Berlin Heidelberg New York
London Paris Tokyo

H. L. J. Makin, Dr.
The London Hospital Medical College
Department of Chemical Pathology
University of London
Turner Street
London El 2AD, GB

R. Newton, Dr.
Biotech Instruments Ltd.
183 Camford Way
Bedfordshire LU 3 3AN, GB

ISBN-13:978-3-642-83469-1 e-ISBN-13:978-3-642-83467-7
DOI: 10.1007/978-3-642-83467-7

Library of Congress Cataloging-in-Publication Data
HPLC/H. L. J. Makin, R. Newton (eds.).
p. cm. -- (Monographs on endocrinology ; vol. 30)
ISBN-13:978-3-642-83469-1 (U.S.)
1. High performance liquid chromatography. 2. Endocrinology--Technique. I. Makin, H. L. J. II. Newton, R. (Raymond), 1947. III. Series: Monographs on endocrinology ; v. 30.
[DNLM: 1. Chromatography, High Pressure Liquid--methods. 2. Endorinology--methods. 3. Hormones-analysis. W1 M057 v. 30/ WK 25 H872]
QP519.9.H53H68 1988
599'.019285--dc 19
DNLM/DLC
for Library of Congress

Softcover reprint of the hardcover 1st edition 1988

2127/3140/543210

Contents

List of Contributors

ALLENMARK, S.
Laboratory of Microbiological Chemistry, Institute of Medical Microbiology, University of Gothenburg, Guldhedsgatan 10 A, 41346 Gothenburg, Sweden

CHADWICK, V. S.
Wellcome Medical Research Institute, University of Otago Medical School, P. O. Box 913, Dunedin, New Zealand

CORBETT, A. D.
Unit for Research on Addictive Drugs, Marischal College, University of Aberdeen, Aberdeen AN9 1AS, Great Britain

DELUCA, H. F.
Department of Biochemistry, University of Wisconsin - Madison, 420 Henry Mall, Madison, WI 53706, USA

GAMES, D. E.
Department of Chemistry, University College, P. O. Box 78, Cardiff CF1 1XL, Great Britain

HEFTMANN, E.
P. O. Box 928, Orinda, CA 94563-0818, USA

JONES, G.
Departments of Biochemistry and Medicine, Queens's University, Kingston, Ontario K7L 3N6, Canada

KOSTERLITZ, H. W.
Unit for Research on Addictive Drugs, Marischal College, University of Aberdeen, Aberdeen AN9 1AS, Great Britain

MAKIN, H. L. J.
Steroid Laboratory, Department of Chemical Pathology, The London Hospital Medical College, Turner Street, London E1 2AD, Great Britain

MATON, P. N.
Department of Health & Human Services, National Institutes of Health, Building 10, Room 9C-103, Bethesda, MD 20205, USA

McKNIGHT, A. T.
Neuroscience Research Centre, Merck Sharp & Dohme Research Laboratories, Terlings Park, Harlow, Essex CM20 2QR, Great Britain

NEWTON, R.
Biotech Instruments Ltd., 183 Camford Way, Bedfordshire LU3 3AN, Great Britain

PATERSON, S. J.
Unit for Research on Addictive Drugs, Marischal College, University of Aberdeen, Aberdeen AN9 1AS, Great Britain

RAMSEY, E. D.
Department of Chemistry, University College, P. O. Box 78, Cardiff CF1 1XL, Great Britain

ROBSON, L. E.
Unit for Research on Addictive Drugs, Marischal College, University of Aberdeen, Aberdeen AN9 1AS, Great Britain

SELDEN, C.
Gastroenterology Unit, Department of Medicine, Royal Postgraduate Medical School, Duncane Road, London W12 ONN, Great Britain

TAVARES, I. A.
Department of Surgery, King's College School of Medicine and Dentistry, The Rayne Institute, 123 Coldharbour Lane, London SE5 9NU, Great Britain

Introduction

This book originates from a symposium held at the London Hospital Medical College under the auspices of Applied Chromatography Systems Ltd. to discuss the place of HPLC in the endocrinology field. Many of the authors of the present book were speakers at this symposium. It seemed to us that many endocrinologists did not at that time fully appreciate the value of HPLC, and this book was designed to publicise the potential value of this technique. A survey of methods used in the steroid field (p 185) confirmed the view that HPLC is not being used as widely, particularly in research described in clinical journals, as might be expected. We hope that this book will illustrate, albeit in a few selected areas of endocrinology, just how versatile and powerful a technique HPLC is, and encourage those who have not yet experienced it to have a go. The beginner does not need to buy expensive instrumentation – all that is required is a pump, injector, column and detector – the rest can come later! All the authors have practical experience in the use of HPLC in the particular area they discuss. All readers who discover apparent errors or who feel that the treatment of a topic of interest can be improved upon are encouraged to contact the editors. All criticism, especially if constructive, is welcomed. We are still learning, and other peoples' experience is always valuable.

As editors we should like to express our thanks to the contributors to this book, to whom we owe a considerable debt of gratitude for their patience and courtesy to us. We should also like to thank Dr. J. Wieczorek and his colleagues at Springer-Verlag and Miss S. Schulmeister, who looked after the proofs. We are also very grateful to Mrs. Linda Harvey who, as ever, typed and re-typed the many drafts of our chapters and the index.

July 1988

H. L. J. Makin
R. Newton

Introduction to the Theory and Practice of High-Performance Liquid Chromatography

R. NEWTON

The success of High Performance Liquid Chromatography (HLPC) as an analytical technique, is obvious to anyone who cares to study the literature. In fact, the rate of growth and acceptance of HPLC as a useful method of analysis has been remarkable. This is true in the field of endocrinology as in any other area of analytical chemistry or biochemistry.

In the light of this success, it is perhaps worth considering the advantages of the technique, its theoretical basis, and how HPLC can be used to solve problems in endocrinology.

The main advantages of HPLC can be summarized in the following brief statements in their approximate order of importance.

1. It is a separation technique.
2. It has wide sample applicability.
3. For compounds with good natural UV absorbance, fluorescence or electrochemical response it is a sensitive technique.
4. Modern apparatus can be relatively inexpensive, reliable and can be readily automated.
5. For many analyses it is rapid when compared to other sequential techniques.

At first glance the first two points on this list are not unique to HPLC. However, their full significance in relation to the technique can only be judged by an examination of the fundamental theory of chromatography and of some basic chromatographic definitions.

Basic Theory

Fundamental accounts of the theory of chromatography have appeared many times (Giddings 1965: Done Knox & Loheac 1974, Snyder & Kirkland 1979) and it is beyond the scope of this chapter to give a detailed description of the concepts involved. However, any user, or potential user of HPLC, must understand the important features of chromatographic theory and basic chromatographic definitions. The following section briefly describes the basic principles of chromatography, and their practical relevance.

When the molecules of a compound are introduced to the top of a chromatographic column, they are subject to two opposing forces: that of the mobile phase tending to force them through the column, and that of the stationary phase

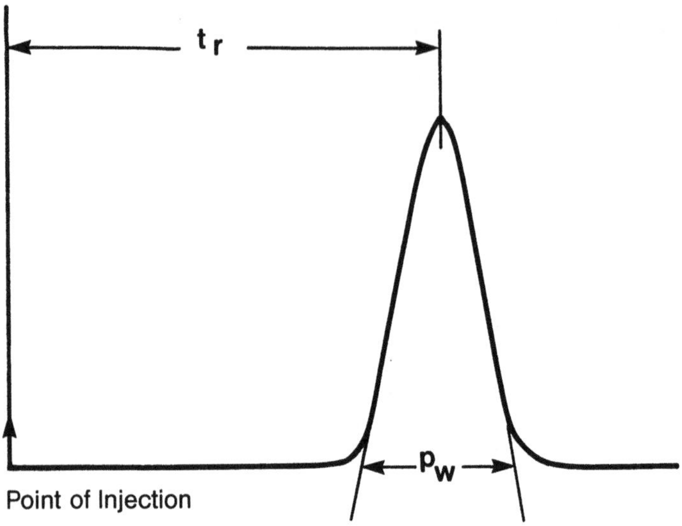

Fig. 1. Theoretical representation of HPLC peak

tending to retain them. A dynamic equilibrium is set up as molecules travel through the column, passing constantly between the mobile and stationary phases, and emerging at the end of the column as a band of molecules. This emerging band can theoretically be represented as a Gaussian peak as shown in Figure 1.

In Figure 1 t_r is defined as the retention time of the compound and P_W is the peak width at the base of the peak.

The column in fact may be assumed to consist of a number of separate elements or theoretical plates in which partition of the solute between the two phases is at equilibrium.

The number of plates (N) in a particular column may be determined from the Gaussian Peak of Figure 1 as:

$$\text{Plate Number (N)} = \left(\frac{tr}{\sigma}\right)^2 \tag{1}$$

where σ is the standard deviation of the Gaussian peak of Fig. 1.

Alternatively, (N) can be described directly in terms of the peak width (W) i.e.

$$N = 16 \left(\frac{tr}{w}\right)^2 \tag{2}$$

In practice, measurement of the peak width at the base can be difficult and the equation

$$N = 5.54 \left(\frac{tr}{w^{1/2}}\right)^2 \tag{3}$$

($W^{1/2}$ = peak width at half the peak height)

is commonly used to calculate the plate number for individual chromatography columns.

In order to compare columns of different lengths (L) the height equivalent to a theoretical plate (H) is often used where:

$$H = \frac{N}{L} \tag{4}$$

H & N are a measure of the column performance or efficiency. Since the chromatographic process involves an equilibrium mechanism' it can also be described thermodynamically in terms of the equilibrium constant (K) which is different for each of the individual components of the sample i.e.

$$K = \frac{\text{Concentration of the component in the stationary phase}}{\text{Concentration of the component in the mobile phase.}} \tag{5}$$

K is also often referred to as the distribution or equilibrium coefficient. The capacity factor, k' is defined in a similar fashion to K but "concentration" is replaced by "total moles" i.e. $k' = V_s/V_m$ where V_s is the volume of the stationary phase in the column (the loading) and V_m is the volume of the mobile phase (i.e. the dead or void volume). The capacity factor is an expression of the relative affinity of the component molecules for the stationary and mobile phases, and as such, is related directly to the component's retention time.

If a component has little affinity for the stationary phase and it will spend most of its time in the mobile phase. Thus, it will pass through the column relatively unretained. In this case, at any time, the concentration of the component in the stationary phase will be low and K will be low. However, if the component has great affinity for the mobile phase then its retention time will be long, its concentration in the mobile phase will be high and K will be high. In fact it can be shown that:

$$K = \frac{(t_r - t_o)}{t_o} \tag{6}$$

where t_r is the component retention time and t_o is the retention time for the an unretained component i.e. a compound with no affinity for the stationary phase.

Resolution

Now consider the case of two compounds (A and B) passing down the column and emerging completely separated at the bottom of the column.

The degree of separation, defined as the resolution (R) can be expressed in terms of the peak widths and retention times of A and B i.e.

$$R = \frac{t_{R_B} - t_{R_A}}{\frac{(W_A + W_B)}{2}} \tag{7}$$

or

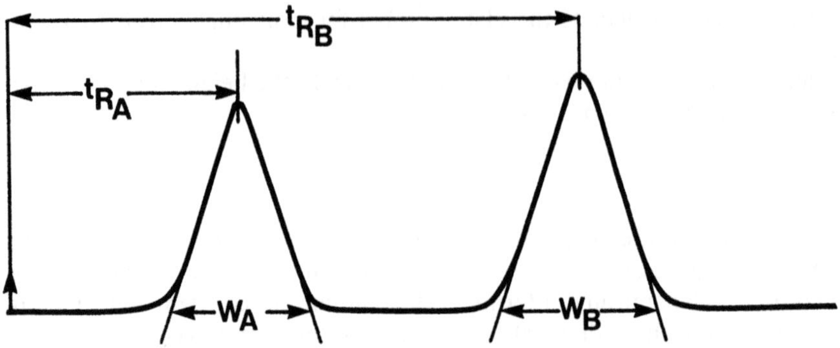

Fig. 2. Separation of two HPLC peaks

$$R = \frac{(2/t_{R_B} - t_{R_A})}{W_A + W_B} \tag{8}$$

or

$$R = \frac{2 \;\Delta\; t_R}{W_A + W_B} \tag{9}$$

When the two compounds emerge partly separated as in Figure 3, this equation is still valid.

In practice Figure 3 represents an unsatisfactory chromatographic separation and the analyst would normally seek to improve this separation.

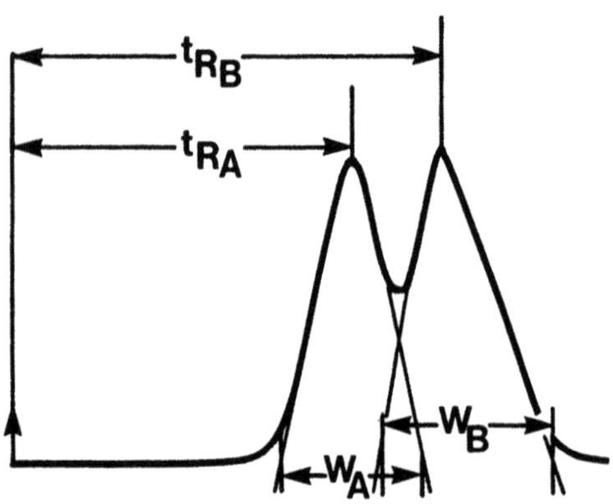

Fig. 3. Resolution of two HPLC peaks

This can be done in two ways:

a) increasing the distance between the two peaks – i.e. changing the selectivity (α).

b) reducing the peak widths W_A and W_B.

In order to change the selectivity, the retention times t_{R_A} and t_{R_B} must be changed. This can only be done by changing the capacity factors of peak A and/or B. In fact selectivity can be defined as:

$$\alpha = \frac{K_B}{K_A} \tag{10}$$

or using equation (6)

$$\alpha = \frac{t_{R_B} - t_o}{t_{R_A} - t_o} \tag{11}$$

This simple conclusion is the most important feature of any chromatographic method in which the mobile phase takes an active part in the partition process occurring between the mobile and stationary phases. This includes most types of liquid chromatography as opposed for example to gas chromatography in which there is no interaction between solute and mobile phase.

Its consequences can be examined if HPLC, or even thin layer chromatography (TLC) are compared to gas chromatography (GC). In the case of GC the mobile phase is an inert gas and the mobile phase an involatile liquid supported on an inert particle, the column being held at an elevated temperature to ensure that the solute is in the vapour phase. If incomplete resolution between two or more peaks occurs then only by either changing the temperature or the chemical nature of the mobile phase can it be improved. Since the latter course of action involves changing the entire column, it is not always convenient and very often a large number of columns packed with different stationary phases must be kept available if a wide variety of sample types is encountered routinely. With HPLC or TLC, however, slight changes in mobile phase composition can drastically change the capacity factor, thus, dramatically changing the resolution. This is far simpler and much more convenient than changing columns. This makes HPLC an extremely flexible technique and is the basic reason why a wide variety of different analyses from different sample types can be performed with the same apparatus, and column system. Furthermore, although it is possible to operate HPLC columns at elevated temperatures, in practice it is rarely necessary since the mobile phase composition has a much greater effect on the capacity factor than changing the temperature. Thus most HPLC separations are carried out at room temperature, and sample decomposition during analysis is less likely to occur. Temperature control may however still be used in HPLC to improve resolution and to maintain the stability of the chromatographic system.

As discussed earlier in this section, a further method of improving the resolution of Figure 3 would be to reduce the peak widths W_A and W_B (i.e. to increase the number of theoretical plates in the column). In order to see how this might be achieved, it is necessary to discuss those factors which affect the column efficiency.

The equation relating resolution (R) to selectivity (α), the number of theoretical plates (N) and the capacity factor (k′) is

$$R = \frac{1}{4}\left[\frac{\alpha-1}{\alpha}\right]\sqrt{N}\left[\frac{k'}{1+k'}\right] \tag{11a}$$

Further Considerations on Efficiency

Peak broadening in chromatography arises almost entirely because of a number of phenomena which occur as a result of diffusion on the column.

A number of workers (Van Deemter 1956, Giddings 1965) have studied these diffusion phenomena in both GC and HPLC and concluded that the total contribution to band broadening can be described by an equation of the type:

$$H = \frac{2\,\lambda \cdot dp}{} + \frac{2\,\gamma \cdot Dm}{u} + \frac{q \cdot r \cdot D^2 \cdot u}{Ds} + \frac{w \cdot dp^2 \cdot u}{Dm} \qquad (12)$$

where H = Height equivalent to a theoretical plate
dp = Average particle diameter of the column packing
u = Linear velocity of the mobile phase
D = Stationary phase thickness
Ds = Diffusion coefficient in the stationary phase
Dm = Diffusion coefficient in the mobile phase
λ, γ, q and r are constants related to the particular column and packing.

This complicated equation has several important practical consequences and, can in fact, be discussed relatively simply in terms of its individual components.

The "eddy diffusion" or "multiple path" term 2 λ dp occurs as a result of the sample taking various paths through the column. The presence of a large number of different mobile phase streams moving at different speeds down the column will obviously cause some portions of the sample to be held up more than others and a broad peak will result. It will be noted that the contribution of this effect to band broadening is dependent on particle diameter, as well as to a constant λ which is related to the type of packing and the way the particles pack together. It would be expected that sperical particles are more likely to produce better and more reproducible eddy diffusion characteristics than irregular particles. This is because spherical particles of a uniform size will tend to adopt an hexagonal close packed configuration, whereas, irregular particles will tend to form more random configurations.

The "molecular diffusion" term $\frac{2\,\gamma \cdot Dm}{u}$ describes the simple diffusion of sample molecules as they pass down the column. This diffusion will occur in all directions i.e. longitudinally in the direction of flow and laterally at right angles to the direction of flow. Since the mobile phase is flowing, longitudinal diffusion will have greatest effect on those components with long retention times hence, H is inversely proportional to the linear velocity v. The effects of radial diffusion can be more complex. Knox (1976) has described the effects of sample molecules diffusing radially to the column walls where it is likely that the mobile phase is travelling more slowly than that travelling down the centre of the column. This led Knox to develop the concept of the infinite diameter column in which the sample passes down the columns without reaching the column walls.

If band broadening is to be low then the sample molecules must be able to move rapidly between the mobile and stationary phases i.e. mass transfer between the two phases must be rapid and unhindered.

The two final terms in equation (12), $q \cdot r \cdot D^2 \cdot u/Ds$ and $w \cdot dp^2 \cdot u/Dm$ are known as the "mass transfer" terms. The former describes the movement of molecules in the stationary phase and the latter, the movement of molecules in the mobile phase. In both cases for high efficiency, diffusion must be rapid, thus it is possible to predict that low viscosity mobile phases will give less band broadening than those of high viscosity.

In the case of the stationary phase, the "film thickness" is an important factor leading to rapid diffusion. In the case of GC this is a simple concept to visualize. In HPLC where the chromatography is likely to occur on a series of single active absorption or ion exchange sites, it is a little more difficult. It is, however, possible to conceive of a large number of similar chromatographic sites with a range of chromatographic activities, so that some sample molecules are retained more than others. Even with this model, however, the band broadening effects of diffusion within the stationary phase can be considered to be small, if not negligible, in HPLC columns.

In both mass transfer terms, H is proportional to the linear velocity of the mobile phase, and in the case of the mobile phase mass transfer term, H is proportional to the square of particle size.

Equation (12) in fact shows that H (which should be low for an efficient column) is dependent directly on the particle diameter, the mobile phase velocity, the diffusion coefficients of the mobile and stationary phases and a number of constants which in turn are related to the packing material and the way that it is packed in the column.

Put even more simply, with a given mobile phase at a given velocity, or flow rate, the efficiency of the column increases with a decrease in particle size, provided that the particles are of a uniform size and that they can be packed effectively in the column.

The history of the early development of HPLC, is in fact, the history of the development of packing materials of low diameter, with a narrow range of particle size distribution. It was only when packing materials of this type were readily available that the full potential of HPLC as a highly efficient and rapid chromatographic method was realized. Current particle diameters are typically 3-5 μm with column lengths of 10–15 cm.

Further practical information may be gained from equation (12) by use of the reduced plate height (h) suggested by Giddings where

$$h = H/dp \tag{13}$$

i.e. the number of particles per theoretical plate.

Furthermore, the linear velocity (v) can be converted to a reduced velocity (V) where:

$$V = vdp/Dm \tag{14}$$

By using these reduced parameters (Giddings 1965) it is possible to make decisions about the quality of packed columns regardless of packing material or particle size e.g. for a good column h should be 2.

Equation (12) can also be written in terms of these reduced parameters i.e.:

$$h = AV^{0.33} + \frac{B}{V} + CV \qquad (15)$$

$AV^{0.33}$	$\frac{B}{V}$	CV
↑	↑	↑
Eddy Diffusion Term	Molecular Diffusion Term	Mass transfer in the mobile phase

The minimal effect of mass transfer in the stationary phase has been ignored here.

Knox (1974) has shown that a plot of log h against log V for a given column produces a characteristic curve of the type shown in Figure 4.

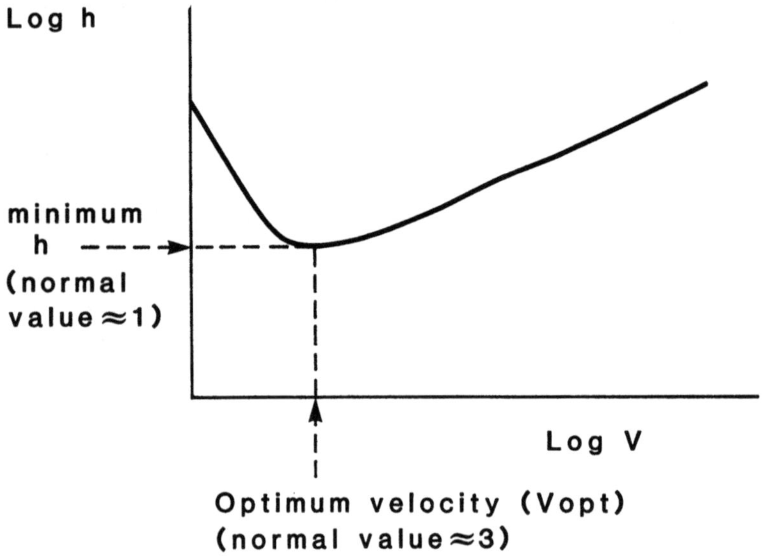

Fig. 4. Reduced HETP (height equivalent to a theoretical plate) plot for a conventionally packed HPLC column

V_{opt} is the optimum velocity to obtain maximum efficiency. In practice higher velocities than V_{opt} are used in order to obtain rapid analysis.

The main value of these curves and the use of reduced plate heights is that they allow rapid evaluation of column performance regardless of packing material or particle size, since all columns should produce curves of the general shape with the typical values for the minimum h and V_{opt} shown in Figure 4.

Translated into practical terms, modern analytical packed columns are typically 10, or 25 cm long, with an internal diameter 4.5 mm to 5.0 mm and are packed with 3 μ, 5 μ or 10 μ particles. Flow rates are typically 1-3 mls/min.

One consequence of using small particles is that the pressure drop across the column might be expected to increase. This is in fact the case, although column back pressure is also dependent on the particle shape, particle size distribution and the mobile phase viscosity. Typical back pressures for 25 cm × 0.45 cm 1.D columns packed with 5 μ sperical particles operating at 1–2 mls/min are between 1000 p.s.i. and 3000 p.s.i. 3 μ particles give very much higher back pressures in 25 cm columns typically > 5000 p.s.i. However, the additional efficiency obtained with 3 μ particles frequently means that excellent separations can be obtained using 10 cm or even 5 cm long columns. The back pressures of these short 3 μ columns are of the same order as those obtained with longer 5 μ particle columns.

In the early 1970s HPLC was used to abbreviate the term "High Pressure Liquid Chromatography". However, when it was realized that by using small particles, high efficiencies could be obtained from short columns at moderate pressures, the phrase "High Performance" was substituted for "High Pressure".

Non - Gaussian peaks

In the previous sections it has been assumed that chromatographic peaks are perfectly Gaussian. In practice this tends not to be the case. The Gaussian nature of a peak can be determined from its P_{skew} value (see Fig. 5).

Fig. 5. Determination of peak skew (non-gaussian peaks)

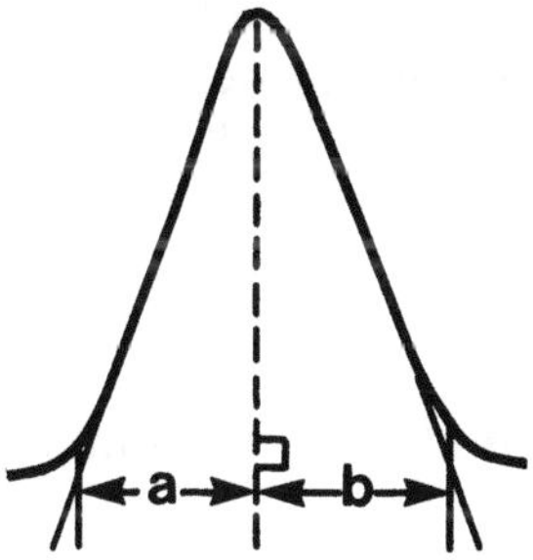

$$P_{skew} = a/b \quad (16)$$

For a perfectly symmetrical peak $P_{skew} = 1$. If $P_{skew} > 1$ a, leading edge is observed as in Figure 6 (a). If $P_{skew} < 1$, then a tailing edge is observed as in Figure 6 (b).

One of the main causes of non-symmetrical peaks is that the capacity factor may change with the amount of sample loaded onto the column. This is normally observed at high sample loadings when both leading and tailing peaks may be seen with consequent loss of resolution. Shifts in retention times are also likely to occur. Simply reducing the sample concentration will remove or reduce this effect (Snyder and Kirkland 1979).

The main reasons for band broadening and the appearance of non-symmetrical peaks, however, are referred to as "extra-column effects". That is, they are a

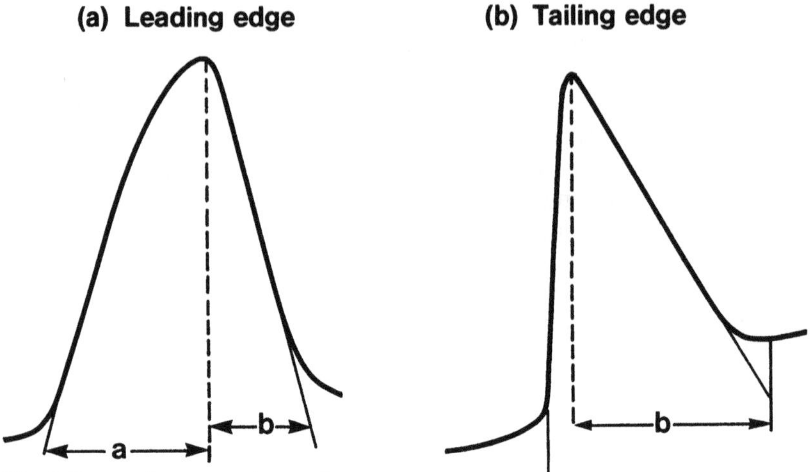

Fig. 6. Types of non-gaussian peaks

result of diffusion which may occur as the sample is introduced onto the column or as the peaks travel from the end of the column into the detector. In order to avoid these effects, columns and fittings must be used with minimum dead volumes as well as using narrow bore, low dead volume tubing throughout the chromatographic system. In addition, injectors and detectors must be designed with low dead volumes and in such a way as to minimize sample diffusion in the mobile phase.

Modes of Separation and Separation Mechanisms

Despite being founded on sound theoretical principles, HPLC is still regarded very largely as being an "art" rather than a "science". This is understandable, since when the analyst is faced with an unknown or new sample type, there is frequently no instant "off the shelf" column/mobile phase combination which will answer the problem. In addition to carrying out literature searches for methods developed for similar samples, it is necessary for the chromatographer to rely on his previous experience, plus a knowledge of the properties of the sample e.g. solubility, stability in various solvents, previous TLC results etc.

In practice, there are four types of separation used in HPLC, which will be discussed in turn in the following sections.

Gel Permeation or Gel Filtration

This is perhaps the simplest type of liquid chromatography to explain and to understand. The particles of the column contain channels, or pores, of a uniform size distribution so that molecules of a larger size than the pores are totally

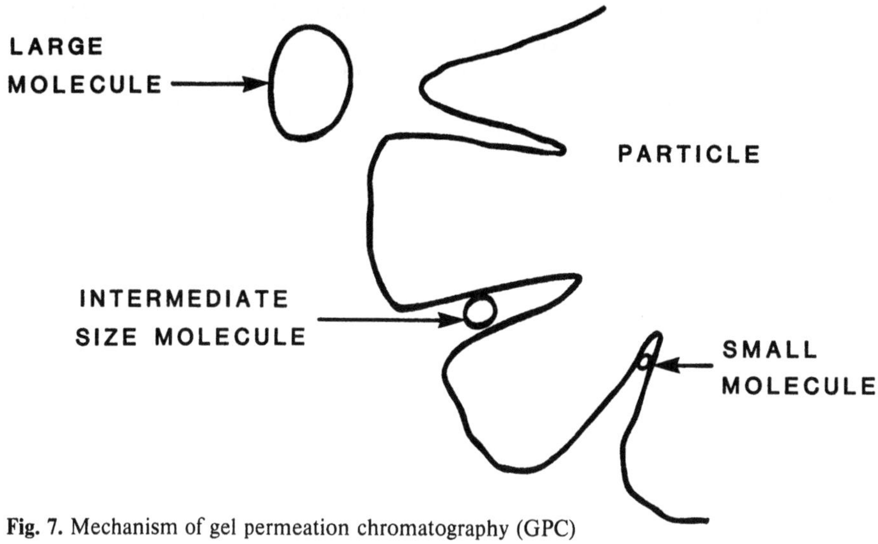

Fig. 7. Mechanism of gel permeation chromatography (GPC)

excluded and pass through the column unretained. Very small molecules permeate deeply into all the pores and are strongly retained and molecules of an intermediate size permeate into those pores large enough to contain them. Thus, the sample is separated according to the molecular size of its components. This process is shown diagramatically in Figure 7.

Gel permeation, or size exclusion chromatography, has been traditionally used for the separation of high molecular weight molecules, although, in fact, the technique can be used in a wide variety of applications involving molecular weights from several hundred upwards to several million. GPC has a number of unique features which can be summarized as:

1. The sample must be totally soluble in the mobile phase.
2. It is the only form of liquid chromatography in which the mobile phase composition does not affect the separation.
3. The order of elution is large molecules first with the smallest molecules eluting last.
4. Provided that the only chromatographic mechanism occurring is gel permeation, then all the components of the sample will elute from the column.
5. GPC columns can be calibrated with compounds of known molecular weight so that molecular weight distributions of unknown samples can be determined.

A number of packing materials have been used in HPLC to achieve GPC separations. The two most common of these are polystyrene/vinyl benzene copolymers, or modified silica gels. The former are normally used in conjunction with organic mobile phases such as tetrahydrofuran for the separation of synthetic organic polymers and associated additives and the latter for the separation of biological macromolucules e.g. proteins or carbohydrates. In this case, the mobile phase is a suitable aqueous buffer.

Although size exclusion chromatography using low pressure cellulosic packing materials has been used for the separation of proteins and peptides for a number of years, it is only in the last five years or so that suitable HPLC packings have been commercially available for rapid and efficient protein separations. As an example, the Toyo Soda Manufacturing Company of Japan have available a wide range of HPLC packings suitable for this type of separation.

Adsorption Chromatography

In the case of adsorption chromatography the packing material is a polar medium such as alumina or much more commonly, silica gel, and the mobile phase is usually a non-polar organic solvent such as hexane, dichloromethane or mixtures of these solvents or others of similar polarity.

The surface of silica gel is structurally complex consisting basically of a random distribution of surface silanol (Si-OH) groups. The particles used in HPLC adsorption chromatography are totally porous with a high surface area, typically 200 m^2/grm.

The porous nature of the particle is important in that each particle must not be thought of as a billiard ball with silanol groups sticking out at angles from a smooth surface. A more accurate picture would be to consider the particles to exist as a "cage" structure with silanol groups projecting randomly at all angles, both inside and outside, the cage structure. In other words most of the chromatography will occur within the pores of the particle since it is at the silanol groups that the chromatographic process occurs.

Separation in adsorption chromatography occurs primarily as a result of hydrogen bonding between the silanol groups of the silica gel and the components of the sample. Thus, the components capable of a high degree of hydrogen bonding will be more strongly retained than those components with a lower capability for hydrogen bonding.

The main features of adsorption chromatography are as follows:

1. In general the order of elution is that the less polar components elute first and the more polar elute last. One point to be borne in mind here, is that some components may be so polar as to be permanently retained on the column. This in turn can have the further consequence that a build up of such permanently retained material can occur, seriously blocking the pores of the particle and, furthermore, modifying its surface and also its chromatographic characteristics. As a result of this, changes of retention times can be frequently observed in conjunction with a gradual loss of column efficiency.
2. The presence of water in the sample can also affect the surface characteristics of the stationary phase, since water molecules have a strong hydrogen bonding affinity for the surface hydroxyl groups of silica gel. This can cause similar retention shifts to those described above, since the chromatographic properties of the silanol groups will be modified by this hydrogen bonded water.

This effect can be reduced by adding small amounts of an alcohol such as isopropanol to the mobile phase, but this may not eliminate the problem. Biological and biochemical samples are complex, containing a wide range of substances of vastly different polaraties as well as varying amounts of water and, therefore, adsorption chromatography is not the ideal method for their analysis. Adsorption columns have been fairly widely used however for the analysis of non-polar compounds such as polynuclear aromatic compounds, fat soluble vitamins and non-polar lipids.

3. The problems of adsorption on the stationary phase by water and polar compounds also means that adsorption columns can take a long time (i.e. several hours) to "condition" when the mobile phase is changed (i.e. any particular solvent is in equilibrium with the adsorbent surface and it takes a finite time for equilibrium to be established even if only slight changes in mobile phase composition are involved).
4. An increase in the polarity of the mobile phase will result in a decrease in retention times.

Bonded Phase Chromatography

Although silica gel is commonly regarded by chemists as an inert material, its surface silanol groups are capable of undergoing a number of interesting, and to the chromatographer, useful reactions. These can be summarized as follows:

a) reaction with alcohols to form silicate esters (Holasz and Sebastian 1969) of the type Si - O - R.
b) reaction with thionyl chloride to form silica chloride, Si - Cl. This can subsequently be treated with a Grignard reagent (Locke et al. 1972) to form an Si - C bond or an amino compound (Brust et al. 1973) to form an Si-N bond.
c) Reaction with organosilane compounds to form silane ethers of the type: Si - O - Si - R (Stewart and Perry 1968).

Reactions of the latter type were originally developed to reduce adsorption effects which can occur on gas chromatography supports. They can also be used, however, to prepare some extremely useful chromatographic materials for HPLC.

A number of suitable organosilane derivatives are readily available for the production of chromatographically useful "bonded phases". The term bonded phase is used to describe HPLC packings in which the silanol groups of silica gel have been chemically reacted in one of the modes described above to produce a surface structure in which organic groups have become chemically bound to the silica surface.

Chromatographically the most useful bonded phases are the silane ethers i.e.

Si - O - Si - R

where R is commonly one of the following:

i) - $(CH_2)_{17}$ - CH_3 Octadecyl
ii) - $(CH_2)_7$ - CH_3 Octyl
iii) - $(CH_2)_3$ - NH_2 Amino-propyl
iv) - $(CH_2)_3$ - CN Cyno-propyl
v) - Phenyl

All these, and a number of others, are all commercially available from several manufacturers. Simpson (1982) has listed eleven different types of octadecylsilane phases which were available from seven different manufacturers in 1982. To the author's knowledge, all of these are still available.

Bonded phases can be reproducibly prepared with a range of surface polarities and some have been found to be extremely useful in specific application areas. The best known of these is probably the use of amino-propyl packings with water/acetonitrile mobile phase mixtures in the separation of carbohydrates. Bonded phases can be broadly separated into two types i.e. the so-called reversed phase materials, and the polar bonded phases.

Reversed Phase HPLC

The term "reversed phase" is used to describe those packings in which the bound organic derivative is a straight chain alkyl group. The most common being the octadecyl silane packings as well as the octyl and hexyl packings. All of these packings have similar retention characteristics, the main difference being that the retention time for a given compound is proportional to the length of the bound alkyl chain.

The separation mechanism in reversed phase liquid chromatography arises out of the fact the stationary phase now consists of a non-polar surface and the mobile phase is now polar, normally a water/methanol, water/tetrahydrofuran or water/acetonitrile mixture. Sometimes ternary or quaternary mixtures of three or four solvents are used. This non-polar surface/polar mobile phase combination gives the method its name in that it is the reverse situation to that of adsorption chromatography.

Molecules of the sample are now partitioned between the bonded alkyl chain and the mobile phase producing a separation in which the polar components elute first followed by the non-polar components. Partition between a liquid mobile phase and a solid stationary phase may be difficult to visualize, however, attraction of the hydrophylic polar components for the mobile phase and attraction of the hydrophobic non-polar components for the stationary phase can be more readily understood. Alternatively, it is possible to conceive of a situation in which the molecules of the mobile phase become "structured" in the region of the bound alkyl groups, i.e. repulsion of water molecules by the hydrophobic bonded phase would lead to regions near the surface which are "rich" in organic solvent and "deficient" in aqueous phase. A corresponding region of mobile phase just beyond this would then exist, which was correspondingly rich in terms of its aqueous phase. It is now possible to conceive of partition of sample molecules between the two layers of water "rich" and water "deficient" solvents.

The overall result is that slight changes in bulk mobile phase composition can result in quite large changes in retention time and resolution.

The main features of reversed phase chromatography are:

1. Polar compounds elute before non-polar compounds.
2. Equilibration times when solvents are changed or during gradient elution are short. This helps in method development.
3. The method is flexible in that a wide range of sample types can be analyzed. (Approximately 80% of publications on HPLC describe the use of reversed phase methods.) It can, for example, be used for the separation of mixtures of ionic and non-ionic components, particularly if buffered aqueous phases or ion suppression techniques are used. It has been found to be useful for water soluble compounds and, in particular, biological and biochemical separations.
4. An increase in the concentration of the organic solvent in the mobile phase will cause a decrease in retention times.

Polar Bonded Phases

Polar bonded phases are those in which the bound organic group is something other than an alkyl chain. The most common examples have been mentioned previously, and the most important of these are the nitrile ($(CH_2)_3 - CN$) and the amino propyl ($(CH_2)_3 - NH_2$) packings.

Polar bonded phases form an interesting group of materials chromatographically in that they can be used in either an adsorption mode with non-polar organic solvents, or in a reversed phase mode with aqueous/organic mobile phases. Their use has been generally limited to specific application areas e.g., the separation of carbohydrates on amino-propyl packings (Jones et al. 1977) and the separation of PTH amino acids on nitrile packings (Hunkapiller 1983), both of which are now in common use.

Ion Exchange Chromatography

The principles of ion exchange chromatography are well known and the technique is well established.

In ion exchange methods the stationary phase is a resin which has been chemically modified to produce a fixed ionic charge permanently bound to the surface. This surface charge can be positive (anion exchange) or negative (cation exchange). The mobile phase is an aqueous buffer, although small amounts of organic solvents can be added to facilitate certain separations e.g. the addition of 5–10% of methanol can improve the separation of threonine and serine in amino acid analysis. Since the sample is also ionic, competition for the ion exchange sites of resin is set up between the sample ions and the counter ions of the mobile phase.

This is shown diagramatically for cation exchange chromatography in Figure 8.

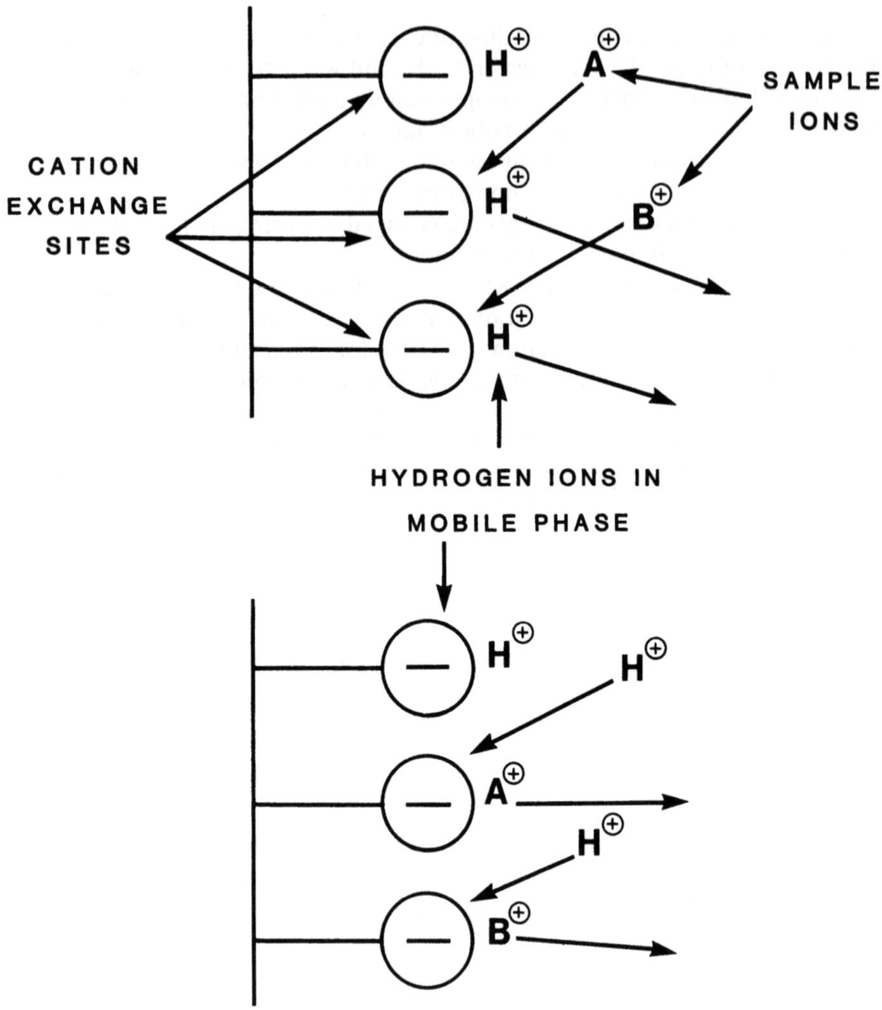

Fig. 8. Mechanism of ion exchange chromatography

The residence times of A^+ and B^+ on the ion-exchange sites at a given mobile phase, ionic strength and pH, will depend on a number of factors e.g., the ionic radius of A^+ and B^+ and the ionic charge, divalent ions being retained longer than monovalent ions.

In the case of cation exchange, the stationary phase is produced by sulphonating the surface of the resin producing bound SO_3^- groups on the surface. In the case of anion exchange materials, the surface ionic species is commonly a quaternary ammonium (R_4N^+) group. A number of materials have been used as the resin substrate including polystryrene/vinyl benzene copolymers, silica and cellulose.

The main features of ion-exchange methods are:

1. They can only be used for the separation of ionic water soluble compounds.
2. An increase in ionic strength of the mobile phase will result in a decrease in retention times.
3. In cation exchange chromatography an increase in the pH of the mobile phase will result in a decrease in retention times. The reverse is true for anion exchange.
4. The use of gradient techniques with ion exchange methods can be extremely useful particularly for compounds such as amino acids, which form Zwitter ions. For a brief discussion of gradient formation please refer to the following section on instrumentation.
5. The nature of the counter ion can also affect the separation e.g., Citrate ions have a greater affinity for anion exchange resins, than do acetate ions. Potassium ions have a greater affinity for cation exchange resins than sodium ions.
6. Unlike other forms of HPLC, ion exchange separations can be affected quite dramatically by changes in column temperature.

Typical applications for ion exchange chromatography would be in the separation of amino acids by cation exchange and the separation of nucleotides by anion exchange chromatography.

Ion exchange chromatography cannot be used to separate mixtures containing both ionic and non-ionic materials. However, the technique of ion-pair chromatography, the development of which is generally attributed to Schill (1973) in his work on ion pair extraction techniques, has proved to be extremely useful for the analysis fo such mixtures.

Ion Pair Chromatography

Ion pair methods are in many respects closely related to ion exchange techniques. Although ion pair chromatography (IPC) can be carried out with either adsorption or reverse phase packings, the use of the latter is far more common.

The technique involves the addition of a small amount (typically 5×10^{-3} M) of a large counter ion of opposite charge to the ions being analysed. Consider the example of a solute ion x^- and a counter ion c^+ in a mixture of an organic solvent such as methanol and water. The solute ion and counter-ion form an ion pair described by the equation.

$$\left[x^- \text{ aqu}\right] + \left[c^+ \text{ aqu}\right] \rightleftharpoons \underset{\text{ion pair}}{\left[c^+ x^- \text{ org}\right]}$$

In the situation which exists in a reverse phase column, this uncharged ion pair can now partition with the hydrophobic alkyl groups of the reversed phase packing as described previously. In fact two mechanisms have been suggested to explain the separations obtained in ion pair chromatography.

a) the ion pair partitions between the reversed phase surface and the mobile phase in exactly the same way that a non-ionic compound would.

b) the counter-ion which is a large molecule such as heptane sulphonate is capable of interacting with the alkyl chains of the reversed phase particle, temporarily forming an ion exchange surface.

There is considerable evidence for (b) since it is well known that the retention characteristics of reversed phase columns change dramatically following their use with ion pair mobile phases. This is true even after washing with large volumes of water. In fact, it is common practice never to mix columns which have been used for IPC with those which are used for normal reverse phase applications.

Typical counter ions used for IPC are quaternary amines e.g., letrabutyl ammonium salts and alkyl sulphonates e.g. heptane sulphonic acid. Applications of IPC include the analysis of biogenic amines, carboxylic acids, ionic drugs and dyestuffs.

The most important feature of IPC is its ability to separate mixtures of ionic and non-ionic compounds. The main features affecting retention are:

1. the type of counter ion and its ion pairing ability
2. the size of the counter ion. In general the larger the counter ion the longer will be the retention time.
3. concentration of counter ion: generally retention will increase with concentration, although the relationship between retention time and counter ion concentration can be complex.
4. pH, this is important as it will dictate the concentration of the ionic form of the solute. This, in turn, will affect the formation of the ion pair.

In all other respects IPC shows the same general features as normal reverse phase chromatography.

Instrumentation

The components of a modern High Performance Liquid Chromatograph are shown schematically in Figure 9. This diagram can be discussed in terms of its individual components.

Solvent Delivery System

This includes the solvent reservoirs, pumps and gradient control system if required. The solvent reservoirs are normally glass bottles of 500 ml or 1 litre capacity. In order to prevent contamination of the mobile phase it is best to have screw caps to the solvent reservoirs and for some solvents it can be advantageous to use amber bottles e.g. to minimize decomposition of tetrahydrofuran THF, in the presence of light. Dissolved air in the mobile phase can also cause problems in the pump heads, particularly, with methanol water mixtures and some manufacturers now fit helium degassing systems as an integral part of the solvent reservoir.

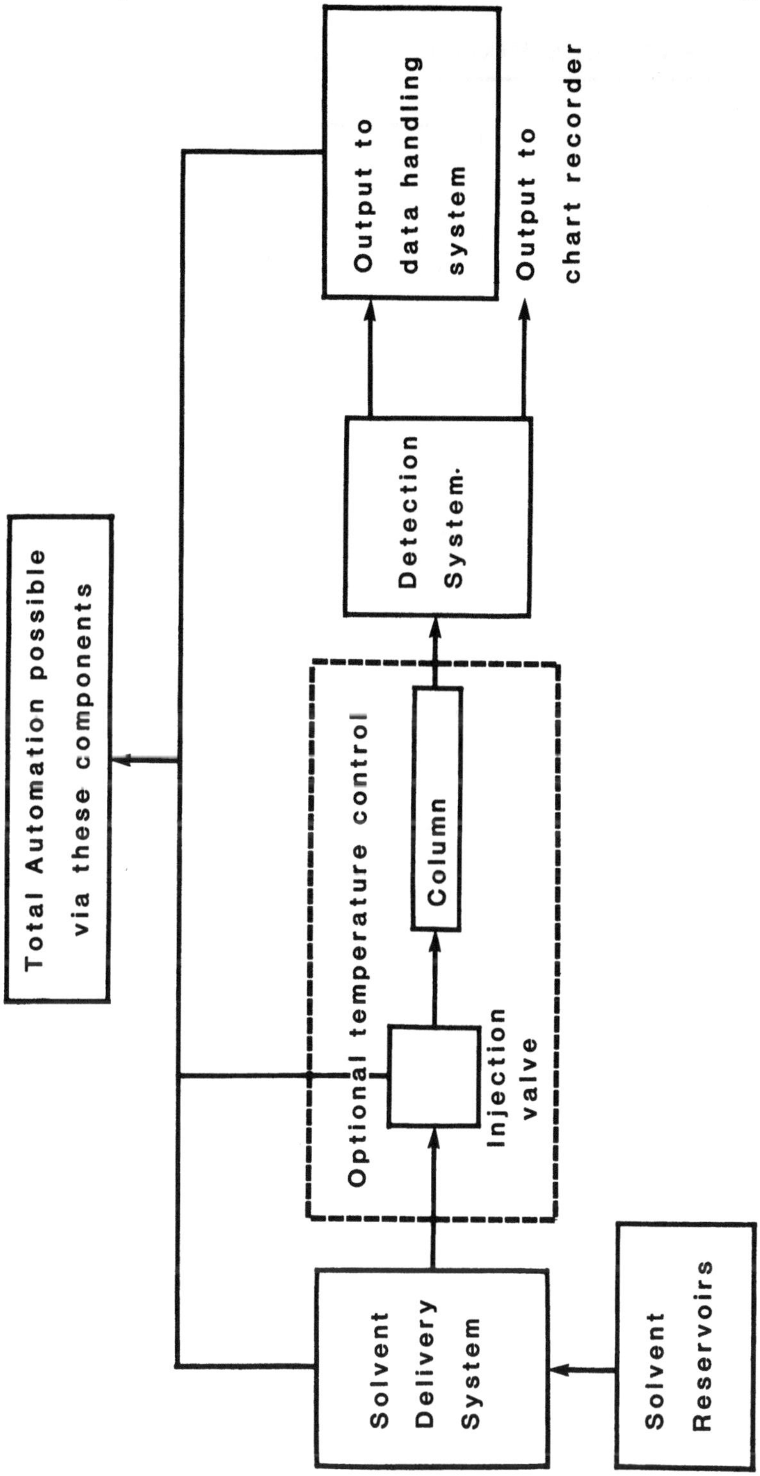

Fig. 9. Schematic diagram of typical HPLC system

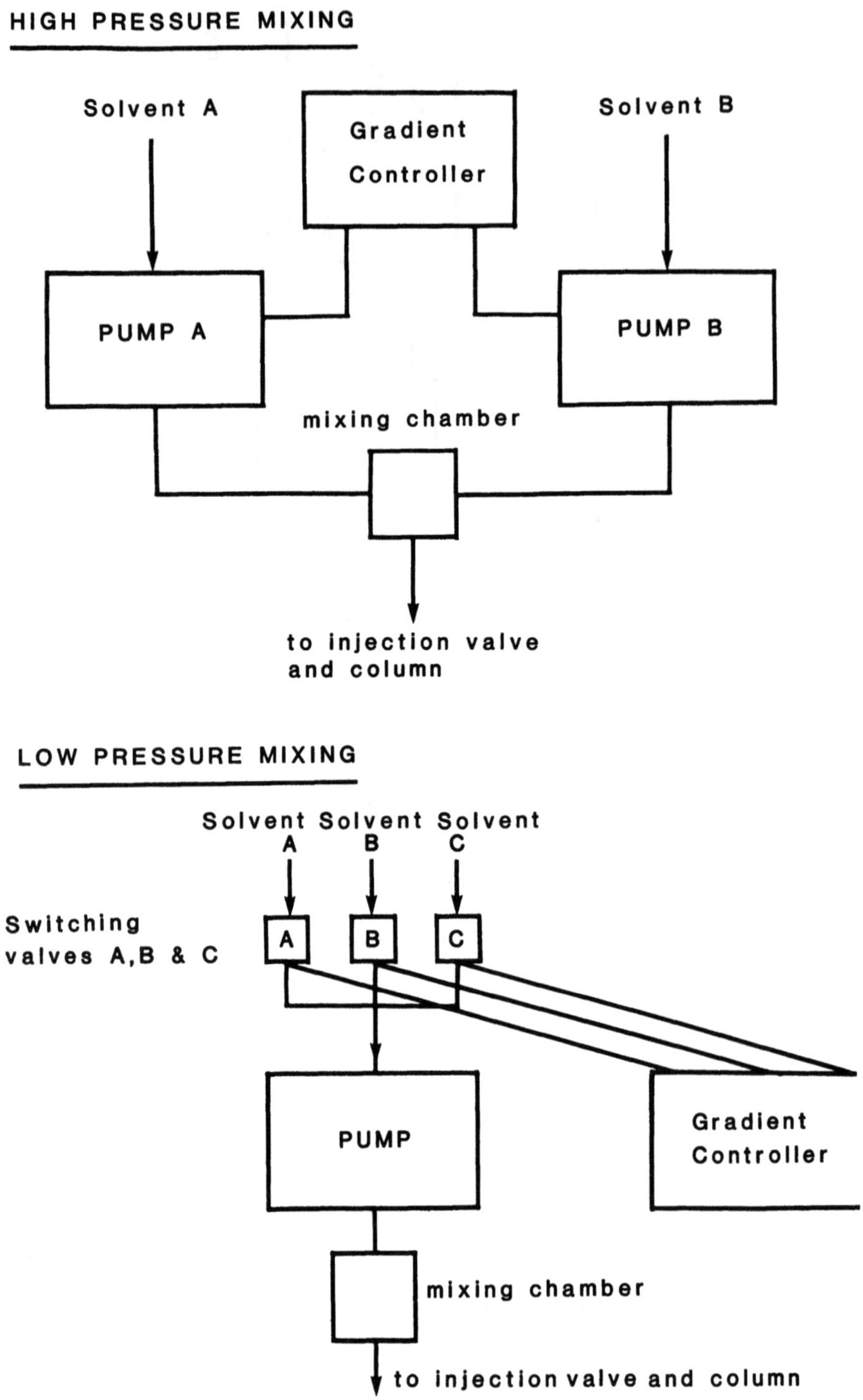

Fig. 10. High- and low pressure mixing of solvents for HPLC

In the early days of HPLC, two types of pump were in common use, large volume (between 60 mls and 500 mls) syringe pumps, driven either via gas actuators or motorized via a worm-gear mechanism, or reciprocating pumps. All the commercial pumps now available for conventional packed column HPLC are reciprocating pumps, although a number of manufacturers are now supplying motorized syringe pumps for capillary chromatography. Capillary HPLC will be discussed briefly in a later section.

Reciprocating pumps fulfill nearly all the requirements necessary for HPLC. They can be operated at pressures up to 6000 p.s.i.; they are simple to use and compact; they can be manufactured from materials which are resistant to all the common mobile phases used in HPLC. In addition, their ability to continue supplying mobile phase is limited only to the volume of the reservoir, which means they can be used in automatic HPLC systems. Finally, modern reciprocating pumps are reliable and inexpensive to service and maintain.

The only disadvantage of reciprocating pumps is that they deliver a pulsating flow of solvent which can be transmitted to the detector as noise. This can reduce the overall sensitivity of the detector. In practice, manufacturers of modern pumps have virtually eliminated the problems associated with pulsations by incorporating electronic pulse damping mechanisms or by using two or three reciprocating heads operating out of phase within the pump. Frequently a combination of both pulse damping methods are used. Furthermore, modern HPLC detectors tend to be less sensitive to pulsations. This is not true, however, for electrochemical detectors which are still pulse sensitive, and it is true that some reciprocating pumps operate better with electrochemical detectors than others.

Gradient Elution

If the sample to be analyzed consists of compounds of similar polarity, or structure, then it is likely that they will all separate satisfactorily using a single mobile phase in a reasonable time period. This is described as an isocratic (single mobile phase) method, and required only one solvent reservoir and a single pump with a simple single solvent inlet. If the sample is more complex, however, it may be necessary to change the composition of the mobile phase during the elution of the sample down the column. This technique is called gradient elution.

Two instrumental methods are available commercially for carrying out gradient elution, high pressure mixing systems and low pressure mixing systems. The two systems are shown schematically in Figure 10.

In high pressure mixing the different mobile phases are pumped via separate pumps and are mixed, as the name suggests, on the high pressure side of the system. An electronic device, the gradient controller, controls the relative amounts of the two solvents “A” and “B” delivered to the column by directly controlling the flow rates of the two pumps.

In low pressure mixing the gradient controller controls the times of opening of a number of switching valves, and thus the composition of the mobile phase reaching the pump. For efficient mixing of the solvents it is general practice to fit a high pressure mixing chamber on the outlet side of the pump, as shown in Figure 10.

The gradient controllers are generally microprocessor controlled, and several manufacturers are currently using computers such as the IBM pc fitted with special interfaces as gradient control devices. Gradient controllers of this type are capable of forming a wide range of sophisticated gradient shapes. In practice it is best to use the simplest gradients possible to perform separations as the more sophisticated the gradient profile, the more difficult it is to reproduce.

Either high pressure or low pressure systems are capable of producing quite satisfactory gradients. Low pressure systems have the advantage that it is simple to carry out gradients using three or four solvents (ternary and quaternary gradients) by simply using 3 or 4 switching valves. This can be useful in method development. High pressure systems which use two pumps can be more flexible since, if more than one detector is available, the high pressure gradient system can, it required, be split into two separate isocratic systems.

Injection Valves

The ideal sample injection system should be capable of injecting a wide range of sample sizes (typically 0.5 μ litres up to 50 μ litres for analytical work) without causing band spreading before the sample reaches the column.

Early injection devices for HPLC were, in fact, septum based systems directly analagous to those used in gas chromatography. These methods when used in HPLC, however, were prone to needle blockage and system leakage and have largely been replaced by injection valves. Modern injection valves, as typified by the Rheodyne model 7125 valve (Rheodyne Corporation, USA) fulfil nearly almost perfectly the requirements of the ideal HPLC sample injection system. Valves of this kind are fitted with external loops with volumes up to several mls. They are designed in such a way that the loop may be totally, or partially, filled using a Hamilton micro syringe. The loop is fitted at atmospheric pressure but by using a system consisting of a rotor and stator the loop can be flushed with solvent from the pump, flushing the sample onto the column. They are designed specifically to have zero or almost zero dead volume, so that sample wastage and cross contamination from sample to sample is minimized. Sample volumes of 1 μ litre or less can be injected with these valves with a precision of better than 1%. Automated versions are also available.

Detectors for HPLC

A vast range of literature exists on the subject of detectors for HPLC, a number of review articles on the subject having appeared in recent years (e.g. White 1984).

It is in fact possible to identify three distinct modes of HPLC detection.

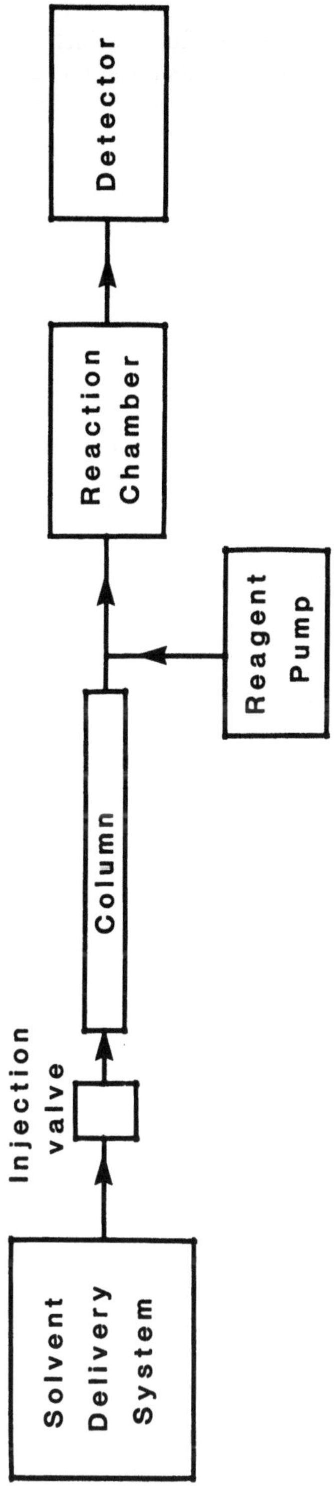

Fig. 11. Schematic of post-column derivatization system

i) solute specific, in which detection depends on monitoring some specific physical or chemical property of the compounds to be analysed
ii) bulk property, in which detection depends on monitoring some bulk physical or chemical property of the mobile phase which changes as a result of the presence of the solute
iii) Desolvation detectors in which the mobile phase is removed, normally by evaporation, prior to detection of the solute by a secondary technique.

Methods (ii) and (iii) above tend to be insensitive with detection limits of the order of a microgram.

The most common bulk property detectors are those based on the monitoring of the refractive index of the column eluent, R.I. detectors. In addition to having poor limits of detection, R.I. detectors are notoriously temperature sensitive and furthermore cannot be used in conjunction with gradient elution. This is because slight changes in mobile phase composition results in a larger change in refractive index than does the presence of a small amount of solute.

One other bulk property detector worth mentioning is the conductivity detector. They are only useful for the detection of ions and are therefore limited in their application to ion exchange separations. Like R.I. detectors, they cannot be used with gradients, have poor detection limits, and are temperature sensitive. These features are, in fact, likely to be shared with any bulk property detector.

Both R.I. and conductivity detectors can be made with small detection cells i.e. dead volumes of 5 μ litre or less. This is an important feature of any detector since it is important that the detector itself does not contribute to band broadening by allowing diffusion of the separated peaks to occur after the columns.

The use of desolvation detectors is now limited almost exclusively to the coupling of HPLC to mass spectrometry HPLC-MS interfacing is discussed in Chapter by Games and Ramsey. This is a far more complex problem than that experienced in g.c. – mass spectrometry, because in HPLC the mobile phase is a liquid, which if it contains an aqueous buffer may not be particularly volatile. Desolvation detectors consist of a continuous e.g. stainless steel, inert moving ribbon or belt to which the column eluent is continuously applied. The moving belt is then heated removing most of the mobile phase before it passes via a series of vacuum locks into the ion chamber (e.g. in a mass spectrometer) where it is again heated to vaporize the sample and any residual mobile phase. The moving belt is then cleaned by further heating before passing again to the column outlet.

Because of the problem of removing large quantities of solvent at flow rates greater than 1 ml. min^{-1}, these systems may be of much greater use in capillary HPLC where flow rates are much lower than with normal packed column HPLC.

Although there is considerable interest (McFaden 1980) in the use of mass spectrometry as a form of HPLC detector, it must be remembered that the technique is always likely to be very expensive.

The most common detectors currently in use for HPLC are the solute specific detectors. Even within this group of detectors only a limited number of solute properties are used, these are listed below.

i) UV detectors which monitor the absorbance of UV light by the sample
ii) Flourescence detectors which monitor the fluorescent characteristics of the sample
iii) Infra-red detectors which monitor the absorbance of infra-red radiation by the sample
iv) Electrochemical detectors which monitor the electrochemical properties of the sample.

All of the above detectors have detection limits in the nanogram range (in the case of electrochemical detectors and fluroescence detectors this can be in the picogram or femtogram range). Furthermore, their measuring cells can be made with dead volumes less than 10 μ litres (EC detector cells are typically < 2 μl).

U.V. Detectors

UV detectors are the most commonly used HPLC detectors. They are easy to use, relatively inexpensive, and can be used in a very wide range of applications.

Three basic designs are currently available commercially. The most flexible design uses a monochromator, in the same manner as a UV spectrophotometer, so that the wavelength of UV radiation which is focussed onto the flow-cell can be selected from wavelengths as low as 200 nm up to 600 nm. It is even possible to use deuterium lamps throughout this range if the detector is carefully designed since there is sufficient emission from these lamps between 300 and 600 nm to allow their use with moderate noise levels.

As an alternative to using monochromators, interference filters can be used and although detectors based on this method of wavelength selection are cheaper, they are considerably less flexible than monochromator based detectors. This is because the monochromator, or continuously variable wavelength detectors, as they are normally called, can be tuned to select virtually any wavelength and, therefore, any compound or group of compounds which absorbs in the UV/Visible spectrum. In a similar way it is possible to detune background absorption from the mobile phase, or interfering peaks, thus increasing the signal to noise ratio. Many continuously variable UV detectors are now available which are microprocessor controlled and allow the user to change wavelengths automatically during the chromatogram. This enables the chromatographer to select the optimum detection conditions for each compound of interest. Such wavelength programming can result in sudden base-line shifts however. These occur as a result of the different background absorption by the mobile phase at different wavelengths.

The third and final type of UV detector uses diode arrays as the photodetection system. Conventional UV detectors are fitted with a single photodetector e.g. a photodiode. These devices can monitor only one wavelength at a time, and are chosen to have as good a sensitivity as possible across the wavelength range. It is in fact largely the recent developments in photodetectors which has meant that detectors using deuterium lamps can be used up to 600 nm without changing lamps.

Within the last five years diode array photodetectors have become available, and these have completely different characteristics to conventional photodiodes. They are relatively small devices similar in appearance to the central processor "chips" of personal computers. They are manufactured in such a way that their sensitivity to UV light varies along their length. Thus, by placing a monochromator behind the flow-cell, the flow-cell is irradiated with all wavelengths simultaneously, and light exiting from the flow-cell is split into its various individual components which can then be individually detected by the diode array. This means that the entire UV spectrum of any material in the flow-cell is being examined at any given moment in time.

Diode array detectors are extremely powerful in that a very large amount of spectral information can be obtained from any given chromatogram or sample. It is, in fact, essential that diode array detectors are operated with their own data handling systems because of the quantity of information with which they have to deal.

Considering their potential analytical power, diode array detectors are relatively inexpensive, although the quality of the results from any given detector can be highly dependent on the quality of the manufacturers' software.

Fluorescence Detectors

Optically fluorescence detectors are similar to fluorescence spectrophotometers. A UV light source is focussed onto a flow-cell and the resultant fluorescent radiation is collected at right angles to the direction of the incident radiation. This is actually quite difficult to achieve in practice since the volume of the flow size must be kept low and, therefore, the quantity of fluorescent radiation collected at the photodetector can be quite low. This can result in a lower sensitivity than might be expected for many fluorescent compounds. One method of overcoming this is to increase the size of the flow-cell. Although this adds to band spreading this tends to be less important with fluorescence detectors since they have two main features.

The first of these is their intrinsic sensitivity to compounds exhibiting fluorescence, and the second is their specificity. By careful selection of the excitation and emission wavelengths it is possible to obtain a response which is highly specific to the solute or solutes of interest. Thus, the chromatograms from fluorescent detectors tend to be very much less complex than those obtained from other detectors i.e. there are fewer problems resulting from the presence of overlapping peaks and, hence, some additional peak broadening can be tolerated. In practice it is possible to obtain satisfactory results from fluorescence detectors with flow-cells as large as 70 μ litres.

Selection of the excitation and emission wavelengths can be carried out by using monochromators or optical filters, or a combination of both.

Electrochemical Detectors

The possibility of using polarography as a method of detection in HPLC has been recognised for sometime. It was not however until the publications of Kissinger (1973) that reliable electrochemical detectors became commercially available. Electrochemical detectors have a number of advantages in theory:
they should be extremely sensitive, inexpensive and should be capable of detecting any compound with measurable redox potential e.g. alcohols, phenols, amines, acids, esters, ketones and aldehydes. Furthermore flow cell volumes should be small.

In construction electrochemical detectors are fitted with three electrodes, a working electrode to which the working potential is applied and at which the redox reactives take place, a reference electrode and an auxillary electrode; the electrodes are normally mounted in a PTFE or kel-F block. The working electrode is kept at a fixed electrical potential equivalent or close to the half-wave potential of the solute under study. When the solute enters the cell a redox reaction occurs at the working electrode surface and current flows between the working and reference electrodes. The auxillary electrode is used to maintain the potential at the reference electrode which must remain fixed, however, the reference electrode potential will change when current flows through it. The auxillary electrode ensures that only a small amount of current passes through the reference electrode thus, maintaining the correct potential difference between working and reference electrodes.

The working electrode is normally glassy carbon, although Perrett (Perrett & Drury, 1982 and Perrett & Rudge, 1985) has used gold amalgam electrodes for some applications.

Electrochemical detectors can only be used with mobile phases which are themselves electrolytes, although Flanagan et al. (1985) has used them for the analysis of certain drugs with quite high methanol concentrations.

Nevertheless the high sensitivity of electrochemical detectors cannot be denied. They are capable of detecting a few picogrammes of catecholamines (see Chapter by Allenmark).

Infrared Detectors

Infrared detectors ore based on similar principles to those used in UV. Many compounds which can be separated on HPLC columns exhibiting strong absorption in the infrared region of the spectrum. Unfortunately, most useful HPLC solvents also absorb infrared radiation and there are serious problems in finding suitable windows in the mobile phase absorption spectra, unless very short (0.1 mm) path lengths are used. This tends to limit sensitivity.

Despite this disadvantage, commercial infrared detectors are available, although they tend to be used only for the analysis of synthetic polymers.

Derivatization Techniques and other Detection Methods

Although UV detectors are capable of being used for the majority of HPLC analyses, there are a number of compounds which can be satisfactorily separated on HPLC, but cannot be detected at useful detection limits.

One way of overcoming the problem is to prepare a derivative of the compound which can be readily detected by one of the methods discussed previously. This can be done in one of two ways, to prepare the derivative prior to injection on the column (pre-column derivatization) or alternatively to separately continuously pump a suitable derivatizing reagent after the column (post-colum derivatization). Post colum derivatization is illustrated in Figure 11. This type of systems has been studied very thoroughly by Frei (1979), and three distinct types of reaction chamber have been described in the literature; coiled tube reactors (Frei 1979), packed bed reactors (Schwedt and Reh, 1981) and segmented flow reactors (Frei, Sholtern and Brinkman 1981).

Although pre-column derivatization is probably simpler experimentally than post-column, the latter has a number of advantages. The post-column method is easier to automate, especially in the case of unstable derivatives, and in addition, the sample is chromatographed in its native chemical state. Derivatization frequently involves reaction with a chemical species which tends to increase the chemical similarity between the components of the sample; this in turn can make them chromatographically and hence more difficult to separate. This is well illustrated in the separation of PTH derivatives of amino-acids which are notoriously sensitive to slight changes of mobile phase composition, column ageing, etc. In the case of precolumn derivatisation reactions, consideration must also be given to the presence of cripurities in the sample/mobile phase or derivatisation reagent which may increase background noise levels.

Before passing onto the next section a number of points are worth mentioning on the subject of HPLC detectors.

The topic is, in fact, so broad and the literature so full of novel detection methods that it merits a book to itself let alone a small section of an introductory chapter.

Brief mention has only been made here to those detection systems which are commercially available. Little mention has been made of the use of combining detectors e.g. monitoring two UV wave-lengths in series, or using a combination of UV and fluorescence detection, techniques which can be extremely useful.

For many biological samples, or where radioactive compounds are being separated, it is frequently possible to operate HPLC systems without detectors at all. Fractions from the eluent can be collected and subsequently examined for biochemical/biological activity using bioassay methods, or in the case of radiolabelled substances the fractions can be analysed using scintillation counting techniques. It must be also pointed out that on-line facilities for the detection of radioactive solutes can also be purchased commercially.

For those serious research workers faced with a novel analytical problem, soluble by HPLC, but apparently lacking a suitable, sensitive detection method, a full literature survey on detection methods is recommended.

Automation and Data Handling

The advent of the microprocessor has revolutionized most aspects of instrumental analysis and HPLC is no exception. For those laboratories with high sample throughputs fully automatic liquid chromatographs are available. In conjunction with these systems a wide variety of data handling devices can be used varying from inexpensive single channel intregrators to sophisticated fully computerized multi-channel data stations.

When data handling is being considered, it is therefore necessary to analyse individual requirements carefully in order to maximize the cost effectiveness of the total system. Some of the factors to be considered are:

> "Is sufficient data really being generated from more than one chromatograph to justify a multi-channel instrument?
> Alternatively, a single channel integrator may be sufficient for current needs, but does the approach fulfil the laboratory's long term requirements?"

Capillary and narrow-bore HPLC

Discussion so far has centred around HPLC columns of length between 5 cm and 25 cm with internal diameters of the order of 4.5 mm, packed with particles of 3, 5 or 10 μm diameter.

In recent years, interest has grown in the use of narrow-bore columns for HPLC. Those most commonly described in the literature are 1–2 mm in diameter in with lengths up to 1 metre packed with similar materials as 4.5 mm bore columns. Such columns appear to have overall efficiencies higher than conventional packed columns although their use does have certain implications with regard to instrument design.

Microbore columns have optimum flow-rates between 5 μ litres min^{-1} and approximately 50 μ litres min^{-1} and in addition their high efficiency (of the order of 150,000 to 200,000 plates per column) means that dead volumes in detectors, injectors etc., are very much more critical than in the case of more conventional columns.

Most reciprocating pumps have some difficulty in operating at flows less than 100 μ litres min^{-1} without major pump head modifications, and often simply changing the piston size and pump head volume is not sufficient. For this reason some manufacturers are re-examining the use of small volume (10–20 ml) motorized syringe pumps.

Many UV detectors are also supplied with the option of having micro flow cells fitted, some of which operate satisfactorily and some of which do not, the major problem being to eliminate dead volume without producing unwanted thermal effects within the flow cell itself.

Few manufacturers supply total HPLC systems dedicated for use with capillary columns. This is largely because many of the technical problems associated with microbore columns remain to be solved, and the majority of HPLC separations are still carried out on wide bore columns. This is despite the higher efficiencies and lower mobile phase consumption of microbore columns.

Summary

For those readers interested in researching the theory of HPLC and the associated instrumentation a brief bibliography is listed below in addition to the references cited.

The reader is also referred to the many good quality International Journals which are dedicated to the techniques of HPLC.

References

Bristow P (1976) Liquid chromatography in practice, (privately published by HETP Components for Liquid Chromatography, Wilmslow, Cheshire). Now out of print but a good guide to the "nuts and bolts" of HPLC.

Brust OE, Sebastian I, Halasz I (1973) Stationäre Phasen mit S-N Bindung für die Flüssigkeits-chromatographie. J Chromatogr:83, 15–24

Done JN, Knox JH, Loheac J (1974) Applications of high speed liquid chromatography. John Wiley, New York

Flanagan RJ, Jane I, McKinnon A (1985) High performance liquid chromatographic analysis of basic drugs on silica columns using non-aqueous ionic eluents. Part II J Chromatogr:323, 191–225

Frei RW (1979) Recent developments in post-column reactions for high performance liquid chromatography. Proc Analyt Div Chem Soc:16, 289–293

Giddings JC (1965) Dynamics of chromatography. Part I Marcel Dekker, New York

Halasz I, Sebastian I (1969) New stationary phase for chromatography. Angew Chem Int Ed:8, 453–454

HPLC of proteins and peptides (1983) Academic Press, London and New York. (Editors Hearn MTW, Regnier FE; Wehr CT)

HPLC in food analysis (1982) Academic Press, London and New York. (Editor Macrae R)

Hunkapiller MW, Hood LE (1983) Analysis of phenylthiohydantoins by ultra-sensitive gradient high performance liquid chromatography. In: Hirs CHW, Timasheff SN (eds) Methods in Enzymology, Enzyme Structure Part I. 91, 486–493

Jones AD, Burns IW, Sellings SG, Cox JA (1977) Preparation, optimisation and slurry packing of an amino bonded phase for the analysis of sugars in food by high performance liquid chromatography. J Chromatogr:144, 169–180

Kissinger PR, Refshauge C, Dreiling R, Adams RN (1973) An electrochemical detector for liquid chromatography with picogram sensitivity. Analyt Letters:6, 465–477

Knox JH, Laird GR, Raven PA (1976) Interaction of radial and axial dispersion in liquid chromatography in relation to the "infinite diameter effect". J Chromatogr:122, 129–145

Locke DC, Schermund JT, Banner B (1972) Bonded stationary phases for chromatography. Anal Chem:44, 90–92

McFadden WH (1980) Liquid chromatography/mass spectrometry systems and applications. J Chromat Sci:18, 87–101

Paris NA (1976) Instrumental to liquid chromatography: A practical manual on HPLC methods. Elsevier, Amsterdam

Perrett D, Rudge SR (1985) The determination of thiols and related compounds using high performance liquid chromatography. J Pharm Biomed Anal:3, 3–27

Schill G (1974) In: Marinskey JA, Marcus Y (eds) Ion exchange and solvent extraction. Marcel Dekker, New York, Volume 6, Chapter 1

Schwedt G, Reh E (1981) Aufbau und Optimierung fluorimetrischer Reaktionsdetektoren mit luftsegmentierten Flüssigkeitsströmen für die Hochdruck-Flüssigkeits-Chromatographie. Teil 2. Reaktormaterialen und Einflüsse variabler dynamischer Größen. Chromatographia:14, 123–126

Scholten AHMT, Brinckmann UAT, Frei RW (1981) Solvent segmentation in liquid chromatography – application to photochemical reaction detectors. J Chromatogr:205, 229–237

Simpson CF (1982) Separation modes in high performance liquid chromatography. In: Macrae R (ed) High performance liquid chromatography in food analysis. Academic Press, New York, pp 79–121

Snyder LR, Kirkland JJ (1979) Introduction to modern liquid chromatography. 2nd Edn, John Wiley, New York

van Deemter JJ, Zuiderweg FJ, Klinkenberg A (1956) Longitudinal diffusion and resistance to mass transfer as causes of non-ideality in chromatography. Chem Eng Sci:5, 271–289

White PC (1984) Recent developments in detection techniques for high-performance liquid chromatography, Part I. Spectroscopic and electrochemical detectors. Analyst:109, 677–697

White PC (1984) Recent developments in detection techniques for high-performance liquid chromatography, Part II. Other detectors. Analyst:109, 973–984

High-Performance Liquid Chromatography of Catecholamines and their Metabolites in Biological Material

S. ALLENMARK

Introduction

The use of high-performance liquid chromatography (HPLC) as an analytical tool for the life sciences has become increasingly important and the techniques and instrumentation involved are continuously developing to higher levels of sophistication. One result of this rapid progress is that a user of HPLC for a certain bioanalytical purpose is often faced with a great problem of selecting from the overwhelming menu offered in the field with respect to choice of column type, mobile phase, elution technique, detector system, degree of computerization etc. To this are then added the particular problems associated with the pre-chromatographic work, i.e. the extraction, prepurification and concentration steps often necessary to produce a sample suitable for injection into an HPLC-system.

It is the purpose of this chapter to give an overview of the methods and techniques available for the determination of catecholamines and their metabolites in biological material by HPLC with selected examples from publications on suitable experimental procedures.

Some earlier reviews in the field have been published; in particular the reader is referred to the papers by Krstulovic (1979, 1982), Krustulovic & Colin (1983), Warsh et al. (1982) and Allenmark (1982).

General Considerations

Chemistry and Metabolism of Catecholamines

The catecholamines (CA) are biosynthesized from the amino acid L-tyrosine as outlined in Scheme 1.

The physiological action of these compounds as transmitters in the adrenergic system has been intensively studied and the interest in a determination of the concentrations of the individual CAs in various body fluids under different conditions, including states of disease, has increased continuously as a result of the development of more and more advanced methods for their analysis.

The known biochemical events leading to metabolism and urinary excretion are summarized in Scheme 2.

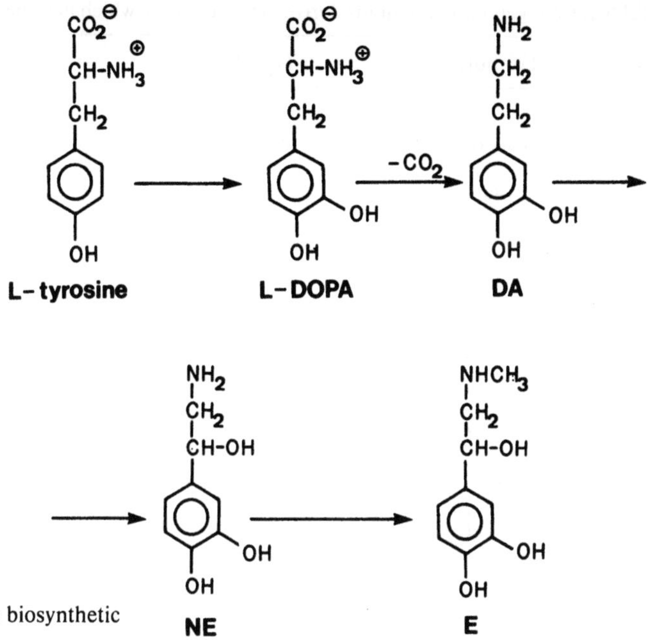

Scheme 1. Catecholamine biosynthetic pathways

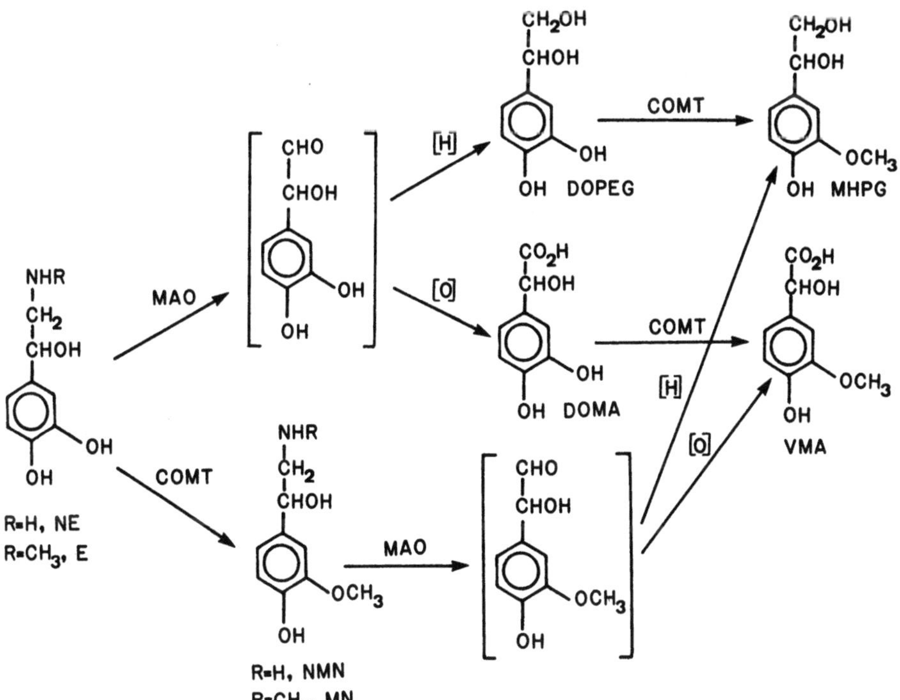

Scheme 2. Transformation of the catecholamines into various metabolites

Table 1. Endogenous compounds derived from catechol which have been determined by HPLC

Basic type	Formula	Trivial name	Abbreviation
	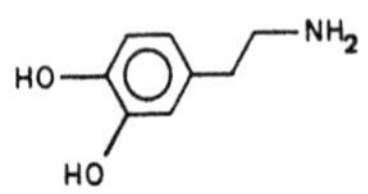	dopamine	DA
	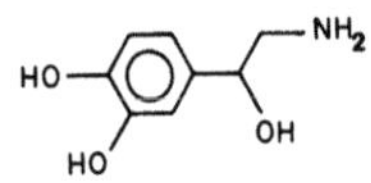	norepinephrine	NE
Amines	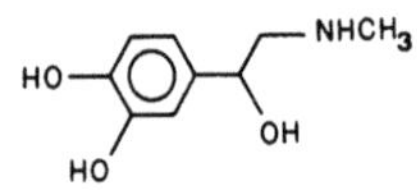	epinephrine	E
	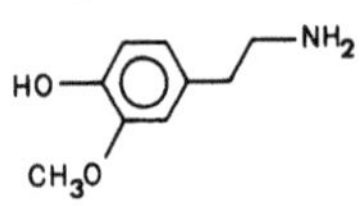	3-methoxytyramine	MT
	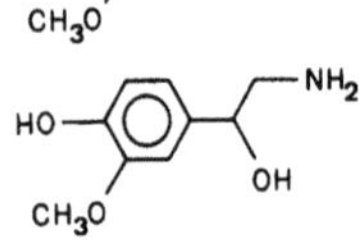	normethanephrine	NMN
	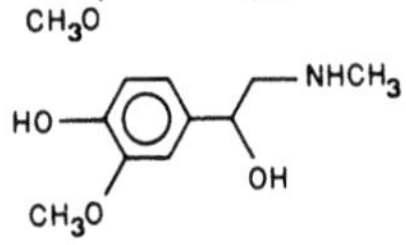	metanephrine	MN
	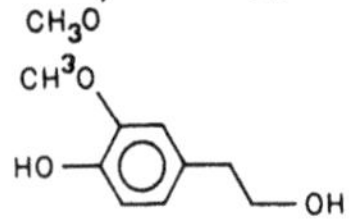	3,4-dihydroxyphenylethanol	DOPET
Alcohols	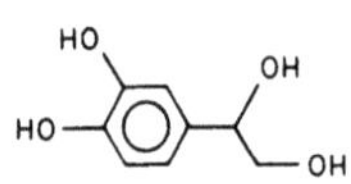	3,4-dihydroxyphenylglycol	DOPEG
	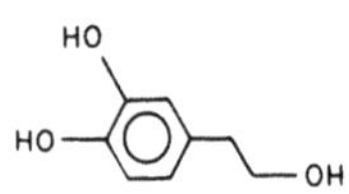	3-methoxy-4-hydroxyphenyl ethanol	MOPET
	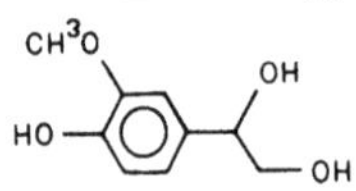	3-methoxy-4-hydroxyphenyl glycol	MHPG (or MOPEG)
	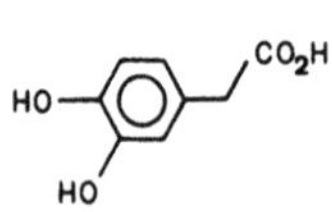	3,4-dihydroxyphenylacetic acid	DOPAC
Carboxylic acids	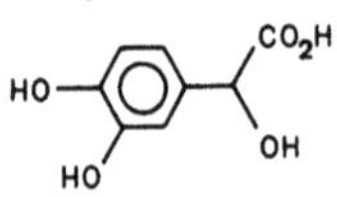	3,4-dihydroxyphenylmandelic acid	DOMA
	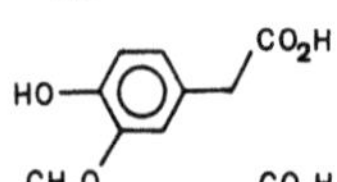	homovanillic acid	HVA
	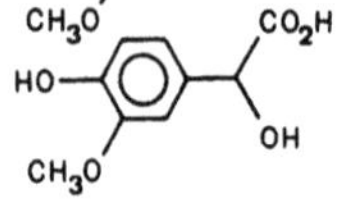	vanillylmandelic acid	VMA

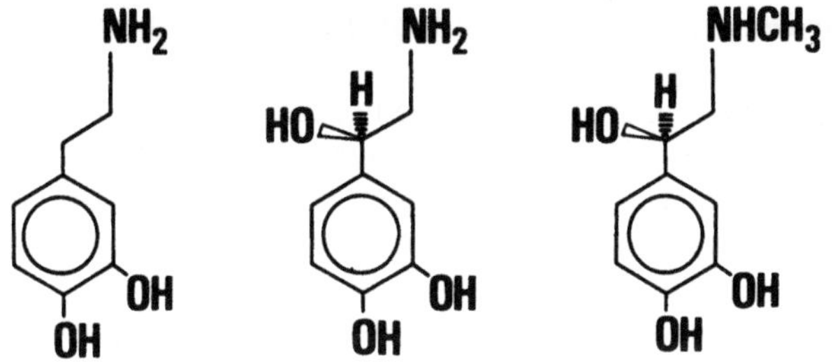

2-(3,4-dihydroxy-phenyl)ethylamine; dopamine (DA)	R-(-)-1-(3,4-dihydroxy-phenyl)-2-aminoethanol; norepinephrine (NE)	R-(-)-1-(3,4-dihydroxy-phenyl)-2-(methylamino)-ethanol; epinephrine (E)

Scheme 3. Structures, including absolute configurations, as well as rational and trivial names of the catecholamines

The CAs all contain a 3,4-dihydroxyphenyl group (catechol unit) as a common structure element, but differ with respect to the 1-substituent. The enzyme dopamine-β-hydroxylase (E.C.1.14.17.1) introduces the hydroxyl group with complete stereospecificity, i.e. converts the achiral DA* into the chiral NE of R-configuration. The same configuration is present in E, as the enzyme phenylethanolamine N-methyltransferase (PNMT; E.C.2.1.1.28) is only involved in the transformation of the primary amino function of NE into a secondary one by methylation.

As immediately evident from their chemical structures (Scheme 3), the CAs possess some common physical and chemical properties. Their highly polar nature and amphoteric character due to the presence of amino- as well as phenolic groups will limit the use of conventional extraction procedures for isolation purposes. Further, the compounds will require derivatization of all polar sites in order to be accessible by GC-methods. Another complicating factor is due to the well-known ease by which all catechol-type compounds oxidize in air, to form o-quinones and secondary products, especially in alkaline media. Thus, these properties have to be considered in all work-up and handling procedures, e.g. by the use of antioxidants as protecting agents and avoidance of high pH or light intensity.

The catechol ring gives rise to UV-absorption maxima at 200–220 nm and at 280–300 nm, respectively. Irradiation at either of these maxima induces a weak native fluorescence (emission maximum at 310–340 nm).

The solution chemistry of the CAs is largely dominated by the nucleophilic character of the amino group which serves as an excellent target for various derivatization reagents. As will be shown later, however, some important reactions used in fluorescence detection techniques require a primary amino group and can therefore not be applied to E. The ease of oxidation of the catechol ring is another property that has been utilized for analytical purposes – the technique of direct electrooxidation at an anodic surface being one of the most powerful and also not restricted to the CAs but applicable to their acid metabolites as well.

The Presence of Catechol-Derived Compounds in Biological Tissue and Fluids

The adrenal medulla acts as the primary source of epinephrine and the extremely low concentration levels of this pressor hormone normally found in blood (plasma) ($\sim 5 \times 10^{-10}$ M) reflect the adrenal medullary secretion. Because of the easily observable physiological effects of this compound, it was the very first hormone to be isolated and chemically characterized. Norepinephrine and its precursor dopamine are both neurotransmitter hormones and accordingly present in nervous tissue. Norepinephrine, which is released from peripheral sympathetic nerves innervating the heart and smooth muscles, represents the principal sympathetic neurotransmitter and is liberated into the blood circulation mainly by exocytosis from nerve endings. The blood (plasma) concentration of NE is normally of the order 2×10^{-9} M, while the figure for (unconjugated) DA is probably less than 5×10^{-10} M.

The enzymes responsible for the two major catabolic pathways, viz. catechol O-methyl transferase (COMT) and monoamine oxidase (MAO), respectively, are widely distributed in the body; however, the highest COMT-activity is located in the liver and that of MAO to the brain and the heart. It is important to note that many factors contribute to the complexity of the problem of quantitation of CAs and metabolites in biological material and, not least, interpretation of data obtained. Thus, CAs as well as metabolites are distributed both in free forms and as conjugates of sulfuric and glucuronic acids. The conjugation is mainly carried out in the liver by certain transferase enzymes. The distribution between free and conjugated forms varies greatly with the type of physiological sample and with the parent catechol derivative. Moreover, the metabolite pattern in urine is derived from both central and peripheral metabolism and accordingly any change in a urinary metabolic profile is often hard to relate directly to a specific area of origin. Only in brain tissue and CSF are the concentration levels not affected by peripheral contributions.

Table 2. Common abbreviations used in the description of various analytical methods

Chromatographic separation techniques	Detection techniques
Gas chromatography: GC	Flame ionization detector: FID Electron capture detector: ECD Mass spectrometry: MS
(High-performance) liquid chromatography: (HP)LC Column type: reversed phase: RP strong cation-exchanger: SCX reversed-phase + ion-pairing additive: RP-IP	Electrochemical detection: EC Fluorimetric detection: F Chemiluminescence: CL
Solvents, reagents, derivatives: Perchloric acid. PCA o-Phthalaldehyde: OPA Fluorescamine: FA Dansyl (derivatives): DNS	

Isolation Procedures and Problems Associated with the Biological Matrix

The selection of a suitable method for sample preparation prior to analysis by HPLC is generally dependent upon a number of factors, such as the biological source, the type of column used and the selectivity of the detection method. In general, however, a minimal sample handling offers many advantages and if the HPLC-conditions are well chosen for a particular analytical problem, very crude samples can often be directly analyzed. Particularly, such "direct-injection" assays reduce the necessity of internal standard addition because recoveries can be assumed to be high and nearly constant. However, in cases where the analyte concentration is very low and/or the analyte is present in a very complex mixture with other compounds, an exhaustive pretreatment may be unavoidable. Thus, Magnusson et al. (1980) reported that DA, DOPAC and HVA in rat brain could be determined simultaneously by electrochemical detection after direct injection of a perchloric acid extract of the brain tissue into a reversed-phase HPLC system. On the other hand sources like plasma or urine are much more complex and will usually require an extractive procedure.

In general, two distinct steps can be found in the work-up of a biological material:

1. a procedure by which the compounds of interest are released into a liquid phase, and
2. a second procedure by which extraction and preconcentration can be achieved.

The first of these involves biological separation techniques, typically tissue homogenization, protein precipitation and centrifugation procedures. The second may involve either liquid or solid phase extraction methods. Usually, acidic and neutral metabolites are easily extracted into organic solvents of medium polarity (ether, ethyl acetate etc.) from acid aqueous extracts ($pH < 3$). Due to their instability at higher pH, extraction of CA's should also in principle be carried out from an acid solution. For reasons already mentioned, however, liquid phase extraction is not applicable. The most obvious technique is the use of a solid cation-exchanger either in a batch or column mode. Such methods have been frequently used for the isolation of biogenic amines in general (i.e. including indoleamines, methoxyphenylethylamines, etc.), but are not selective to CAs.

The most widely used CA-selective solid-phase extraction technique is based on the fundamental investigations by Anton and Sayre (1962, 1964), who found that CAs were selectively adsorbed onto acid-washed alumina at pH 8.6 and could then be readily desorbed by acid. The adsorption mechanism involves the o-dihydroxyphenyl moiety which exhibits a pH-dependent affinity. A quite analogous selective extraction method was introduced more recently by Higa et al. (1977) and makes use of adsorption to a boric acid affinity gel. The adsorption process, based on the reversible formation of a cyclic boronate (or aluminate), is outlined below.

A summary of various well-documented isolation techniques is given in Table 3.

Table 3. Selected examples of different "work-up" procedures used prior to quantitation of catechol-derived compounds in biological material

Compound(s) assayed	Source	Procedure (recovery)	Chromato-graphic method	Reference
DA, DOPAC, HVA	brain	tissue homogenization in perchloric acid, supernatant after centrifugation injected directly	RP-HPLC-ECD	Magnusson et al. (1980)
NE, E	plasma	deproteinization by perchloric acid, pH-adjustment to 8.4, alumina extraction, desorption by acetic acid/methanol (66–68%)	SCX-HPLC-F	Yui et al. (1980)
HVA	urine	none (100%)	RP-HPLC-ECD	Morrisey and Shibabi (1979)
MHPG, DOPAC, HVA	brain	tissue homogenization at pH 3, supernatant after centrifugation extracted with ethyl acetate (MHPG: 64%, DOPAC: 86%, HVA: 98%)	RP-HPLC-F-ECD (detectors in series)	Cross and Joseph (1981)
NE	plasma	undeproteinized plasma, mixed with phosphate buffer (pH 7.5), applied to a boric acid affinity gel column, washing, desorption by acetic acid/methanol (81%)	GC-MS	Yoshida et al. (1980)

Analytical Methods in use other than HPLC

Due to the often extremely small quantities that have to be determined, the analytical method of choice must fulfill high requirements with respect to both selectivity and sensitivity. Therefore, in all truly analytical techniques of practical value, a chromatographic separation is used in combination with a highly sensitive method of detection. The complexity of the method used is often associated with the need of derivatization, which may be performed either pre- or post-chromatographically, and the type of chromatography and detection process used.

A comprehensive review, covering the methods used for the determination of CAs in plasma, was published recently by Holly and Makin (1983).

a) Thin-Layer Chromatographic (TLC) Techniques

While being an excellent technique for qualitative analysis and purity control, TLC has so far not been widely used for quantitative purposes. The reason for this is due to the fact that it is not a procedure of continuous operation. Thus, the chromatographic result has to be evaluated either by densitometric scanning or by removal of the spots after development for separate analysis. The advantage, however, is that many samples can be run in parallel on a TLC-plate. Therefore,

TLC has become an important separation method in combination with radioisotope labelling techniques and quantitation by means of radiation counting, such as liquid scintillation counting of 3H or 14C.

Especially for the radioenzymatic methods, which are based on enzyme-catalyzed introduction of a radiolabel into the compounds to be determined, TLC has become the tool of choice for separation. The strength of these techniques lie in their unsurpassed sensitivity, which has been clearly demonstrated in the case of determination of plasma CAs. A reliable assay method has been described by Peuler and Johnson (1977).

A number of other techniques based on TLC-separation have been used, but rather few seem to have been thoroughly validated. A relatively simple method was described recently by Gelijkens and de Leenheer (1980), by which free urinary CA's were isolated via a boric acid gel column, converted into acetyl derivatives which were separated on a HPTLC-plate, made fluorescent by reaction with ethylenediamine/potassium ferricyanide and assayed by scanning of fluorescence.

b) Gas Chromatographic (GC) Techniques

Many approaches have been made to find suitable methods for GC-derivatization of CAs. In the majority of cases perfluoroacylation has been used which generally gives stable derivatives suitable for mass-pectrometric (MS) or electron capture (EC) detection. The use of pentafluoropropionic anhydride for derivatization of CAs was introduced by Wong et al. (1973) who used the technique to determine urinary CAs by GC-EC. Following an improved isolation procedure the same principle was recently used by Yoshida et al. (1980) for the quantitation of plasma NE by GC-MS using selected ion monitoring with deuterated NE as internal standard.

General Analytical Techniques for the Determination of Catecholamines and Metabolites by HPLC

The main advantage of HPLC is that separation takes place in a liquid phase at room temperature which, unlike GC, poses virtually no restrictions on the physical properties of the compounds to be separated. Therefore, separation efficiency is usually only a question of selecting the right combination of column and mobile phase and derivatization of the compounds is only necessary if required for detection purposes.

Outline of the Main Techniques Used

Because of the often minute amounts of material present in the sample, methods of detection generally have to be very sensitive. Ideally, this can be achieved without derivatization; in certain cases, however, this may not be possible. Derivatization then serves as a means to reduce the limit of detection by introduction of a label in the compounds of interest and utilization of the specific properties of this label. The two most common techniques are radiolabelling, usually by ^3H or ^{14}C, and fluorescence labelling. Such derivatization techniques can, in principle, be carried out either before or after the HPLC-separation (pre- and post-column derivatization, respectively). In all precolumn derivatization methods it is essential that the derivatives are stable enough to be recovered unchanged after chromatography and further that the derivatization yields can be determined for all compounds of interest and show high reproducibility. Postcolumn reactions, on the other hand, should be rapid in order to suppress the band broadening caused by the large postcolumn reactor volumes needed at long reaction times.

Table 4 gives a summary of the various techniques used for HPLC of CAs and metabolites.

Table 4. Different approaches to the problem of quantitation of catecholamines and metabolites by HPLC

Derivatization mode	Labelling technique	Application	Detection mode	Reference
Precolumn	3H/RE	NE, E, (DA)	UV + LSC	Müller et al. (1979), Endert (1979)
	OPA	DA, NE	fluroescence	Davis et al. (1978) Mell et al. (1978)
	DNS	DA, NE	fluorescence	Frei et al. (1979)
	FA	DA, NE	fluorescence	Imai and Zamura (1978)
Postcolumn	OPA	DA, NE	fluorescence	Okamoto et al. (1978)
	THI	NE, E	fluorescence	Yui et al. (1980 a,b)
No derivatization	-	DA, NE, E, DOPA	(native) fluorescence	Krstulovic et al. (1979) Anderson and Young (1981)
	-	DA, NE, E, DOPA, M, NM, MT, VMA, HVA, DOPAC, MHPG	electrooxidation	Kissinger et al. (1981)

Column and Mobile Phase Selection

Due to the rapid development in the field, nearly all commercial HPLC-columns are now based on chemically bound stationary phases. A majority of these columns contain alkyl-bonded silicas which are used for reversed-phase HPLC.

A number of different separation modes for such columns have been developed due to the fact that these stationary phases are readily dynamically coated by certain additives to the mobile phase. These techniques have extended the possibilities of regulating retention in RP-HPLC considerably. One of the main reasons why RP-columns are so attractive is due to the very high column efficiencies that can easily be obtained. The same efficiencies are rarely achieved with ion-exchange columns.

Alkyl-silica columns retain solutes essentially by virtue of hydrophobic interaction in aqueous mobile phases. Therefore, the retention of protolytic solutes (like amines or carboxylic acids) will be very much pH-dependent (Molnar and Horvath, 1976, 1978). With decreasing pH, amines will be less retained whereas the opposite will hold for carboxylic acids. Too high pH in the mobile phase has several disadvantages, one being the harmful effect it exerts on the silica support. Of particular importance in the present context is the increased rate of oxidation of catechols with increasing pH. Therefore, the chromatographic conditions have to be those under which the basic catechols, i.e. the CAs, exist essentially in their protonated forms. Thus, it is evident that chromatography of CAs under such conditions should be based on ionic (coulomb) attraction of some kind and accordingly there are two ways to achieve this, viz. a) by the use of a cation-exchange column, or b) by changing the mode of separation on an alkyl-silica column via addition of ion-pairing agents to the mobile phase (ion-pair reversed-phase chromatography). For CAs, anionic surfactants, particularly short-chain alkyl sulphonates, have been used with great success as ion-pairing agents ("soap chromatography", introduced by Knox and Jurand, 1976). The ion-pair technique in RP-HPLC was further developed by Schill and his coworkers and applied to CAs and related bioactive compounds (Schill, 1974; Johansson, 1981; Johansson et al., 1978). In these techniques, an organic solvent modifier like 1-pentanol is added in a few per cent to the aqueous mobile phase containing the alkylsulphonate or other suitable counterion. It has been suggested that retention is caused by ion-pair partitioning between a stationary liquid organic phase formed on the alkyl-silica and the aqueous mobile phase.

From the point of view of chromatographic theory, however, different models regarding the equilibria taking place at the alkyl-silica surface have been proposed. In summary, these can be described as

1. the dynamic ion-exchange model (Knox and Hartwick, 1981): the anionic surfactant is adsorbed to the alkyl-silica and converts the column dynamically into a cation-exchanger,
2. the ion-pair distribution model (Terweij-Groen et al., 1978): ion-pair formation occurs in the mobile phase and the ion-pairs are equilibrated between the mobile and stationary phases,
3. the dynamic complex exchange model (Melander and Horvath, 1980): equilibria are established where the ion-pair formed in the mobile phase will displace the anionic surfactant bound to the alkyl silica.

In practice then, C-18- or C-8-silica RP-HPLC columns are by far the most widely used for these types of separations. However, for certain CA-determinations SCX-HPLC-columns still represent a good alternative. The choice of separation system will also to some extent depend on the method of detection. Especially in combination with postcolumn derivatization, where extreme column efficiency is not essential and certain mobile phase additives may cause fluorescence quenching, SCX-columns have been frequently used.

Detector Systems

a) UV Detection

Variable wavelength UV detectors are easy to operate and represent the most versatile type of detector in HPLC. Their use for detection of catechol-derived compounds at the 285 absorption band is only limited by the requirement of sensitivity and selectivity. The practical limit of detection under optimal conditions can be roughly estimated to 0.01 nmol (S/N = 2). For purposes of urinary metabolic profiling with respect to catecholamine catabolites this may be sufficient, especially as peak identification is possible via stopped-flow UV-scanning techniques (Krstulovic et al. 1979) or by use of the photodiode array detector system.

b) Fluorimetric Detection

The fluorescence from CAs in their native state on excitation at the 285 nm absorption band has been used (Krstulovic and Powell, 1979) but although rather high sensitivities have been reported with the use of other excitation/emission wavelength combinations (Schüsler-van Hees et al., 1979; Jackman et al., 1980), this technique has not yet found real use in bioanalytical work. Stopped-flow fluorescence scanning techniques has also been used for identification purposes (Krstulovic and Powell, 1979).

An increased sensitivity will result if specific derivatization techniques are used (cf. Table 4). The chemistry involved in these derivatization reactions is outlined in Scheme 4. Of these derivatives, all except THIs have been used in the precolumn mode.

Precolumn derivatization techniques:

1. OPA (according to Mell et al., 1978):

The derivatizing reagent is prepared by dissolving 160 mg of o-phthalaldehyde in 3 ml of ethanol and 0.2 ml of 2-mercaptoethanol. This solution is added to 100 ml of 0.4 M boric acid, adjusted to pH 9.5 with potassium hydroxide.

A 100 μl volume of effluent from an alumina extraction of CA's is then reacted with a 100 μl aliquot of the derivatization reagent for two minutes at room temperature. DHBA is used as an internal standard for determination of recovery.

The OPA-derivatives may be separated on a μBondapak C18 column with methanol/80 mM acetic acid (1/1 by vol.), pH 2.9, and detected at 340/418 nm.

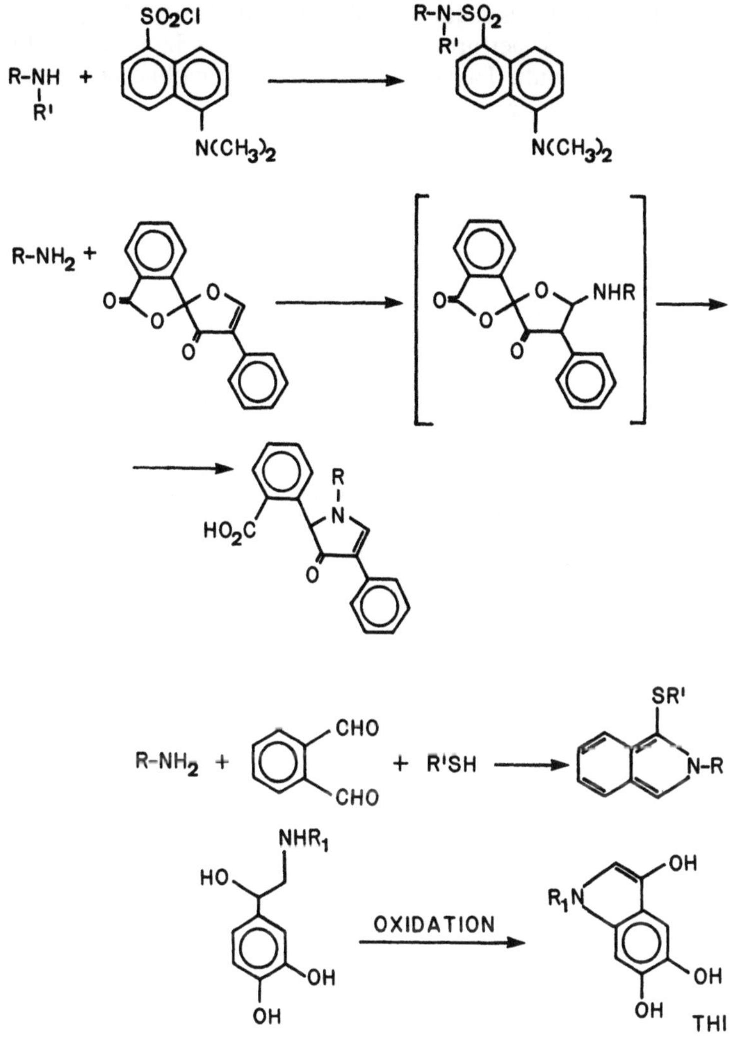

Scheme 4. Commonly employed reactions for the transformation of catecholamines into fluorescent derivatives. a) DNS, b) FA, c) OPA, d) THI

2. FA (according to Imai et al., 1977):

After extraction of CA's onto alumina (0.4 g), desorption is effected with 2 ml of 0.4 M acetic acid in methanol. The eluate is evaporated at 30° C under reduced pressure, dried in a desiccator and the residue dissolved in 70 μl of a 0.1 M sodium borate buffer (pH 8.5) containing EDTA (10 μg/ml) and ascorbic acid (20 μg/ml). Then, 30 μl of a 0.02% solution of fluorescamine in acetone is immediately added. DHBA is used as an internal standard for determination of recovery.

The FA-derivatives require a polar bonded phase of the glycol-type to be well separated and best performance was obtained (at 52° C) with a 10 μ TSK gel 160 (Toyo Soda Manuf., Tokyo) combined with a mobile phase consisting of tris-HCl/methanol (3/1 by vol.), pH 8.0.

3. DNS (according to Frei et al., 1978):

(The method was applied to plasma spiked with epinephrine).

To 1 ml plasma sample 1.2 ml 0.01 M hydrochloric acid and 18 ml dansyl chloride solution (1.87 mg/ml acetone) are added. The pH is adjusted to 8.5 by means of a sodium carbonate solution. The reaction is carried out at 40° C for 20 min. and the acetone is removed in a rotary evaporator. The excess of reagent is hydrolyzed by adding 2 drops of 2 M sodium hydroxide solution and 1 ml of a solution of sodium chloride. Extraction is carried out with benzene (2 × 30 ml) which is evaporated to dryness. The residue is taken up in the mobile phase for HPLC.

For postcolumn derivatization a reactor placed between the column outlet and the detector is used. In this case the reagent, which must not be fluorescent in itself, is intimately mixed with the column eluate before entering the reactor (usually a PFTE coil or a glass bead-packed column). For CA- determination two different techniques are mainly used, viz. the OPA and THI reactions. Reactions with ethylene diamine (Seki, 1978) or alkaline borate buffer (Nimura et al., 1980) have also been employed giving fluorescent products of unknown structure. The latter techniques are not as sensitive as the two first mentioned. Postcolumn techniques involving air segmented reactors are suitable for automatization and fully automated HPLC-systems based on the THI-method were recently described (Mori, 1981; Yamatodani and Wada, 1981). Fig. 1 shows the principle of the automated CA-analyzer described in Mori's paper as well as an example of a chromatogram of plasma NE and E.

c) Electrochemical Detection

An electrochemical detector can be regarded as a type of post-column reactor because the compounds in the column effluent are partly or fully transformed into reaction products on detection. Therefore, the detector response is also highly dependent upon the reaction conditions in the cell, like the electrode potential, mobile phase composition, pH, ionic strength etc.

Catechol-derived compounds are ideally suited for electrochemical detection in the anodic oxidation mode because of the ease by which the catechol ring system will lose two electrons to form, primarily, an o-quinoiod structure. EC detectors in HPLC are therefore more generally applicable in the catecholamine (and indoleamine) metabolism field as compared with detection based on fluorescence labelling techniques (cf. Table 4).

Various types of detector cells are presently in use and the field is covered by very active research. New glassy carbon cells with improved electrode performance have recently been constructed (Hoogvliet et al., 1983) as well as cells

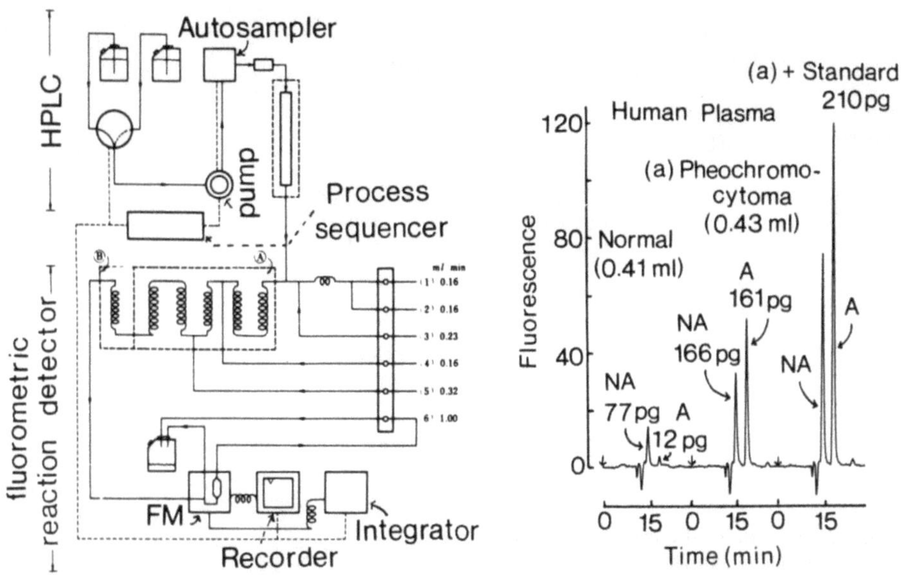

Fig. 1. a) Principle of a microcomputerized automated catecholamine analyzer based on postcolumn THI-derivatization. Reagents: 1 = phosphate buffer (1 M); 2 = potassium hexacyanoferrate(III) (0.1%); 3 = air; 4 = ascorbic acid (0.1%), 5 = sodium hydroxide (4 M). FM = fluoromonitor. b) Example of total system performance. (From: Mori, 1981. Reproduced with permission from the author and the publisher)

housing two electrodes working either in a parallel or in a consecutive configuration, thus improving the selectivity of detection (Mayer & Shoup, 1983). While most of the analytical work until now has been performed with the use of amperometric cells housing impermeable carbon electrodes of the thin-layer or wall-jet type at which only a fraction of the analyte is electrochemically converted, recent advances in electrode technology have now produced cells containing permeable, porous carbon electrodes capable of 100% analyte conversion (Fig. 2).

d) Other Methods of Detection

While most of the techniques for CA-determination based on radioenzymatic (RE) labelling employ TLC for the separation step, a variant of the RE method makes use of HPLC. After collection of the HPLC-separated ^{3}H-methylated CAs, these are oxidized and determined by liquid scintillation counting. Müller et al. (1979), using a reversed-phase HPLC system, claims a detection limit of ca. 1 pg.

Chemiluminescence, based on chemical excitation of a fluorophore, is still a relatively unexplored area of luminescence detection possibilities. DNS-derivatives of amines and amino acids react with 1,2-dioxethane-dione under light emission at 495 nm (Kobayashi and Imai, 1980). An HPLC method for the assay

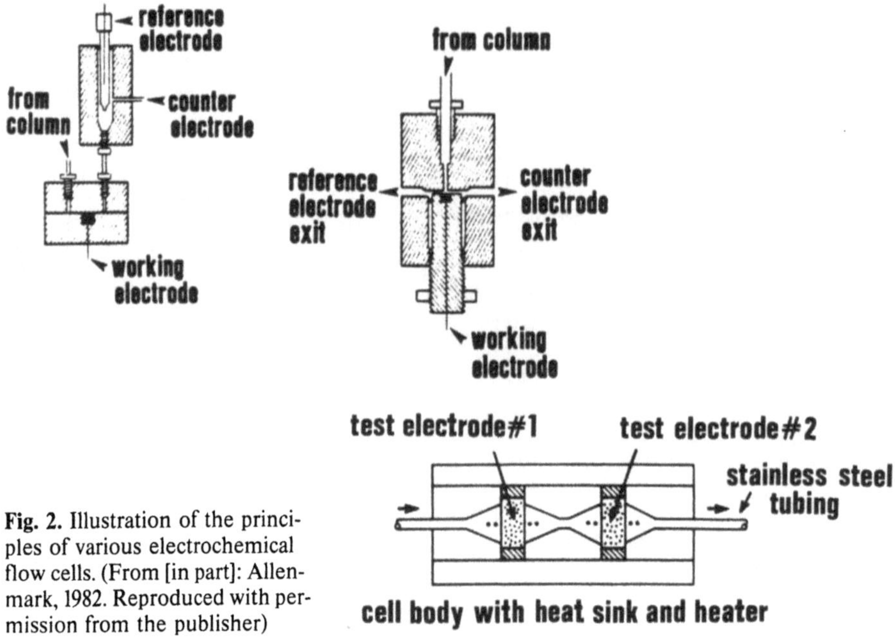

Fig. 2. Illustration of the principles of various electrochemical flow cells. (From [in part]: Allenmark, 1982. Reproduced with permission from the publisher)

of urinary CAs, based on this technique and with a detection limit of 5–10 pg, was recently described by Kobayashi et al. (1981).

A summary of estimated limits of detection for CAs with the most common methods for column effluent analysis is given in Table 5.

Problems Associated with Quantitative Analysis at Low Concentration Levels

It is essential to recognize that a high sensitivity of a method, i.e. a low limit of detection, does not imply a high precision or accuracy. Particularly when derivatization techniques and/or multistep work-up procedures are used, recoveries may differ significantly between the internal standard and an analyte as well as between the analytes of interest. Still worse, inter-assay recoveries may be

Table 5. Estimated Limits of Detection for CA's in HPLC with Different Methods for Column Effluent Analysis. (From: Yui and Kawai, 1981. Reproduced with permission from the authors and the publisher)

Method	Limits of detection (pg)		
	DA	NE	E
1. Ultraviolet	1250	1400	1200
2. Native fluorescence	300	300	300
3. OPA (postcolumn derivatization)	130	75	-
4. THI	800	1	1
5. Electrochemical	25	25	25

significantly different. In fact, in many cases when samples of low concentration are analyzed, inter-assay recovery variations of a given analyte are found to be greater than the intra-assay differences between the individual analytes. Usually such problems, which are amplified in trace analysis, will be associated to a large extent with the sample preparation technique. Among the factors of importance to control are decomposition during storage, contamination produced by unclean glassware, losses due to adsorption to surfaces, etc.

A method used for quantitatie analysis is usually validated in terms of relative standard deviation or coefficient of variation for a series of repeated analycing the same sample, which will give the over-all precision of the method. Inaccurate results may be produced, however, by large systematic errors even though the precision of a method is high. Therefore, it is very valuable if the same material can be repeatedly analyzed by means of two or more quite different methods. Results from an inter-laboratory investigation on plasma CAs published recently (Hjemdahl, 1984) reveal the complexity of the field as well as the methodological difficulties still present and the great care that has to be exerted in the interpretation of results.

Finally, what concerns the use of HPLC in trace organic analysis, a common problem, is often to increase the signal-to-noise ratio by decreasing the background noise level. To do so, it is often necessary to lower the background current of the detector. Very great demands are therefore placed on the chromatographic system with respect to solvent purity, stability of the stationary phase, etc. as well as constant flow rate, electronic shielding etc. It may also be advantageous to keep the HPLC-system stabilized by continuous mobile phase recycling with the detector on over long periods of time.

Selected Methods for Quantitation of Catecholamines and Metabolites in Biological Material by HPLC

Introduction

In the following some recent methods for determination of CAs or CA-metabolites in various biological material by HPLC are presented in experimental detail. The selection has been somewhat arbitrary; however, alternative methods are given in brief comments. Most methods given have been evaluated in long-term runs and found clinically important.

In Urine

a) Catecholamines

The CAs are to a large extent present in conjugated form in human urine. From experiments with L-DOPA administration it is now known that urinary conjugated DA exists mainly as DA-3-O-sulphate with the 4-O-isomer present as a minor component (Arakawa et al., 1979). By the use of a simple hydrolysis procedure

Table 6. Free and Conjugated Catecholamines in Human Urine Determined by HPLC. (From: Elchisak and Carlson, 1982. Reprinted with permission of the authors and the publisher)

	Excretion rate (nmoles/day) Free		Conjugated		Total		Percentage conjugated	
NE	653	(230)	727	(418)	1380	(623)	50.4	(7.4)
E	421	(139)	693	(594)	1114	(703)	55.9	(14)
DA	1161	(500)	2671	(1571)	3832	(2025)	66.8	(7.7)

n = 6, S.D. in parenthesis

prior to the extraction of the free CAs with alumina, it is possible to estimate the degree of conjugation (Elchisak and Carlson, 1982). In this way it was verified that all CAs are present to more than 50% in conjugated form in human urine (Table 6). The sample processing procedure will be dependent upon the HPLC-technique used. Mori (1981), using a SCX-column, postcolumn THI-derivatization and fluorimetric detection in an automated system, determined all three CAs in hydrolyzed and unhydrolyzed urine after a single alumina extraction step. Riggin and Kissinger (1977) and Anderson et al. (1981) used a cation-exchange column extraction procedure prior to the alumina extraction in order to eliminate interfering acid metabolites from the sample. The following procedure has been used in combination with RP-IP-HPLC and native fluorescence detection as well as ECD.

Method used by Anderson et al. (1981):

The over-all recovery of DHBA used as the internal standard was reported to be 38.5%, with similar values for the CA's. The between-day CV:s range between 6.6 and 9.8%.

Flow-sheet 1. Determination of free CAs (DA, NE and E) in urine:

To a 20 ml urine aliquot is added 40–100 μl of a DHBA-solution (100 g/l), 100 μl of sodium metabisulfite (5 g/l) and 100 μl of EDTA (100 g/l).

|

Mix 5 ml of this urine sample with 15 ml of a 0.05 M phosphate buffer (containing 10 g EDTA/l) pH 7.0 and adjust the final pH to within 6.5–7.0.

|

Pour the solution onto a Bio-Rad Bio-Rex 70 CA-isolation column previously equilibrated with 0.1 M potassium acetate buffer pH 6.5 Add 10 ml of distilled water followed by 1.5 ml 0.7 M sulfuric acid; discard the eluate.

|

Elute the CAs into a 5 ml test tube with 4 ml of 2 M ammonium sulfate.

|

Add 100 μl of the metabisulfite solution and 0.5 ml of tris-HCl buffer (3 M, pH 8.6), adjust the pH to 8.4–8.7, add 50 mg of acid-washed alumina and shake for 10 min. After washing the alumina twice with ca. 2 ml of distilled water, the CA's are eluted with 250 μl of 1 M acetic acid (5 min's shaking followed by collection of the supernatant).

|

Inject 20–50 μl into the HPLC system.

|

HPLC: Solvent delivery system: Constant flow solvent pump. Injector: Injection valve with a 20 or 50 μl loop. Column: 300 × 3.9 mm μBondapak-C18. Detector: a) Flowcell-equipped fluorimeter (Perkin-Elmer mod. 650-10 LC) operating at 285/305 nm, b) thin-layer flow-cell with a glassy carbon anode operating at + 0.60 V vs. an Ag/AgCl reference electrode. Mobile phase: 0.01 M sodium phosphate buffer/methanol (95/5) of pH 4.0, containing sodium octyl sulfonate (100 mg/l) and EDTA (50 mg/l). Flow rate: 2.0 ml/min.

An alternative sample pre-treatment procedure is given by Speek et al. (1983) leading to essentially complete recovery and better between-assay coefficients of variation (2.1–5.0%). In this procedure urine, buffered to pH 7 is directly applied to a boric acid column (Affi-Gel 601, Bio-Rad, Inc.). After a washing step, the CAs are eluted with the mobile phase used for HPLC.

The determination of total urinary CAs is carried out via acid hydrolysis of the urine sample. A recently described sample pre-treatment technique is described below (Schleicher et al., 1983):

Flow-sheet 2. Work-up procedure prior to HPLC for the determination of total CAs in urine:

Urine is collected over a 24 h period in a plastic container with 20 ml of 6 M hydrochloric acid

|

Acidify 5 ml of the urine to pH 0.5–1.0 with 6 M hydrochloric acid

|

Boil this sample on a water bath for 20 min

|

Adjust the pH to 6.5 with 0.5 M sodium hydroxide and apply to a weak cation-exchange column (Bio-Rad, Inc.)

|

Wash twice with 10 ml of distilled water

Elute the CAs with 8 ml of a boric acid solution (40 g/l)

b) Metabolites

The determination of the major CA-metabolites (VMA, HVA, MHPG) in urine generally represents (at least from the point of view of sensitivity) a less demanding analytical problem since the concentration levels are higher than those of the parent amines. Relatively simple extraction methods have been shown to be sufficient for sample purification prior to HPLC, provided an electrochemical detection technique is used.

1. VMA:

This urinary metabolite of NE and E has been used (together with NMN and MN) as a diagnostic marker for neural crest tumors (Krstulovic et al., 1979). The method given below is taken from the latter publication. A recovery of 95–99% is claimed.

Flow-sheet 3. Determination of VMA in urine:

Urine (3 ml) is acidified to pH 1 with 6 M hydrochloric acid and extracted with three portions of ethyl acetate (6, 3 and 3 ml)

|

Pool the extracts and evaporate to dryness under dry nitrogen

|

Redissolve the residue in 1.0 ml of distilled water

|

Inject 5 μl directly into the HPLC system.

|

HPLC: Solvent delivery system: Dual constant flow solvent pump system equipped with a solvent programmer. Injector: Injection valve with a 5 or 10 μl loop. Column: μ-Bondapak-C-18 (10 μ). Detector: Thin-layer electrochemical cell with a glassy carbon anode operating at +1.00 V vs. an Ag/AgCl-reference. Mobile phase: A 45 min linear gradient from 0 to 60% of high-strength eluent. Low-strength eluent: 0.1 M potassium dihydrogen phosphate (pH 2.50); high-strength eluent: acetonitrile-water (3:2). Flow rate 1.4 ml/min.

2. HVA:

In addition to the method given under 1. by which also HVA is monitored, a very simple method has been described by Morrisey and Shibabi (1979), who used RP-HPLC-EC and direct injection of untreated urine. Between-assay CV was reported to 7.6%.

HPLC: Solvent delivery system: Constant flow solvent pump. Injector: Injection valve with a 20 μl loop. Column: 250 × 4.6 mm Lichrosorb RP 18 (10 μ). Detector: Thin-layer transducer with a glassy carbon anode operated at +1.2 V (vs. an Ag/AgCl-reference). Mobile phase: Phosphate buffer (60 mM) – methanol (77:23), pH 3.0.

3. MHPG:

This metabolite of NE exists in urine as free MHPG, as a sulfate and as a β-glucuronide. The conjugated forms can be enzymatically hydrolyzed by an arylsulfatase and a β-glucuronidase, respectively. For this purpose enzymes from Aerobacter aerogenes (sulfatase) and from Helix pomatia (glusulase, containing both the sulfatase and the β-glucuronidase) have been used. In this way all existing forms of MHPG may be quantitated. Iso-MHPG has been used as an internal standard.

Method used by Santagostino et al. (1982):

Free MHPG is extracted (from enzymatic hydrolysates) with ethyl acetate – borate buffer – ethyl acetate. Between assay CVs are 3%.

Flow-sheet 4. Determination of free and conjugated MHPG in urine:

To urine (24 hr sample), collected in a glass bottle, is added 0.5 mg sodium metabisulphite/ml

|

From a 20 ml aliquot (stored at −20° C) is taken samples according to the following:

Sample 1 (free MHPG)	Sample 2 (free MHPG + sulfoconjugate)	Sample 3 (total MHPG)
0.5 ml of urine + 0.2 ml 1 M sodium acetate, pH 6.5 + 0.1 ml EDTA (2%)	As sample 1 + 25 μl of arylsulfatase (31 units/ml)	As sample 1 + 10 mg of glusulase (19.5 and 420 units/mg, respectively)

Adjust the pH of the samples to 5.2

Add 100 μl of iso-MHPG (20 ng/μl)/sample

Incubate at 37° C for 20 h

Add ca. 0.7 ml of 1.0 M sodium acetate buffer to pH 6.5 and saturate with sodium chloride

Extract each sample with 3 × 2 ml of ethyl acetate

Reextract each ethyl acetate pool with 2 × 0.5 ml of 0.1 M boric acid

Wash each boric acid pool with 2 × 2 ml of diethyl ether and adjust the pH to 8.7

Extract with 2 × 2 ml of ethyl acetate and evaporate to dryness

Dissolve each residue in 500 μl of distilled water

Inject 20 μl into the HPLC

HPLC: Solvent delivery system: Constant flow solvent pump. Injector: Injection valve with a 20 μl loop. Column: 3.9 × 300 mm μBondapak C-18 (10 μ). Detector: Thin-layer transducer with a graphite paste anode (CPO, BAS) operating at +0.8 V (vs. an Ag/AgCl-reference). Mobile phase: 9 mM citric acid, 89 mM sodium acetate, 10% methanol, pH 5.1.

Fig. 3 shows the increase in MHPG obtained with enzymatic deconjugation.

In Blood (Plasma or Serum)

a) Catecholamines

As the concentrations of CAs in plasma or serum of healthy individuals at rest are extremely low (< 2 pmol/ml), very high demands are placed on the operation

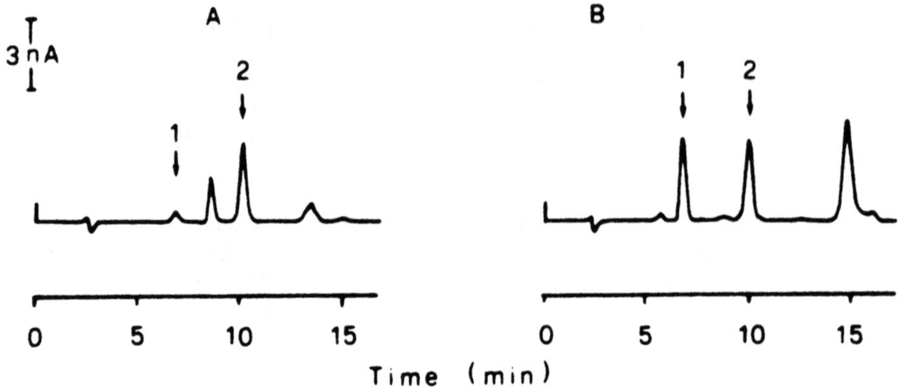

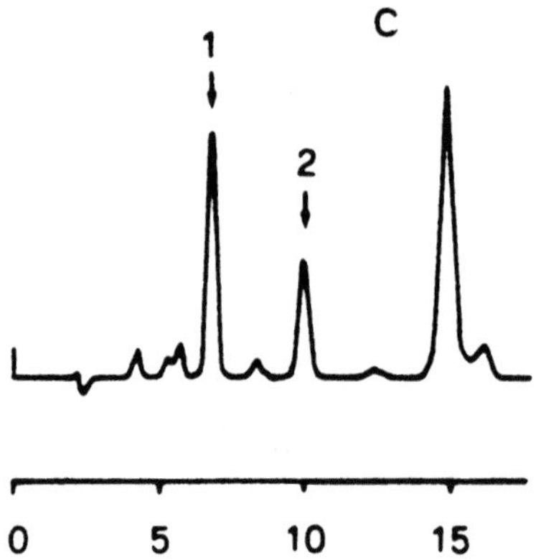

Fig. 3. Typical chromatograms of MHPG extracted from **a)** untreated urine (free MHPG, 0.19 ng/μl), **b)** urine hydrolyzed with arylsulfatase (free MHPG and sulfated MHPG, 1.61 ng/μl), **c)** urine hydrolyzed with glusulase (total MHPG, 3.46 ng/μl). Peaks: 1 = MHPG, 2 = iso-MHPG. (From: Santagostino et al., 1982. Reproduced with permission from the author and the publisher)

of the total HPLC-system as well as on the performance of the extraction procedure for sample preparation. Elevated CA-levels associated with many pathological conditions are readily determined, however.

Method used by Allenmark, Hedman and Söderberg (1980):

The basic analytical principle is SXC-HPLC-ECD. The sample preparation procedure is based on extraction onto acid-washed alumina from deproteinized plasma at pH 8.6. Essentially the same method was used by Hjemdahl et al. (1979) and by Allenmark and Hedman (1979). Previous LC-ECD methods utilizing glass columns were described by Kissinger et al. (1973), Refshauge et al. (1974) and by Hallman et al. (1978).

Flow-sheet 5. Determination of CAs in plasma (or serum).

Centrifugation (ca. 2000 g, 10 min) of blood (10–20 ml) collected in heparinized Vacutainer tubes.

|

Plasma (clear) is spiked with DHBA (0.5 μl 10 μM soln./ml plasma) and sodium metabisulphite (75 μl 2.5 mM soln./ml plasma).

|

6.0 ml of this plasma is diluted with tris-HCl buffer of pH 8.6 (2.25 ml 1 M soln. containing 20 mg EDTA/ml) and water (4 ml).

|

Add acid-washed alumina (50 mg), stir for 15 min, collect the alumina on glass filter (use PFTE-stoppered 100 × 10 mm i.d. siliconized glass filter tubes).

|

Wash the alumina with four 2 ml portions of water. Suck the alumina as dry as possible.

|

Desorb the CAs by shaking with PCA (500 μl, 0.5 M soln.) for 15 min.

|

Inject 100 μl immediately.

|

HPLC: Solvent delivery system: Constant-flow solvent pump in combination with an external pulse-dampening device. Injector: Injection valve with 100 μl loop. Column: 200 × 4.6 mm i.d. stainless steel column packed with Nucleosil SA 5 μ. Detector: Electrochemical cell of thin-layer type with a carbon paste (CPO) anode, operating at +0.65 V vs. an Ag/AgCl reference electrode. Injector, column and detector parts shielded by a Faraday cage. Mobile phase: Citrate-acetate buffer pH 5.2 (prepared according to Keller et al. [1976]; see also Allenmark and Hedman [1979]). Flow rate 0.6 ml/min.

The drastic change in human plasma CA-levels under certain states of disease is readily found from chromatograms obtained by the method given (Fig. 4).

By the use of reversed-phase columns in the "dynamic cation-exchange" mode, i.e. together with an anionic surfactant, there has sometimes been difficulties in obtaining sufficient retention of NE to avoid chromatographic overlap with uric acid (Davis et al., 1981). Reversed-phase systems have recently been very successfully used, however; also combined with electrochemical detector cells housing "glass carbon" anode material of high solvent and mechanical resistance (Davies and Molyneux, 1982).

The alternative detection method in HPLC of plasma CAs is the THI postcolumn reaction with fluorimetric detection.

b) Metabolites

Plasma VMA and MHPG can be determined simultaneously by HPLC-ECD after an isolation procedure involving gel filtration and extraction with ethyl acetate.

Method used by Ong et al. (1982):

Plasma is deproteinized on a Bio-Gel P-10 column (made from a 10 ml disposable syringe containing 5 ml of a gel prepared from 1 g of dry powder) connected to a peristaltic pump and a fraction collector . Hydroquinone (HQ) is used as the internal standard. The acid metabolites are extracted at pH 3 from pooled fractions. The average total recoveries are 59 and 53% for VMA and MHPG, respectively.

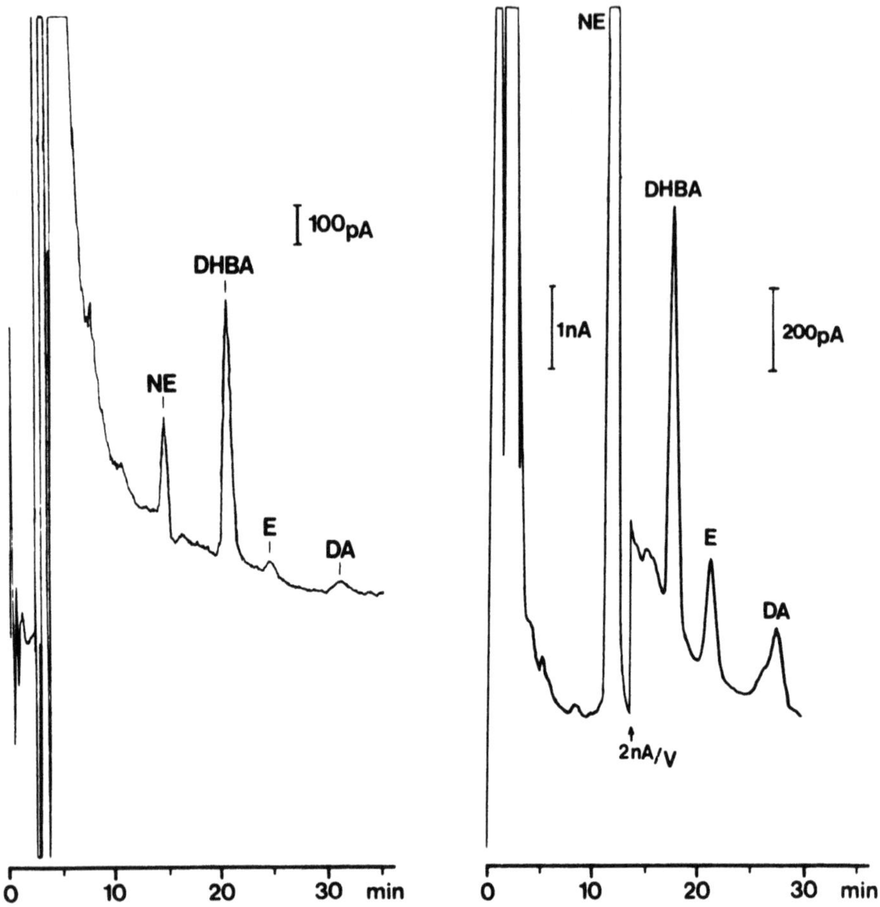

Fig. 4 a, b. Elevated plasma CA-levels due to a pheochromocytoma. **a)** Normal subject, **b)** Patient with a pheochromocytoma. (From: Allenmark, 1982. Reproduced with permission from the publisher)

Flow-sheet 6. Determination of VMA and MHPG in plasma:

Apply 1 ml of plasma mixed with 10 μl of a solution of HQ (0.8 mg/l) to the Bio-Gel column.

|

Having finished the usual sample application procedure, elute with the buffer used for column equilibration (0.05 M sodium phosphate, pH 6.0, containing 2.7 mM EDTA and 63 μM ascorbic acid) at a flow rate of 30 ml/cm 2/h. Collect 0.5 ml fractions. Discard fractions 1–8 and pool fractions 9–15 (3.5 ml).

|

Place the pooled fractions in a 20 ml screw-cap test tube, add 2 g of sodium chloride and adjust the pH to 3 with phosphoric acid (30% by vol.).

|

Extract (5 min) with ethyl acetate (10 ml), centrifuge (5 min, 2000 g), collect the upper (organic) layer. Repeat the extraction procedure twice with 5 ml portions of ethyl acetate. Pool the organic phases and evaporate to dryness (vacuum, 30° C). Rinse the tube with 1.5 ml of methanol and reevaporate.

|

Dissolve the residue in 150 μl of the mobile phase used for HPLC. Inject 100 μl

|

HPLC: Solvent delivery system: Constant flow solvent pump. Injector: Injection valve with 100 μl loop. Column: 250 × 3 mm i.d. SS-column packed with Spherisorb ODS 5 μ, protected by a small guard column (MPLC, Brownlee). Detector: Thin-layer flow-through electrochemical cell with a glassy carbon anode operating at +0.80 V vs. an Ag/AgCl reference. Mobile phase: Phosphate buffer (0.07 M)/methanol (97/3 by vol.) containing 0.01% EDTA and pH-adjusted (phosphoric acid) to 2.5. Flow rate 1.2 ml/min.

An alternative method for the determination of MHPG was recently described by Scheinin et al. (1983a).

The determination of free plasma HVA has been achieved by direct injection of deproteinized plasma into a RP-HPLC-EC-system using vanillic acid (VA) as an internal standard. Method used by Javaid et al. (1983):

Flow-sheet 7. Determination of HVA in plasma:

Add 100 ng of VA to a 1 ml cold plasma sample. Precipitate the proteins by adding 35 μl of concentrated PCA.

|

Centrifuge the sample

|

Transfer the supernatant to a clean glass vial and inject 50–100 μl

|

HPLC: Solvent delivery system: Constant flow solvent pump. Injector: Injection valve with a 50 or 100 μl loop. Column: 250 × 4.5 mm i.d. packed with octylsilica (Analytical C-8, 10 μ, Perkin-Elmer). Detector: Thin-layer flow-through electrochemical cell with a glassy carbon anode operating at +0.95 V vs. Ag/AgCl. Mobile phase: 0.1 m sodium acetate, pH 4.5 with EDTA (1 mM) and 10% (v/v) of methanol. Flow rate: 1.0 ml/min.

In Brain Tissue

Numerous neurochemical studies have been performed using RP-IP-HPLC since the demonstration of the utility of the technique for studies of brain CA's (Felice et al., 1978). Most of these have involved a selective extraction onto alumina, permitting chromatographic determination of NE and DA. Recently, it was demonstrated that PCA-extracts of brain tissue may be used for direct injection, permitting studies of acid metabolites as well (Magnusson et al., 1980). The method given below is taken from their paper.

Flow-sheet 8. Determination of DA, DOPAC and HVA in brain tissue:
The brain tissue (rat) is immediately dissected on ice.

|

An appropriate tissue sample (4–50 mg) is placed in a test tube with a solution (300–500 μl) consisting of 0.1 m perchloric acid, 4×10^{-5} M sodium bisulphite and epinine (2.5×10^{-6} M; int. std.).

|

Sonicate (200 W, 5 sec) and centrifuge the homogenate (8000 × g, 2–3 min).

|

Inject 20 μl of the supernatant into the HPLC.

|

HPLC: Solvent delivery system: Constant flow solvent pump. Injector: Injection valve with a 20 μl loop. Column: 100 × 3 mm i.d. glass-lined steel column (SGE) packed with Nucleosil C-18 5 μ. Detector: Thin-layer electrochemical cell with a graphite paste (CPO, BAS) anode operating at +0.60 V (vs. an Ag/AgCl-reference). Mobile phase: Citrate buffer (ionic strength 0.1, pH 4.25) - methanol (92:8), containing hexyl sulfate (1.7 mM).

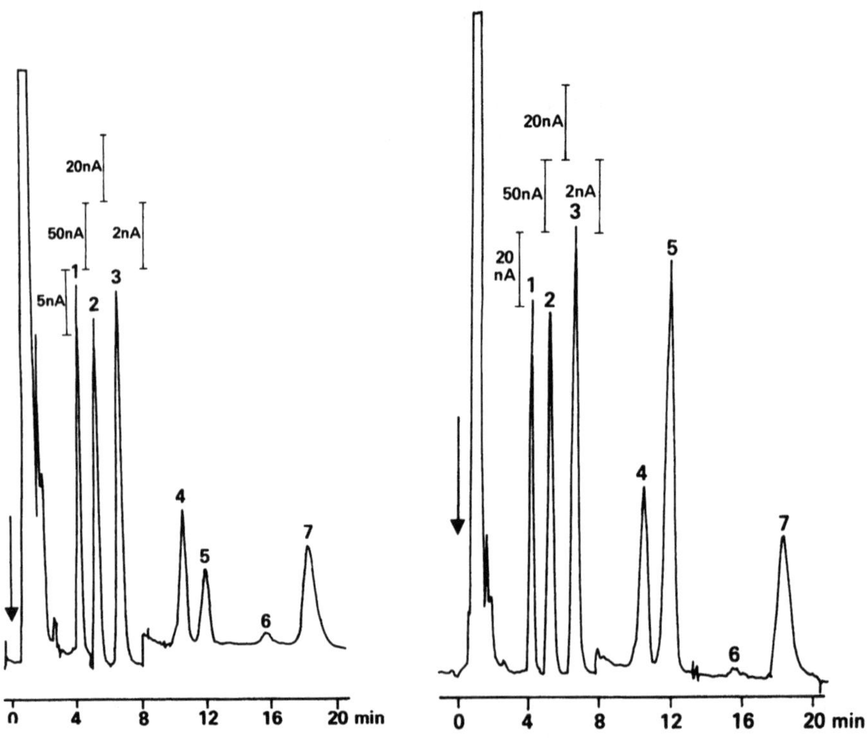

Fig. 5. Illustration of the effect of haloperidol on the levels of DOPAC and HVA in rat brain striatum. Peaks: 1 = DOPAC, 2 = DA, 3 = epinine (int. std.), 4 = 5-HIAA, 5 = HVA, 6 = MT, 7 = 5-HT. **a)** No treatment: 1, 2, 3 and 4 correspond to 7.5, 99, 50 and 7.8 pmol, respectively; **b)** 2 h after haloperidol treatment: 1, 2, 3 and 5 correspond to 22, 67, 50 and 22 pmol, respectively. (From: Magnusson et al., 1980. Reproduced with permission from the authors and the publisher)

Fig. 5 shows some typical chromatograms obtained with the technique.

Recently, a method was described permitting simultaneous observation of DOPEG, DOPAC and all three CA's in rat brain tissue (Howes et al., 1983). The authors claim a detection limit (S/N = 2) of < 5 pg for each compound and recoveries of 50–75%. An alumina extraction procedure is used. The method is given briefly below.

Flow-sheet 9: Determination of DOPEG, DOPAC and CAs in brain tissue:
Tissue pieces are homogenized in 10 vol. of cold 0.1 M PCA containing 0.05% EDTA, 0.05% sodium metabisulfite and DHBA (int. std., 100 ng/ml)

|

Centrifuge the homogenate at 8000 × g for 8 min

|

Mix the supernatant (150–500 μl) with 50 mg acid-washed alumina and an equal volume of 1 M Tris-buffer (pH 8.6) in the upper compartment of a microfilter tube (MF-1, Bioanalytical Systems Inc.) on a rotary mixer for 15 min. After centrifugation (1000 × g, 5 min) the alumina in the filter tube is washed with 3 × 0.5 ml of 0.1 M Tris-buffer (pH 8.6).

|

Elute the catechols with 0.5 ml of 0.1 M PCA containing 0.5% EDTA.

|

Inject a 150 μl aliquot into the HPLC system.

|

HPLC: Solvent delivery system: Constant flow solvent pump (Waters M 45). Injector: WISP 701 automatic sample injector (Waters). Column: μBondapak C-18 (300 mm length) with small precolumn. Detector: Thin-layer electrochemical cell (TL-3, BAS) with a carbon paste anode (+0.70 V vs Ag/AgCl) combined with an LC-4 (BAS) amperometric detector unit. Mobile phase: Sodium dihydrogen phosphate (pH 3.8) with 10% methanol, 2 mM heptane sulfonic acid and 0.05% EDTA. Flow rate: 1.0 ml/min.

In Cerebrospinal Fluid (CSF)

The acid and neutral metabolites of DA and NE present in CSF, i.e. specifically DOPAC, HVA, VMA and free and conjugated MHPG, are readily accessible for HPLC after a simple ethyl acetate extraction procedure from acidified CSF. As the levels of these metabolites are related to central CA turnover and human lumbar CSF is readily available, such determinations are of great clinical interest.

Method used by Krstulovic et al. (1982):

Because recoveries between 94.5 and 98% are reported in this procedure, no internal standard was used. The between-assay CV's were around 1.1%.

Flow-sheet 10. Determination of HVA, VMA, DOPAC and (free) MHPG in CSF:
Human lumbar CSF is collected and stored at −20° C.

|

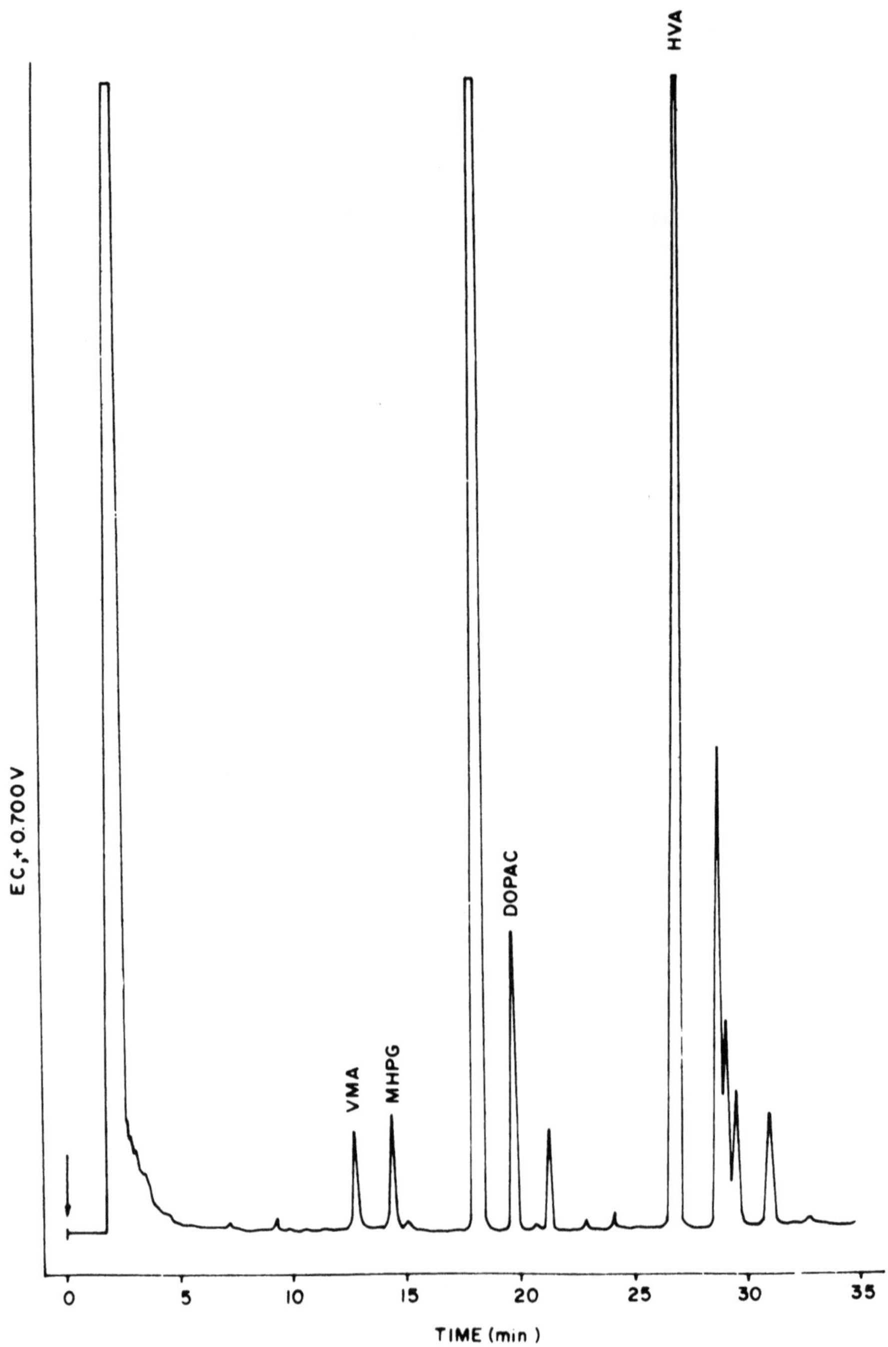

Fig. 6. Chromatogram of the ethyl acetate extract of a sample of human lumbar CSF (corresponding to a volume of 190 μl CSF). (From: Krstulovic et al. 1982., Reproduced with permission from the authors and the publisher)

An aliquot (0.5–1.0 ml) is acidified with 6 M hydrochloric acid to pH < 1

|

Extract with 2 × 3 ml + 1 × 6 ml of ethyl acetate and pool the extracts

|

Evaporate to dryness under dry nitrogen and reconstitute with distilled water to a volume smaller than the original sample volume

|

Inject 50–100 μl into the HPLC

|

HPLC: Solvent delivery system: Constant flow solvent pumps equipped with a linear gradient programmer. Injector: Injection valve with a 50 or 100 μl loop. Column: 250 × 4.6 mm i.d. Ultrasphere ODS 5 μ. Detector: Electrochemical flow cell of the wall-jet type employing a glassy carbon anode operating at +0.70 V (vs. an Ag/AgCl-reference). Mobile phase: A 35 min gradient from 0 to 100% of the high-strength eluent. Low-strength eluent: 0.1 M potassium dihydrogen phosphate, pH 2.50. High-strength eluent: Methanol-water (3:2). Flow rate: 1.2 ml/min.

The appearence of the metabolites in a chromatogram of an extract of CSF is shown in Fig. 6.

By pH-adjustment to 5.2 of the aqueous layer obtained from the extraction with ethyl acetate, followed by incubation with sulfatase for 16 h, acidification and re-extraction with ethyl acetate, the sulfoconjugate of MHPG may be determined as well (Krstulovic et al., 1981).

An alternative method for the determination of MHPG in CSF has been given by Scheinin et al. (1983b).

In Heart Tissue

In connection with studies of ischemia-induced changes of catecholamine levels in rat myocardium (Holmgren et al., 1983), a procedure for the isolation of CA's from heart tissue as well as their quantitation by RP-IP-HPLC was developed.

Flow-sheet 11: Determination of CAs in heart tissue:

Frozen tissue is cut into small pieces, placed in a tissue grinder and weighed per 0.1 g of tissue and the contents in homogenized

|

Centrifuge the homogenate (4° C, 10 min, 25000 × g)

|

The supernatant (1.0 ml) is transferred to a 4 ml centrifuge tube with the addition of antioxidant (GSH, 0.05 m, 50 μl), EDTA (0.3 M, 50 μl) and alumina (20 mg)

|

Add 1.0 ml of tris-buffer (1 M, pH 8.6) to the tube and rotate the tube for 10–30 min

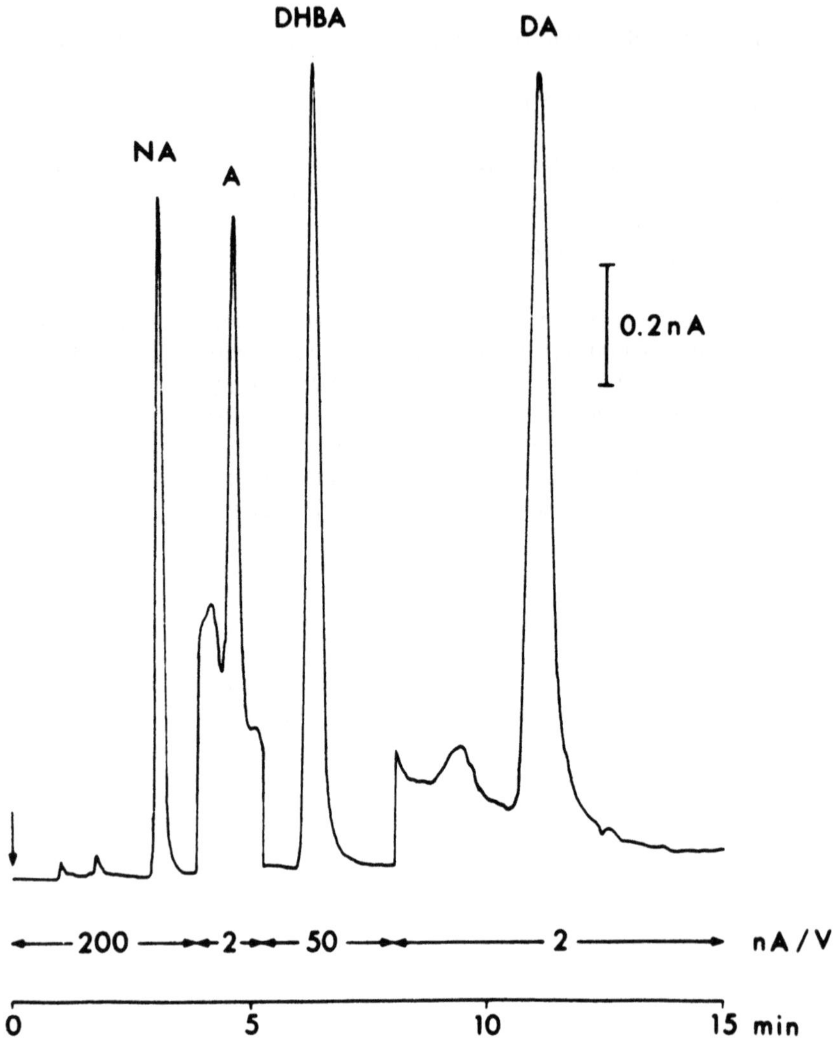

Fig. 7. Chromatogram of CAs obtained from alumina extraction of 100 mg or rat heart tissue. NA (NE) = 608 pmol, A (E) = 5.2 pmol, DHBA (int. std.) = 250 pmol, DA = 11.6 pmol. (From: Eriksson et al., 1982. Reproduced with permission from the authors and the publisher)

|

Discard the supernatant and wash the alumina briefly three times with an EDTA solution (3 mM, pH 7)

|

Remove any excess of liquid by centrifugation

|

Elute the CA's by vortexing for 1 min with 150 μl of PCA (0.2 M)

|

Inject 50 μl into the HPLC

|

HPLC: Solvent delivery system: Constant flow solvent pump equipped with extra pulse dampening devices. Injector: Injection valve with a 100 μl loop. Column: 150 × 4.5 mm i.d. Lichrosorb RP-18 5 μ. Detector: Thin-layer electrochemical flow cell with a carbon paste anode operating at +0.70 V (vs. an Ag/AgCl-reference). Mobile phase: Acetate - citrate buffer (100 mM sodium acetate, 40 mM citric acid, 60 mM sodium hydroxide) of pH 5.2 containing 10% of methanol and 3.5-dimethylcyclohexyl sulfate (3 mM) as an ion pairing agent.

Fig. 7 shows a chromatogram obtained from 100 mg of rat heart tissue.

Conclusion

From this survey of HPLC methodology as applied to the problem of determination of CAs and metabolites in biological material, it should be evident that there are now many highly reliable quantitation techniques available. Generally speaking, factors influencing the reproducibility tend to be associated with the sample purification procedure and not with the HPLC step. There is a definite trend towards the use of the analytes in underivatized form for HPLC (which is actually one of the main advantages of HPLC) and to avoid complicated work-up procedures. Therefore, highly selective detection techniques are favourable, which explains the frequent use of fluorimetry with or without postcolumn derivatization and particularly electrochemical detection. With simple work-up techniques recoveries may be close to 100% and an internal standard may be unnecessary. However, this is generally possible only when relatively high concentrations of analytes are present and/or the biological matrix is not of a too complex nature. Future progress in the field will very likely be associated with the use of microcolumns (Goto et al., 1983), precolumn sample enrichment/column switching techniques (Lachâtre et al., 1982) as well as the development of new highly sensitive and selective detector systems, like those based on chemiluminescence phenomena (Yamaguchi et al., 1982).

References

Allenmark S (1982) Analysis of catecholamines by HPLC. J Liquid Chromatogr 5 (Suppl 1):1–41

Allenmark S, Hedman L (1979) Cation exchange liquid chromatography with amperometric detection as a method for the analysis of endogeneous catecholamine concentrations in plasma or serum. J Liquid Chromatogr 2:277–286

Allenmark S, Hedman L, Söderberg A (1980) Microanalysis of catecholamines in human plasma by high-performance liquid chromatography with amperometric detection as compared with a radioenzymatic method. Microchem J 25:567–575

Anderson GM, Young JG (1981) Applications of liquid chromatographic – fluorometric systems in neurochemistry. Life Sci 28:507–517

Anderson GM, Young JG, Jatlow PI, Cohen DJ (1981) Urinary free catecholamines determined by liquid chromatography – fluorometry. Clin Chem 27:2060–2063

Anton AH, Sayre DF (1962) A study of factors affecting the aluminium oxide - trihydroxyindole procedure for the analysis of catecholamines. J Pharmacol Exp Ther 138:360–374
Anton AH, Sayre DF (1964) The distribution of dopamine and dopa in various animals and a method for their determination in diverse biological material. J Pharmacol Exp Ther 145:326–336
Arakawa Y, Imai K, Tamura Z (1979) High-performance liquid chromatographic determination of dopamine sulfoconjugates in urine after L-DOPA administration. J Chromatogr 162:311–318
Cross AJ, Joseph MH (1981) The concurrent estimation of the major monoamine metabolites in human and non-human primate brain by HPLC with fluorescence and electrochemical detection. Life Sci 28:499–505
Davies CL, Molyneux SG (1982) Routine determination of plasma catecholamines using reversed-phase, ion-pair high-performance liquid chromatography with electrochemical detection. J Chromatogr 231:41–51
Davis TP, Gehrke CW, Gehrke Jr CW, Cunningham TD, Kuo KC, Gerhardt KO, Johnson HD, Williams CH (1978) High-performance liquid-chromatographic separation and fluorescence measurement of biogenic amines in plasma, urine and tissue. Clin Chem 24:1317–1324
Davis GC, Kissinger PT, Shoup RE (1981) Strategies for determination of serum or plasma norepinephrine by reversed-phase liquid chromatography. Anal Chem 53:156–159
Elchisak MA, Carlson JH (1982) Assay of free and conjugated catecholamines by high-performance liquid chromatography with electrochemical detection. J Chromatogr 233:79–88
Endert E (1979) Determination of noradrenaline and adrenaline in plasma by a radioenzymatic assay using high pressure liquid chromatography for separation of the radiochemical products. Clin Chim Acta 96:233–239
Eriksson BM, Persson BA (1982) Determination of catecholamines in rat heart tissue and plasma samples by liquid chromatography with electrochemical detection. J Chromatogr 228:143–154
Felice LJ, Felice JD, Kissinger PT (1978) Determination of catecholamines in rat brain parts by reversed-phase ion-pair liquid chromatography. J Neurochem 31:1461–1465
Frei RW, Thomas M, Frei I (1978) A study of adrenaline-dansyl derivatives for separation by liquid chromatography. Application to blood plasma analysis. J Liquid Chromatogr 1:443–455
Gelijkens CF, de Leenheer AP (1980) Simple method for isolating free urinary catecholamines by boric acid gel chromatography. J Chromatogr 183:78–82
Goto M, Zou G, Ishii D (1983) Current amplification in dual electrochemical detection for micro high-performance liquid chromatography. J Chromatogr 268:157–167
Goto M, Zou G, Ishii D (1983) Determination of catecholamines in human serum by micro high-performance liquid chromatography with micro precolumn and electrochemical detection. J Chromatogr 275:271–281
Hallman H, Farnebo LO, Hamberger B, Jonsson G (1978): A sensitive method for the determination of plasma catecholamines using liquid chromatography with electrochemical detection. Life Sci 23:1049–1052
Higa S, Suzuki T, Hayashi A, Tsuge I, Yamamura Y (1977) Isolation of catecholamines in biological fluids by boric acid gel. Anal Biochem 77:18–24
Hjemdahl P (1984) Inter-laboratory comparison of plasma catecholamine determinations using several different assays. Acta Physiol Scand Suppl 527:43–54
Hjemdahl P, Daleskog M, Kahan T (1979) Determination of plasma catecholamines by high performance liquid chromatography with electrochemical detection: Comparison with a radioenzymatic method. Life Sci 25:131–138
Holly JMP, Makin HLJ (1983) The estimation of catecholamines in human plasma. Anal Biochem 128:257–274
Holmgren S, Abrahamsson T, Almgren O, Eriksson BM (1981) Effect of ischemia on the adrenergic neurons of the rat heart: A fluorescence histochemical and biochemical study. Cardiovasc Res 15:680–689
Hoogvliet JC, van Beezoyen JAG, Hermans-Lokkerbol AJC, van der Poel CJ, van Bennekom WP (1983) Fabrication and characterization of leak-tight glassy carbon electrodes, sealed in glass employing silicon coating, for use in electrochemical detection. J Liquid Chromatogr 6:1849–1861

Howes LG, Summers RJ, Rowe PR, Louis WJ (1983) The simultaneous determination of 3,4-dihydroxyphenylethylene glycol, 3,4-dihydroxyphenylacetic acid and catecholamines in brain tissue by high-performance liquid chromatography with electrochemical detection. Neurosci Lett 38:327–332
Imai K, Tamura Z (1978) Liquid chromatographic determination of urinary dopamine and norepinephrine as fluorescamine derivatives. Clin Chim Acta 85:1–6
Jackman GP, Carson VJ, Bobik A, Skews H (1980) Simple and sensitive procedure for the assay of serotonin and catecholamines in brain by high-performance liquid chromatography using fluorescence detection. J Chromatogr 182:277–284
Javaid JI, Liu TS, Maas JW, Davis JM (1983) Measurement of 3-methoxy-4-hydroxyphenylacetic acid (HVA) in plasma by high-performance liquid chromatography with electrochemical detection (HPLC-ECD). Anal Biochem 135:326–331
Johansson IM (1981) Retention in reversed-phase ion-pair chromatography of amines on alkyl-bonded phases. J Liquid Chromatogr 4:1435–1457
Johansson IM (1981) Retention behaviour of amines in reversed-phase ion-pair chromatography with octyl sulfate and 1-pentanol in the eluent. Acta Pharm Suec 18:9–24
Johansson IM, Wahlund KG, Schill G (1978) Reversed-phase ion-pair chromatography of drugs and related organic compounds. J Chromatogr 149:281–296
Keller R, Oke A, Mefford J, Adams RN (1976) Liquid chromatographic analysis of catecholamines. Routine assay for regional brain mapping. Life Sci 19:995–1004
Kissinger PT, Bruntlett CS, Shoup RS (1981) Neurochemical applications of liquid chromatography with electrochemical detection. Life Sci 28:455–465
Kissinger PT, Refshauge CJ, Dreiling R, Adams RN (1973) An electrochemical detector for liquid chromatography with pictogram sensitivity. Anal Lett 6:465–477
Knox JH, Hartwick RA (1981) Mechanism of ion-pair liquid chromatography of amines, neutrals, zwitterions and acids using anionic hetaerons. J Chromatogr 204:3–21
Knox JH, Jurand J (1976) Separation of catecholamines and their metabolites by adsorption, ion-pair and soap chromatography. J Chromatogr 125:89–101
Kobayashi S, Imai K (1980) Determination of fluorescent compounds by high-performance liquid chromatography with chemiluminescence detection. Anal Chem 52:424–427
Kobayashi S, Sekino J, Honda K, Imai K (1981) Application of high-performance liquid chromatography with a chemiluminescence detection system to determine catecholamines in urine. Anal Biochem 112:99–104
Krstulovic AM (1979) The current state of the art in the analysis of catecholamines. Adv Chromatogr 17:279–309
Krstulovic AM (1982) Investigations of catecholamine metabolism using high-performance chromatography. Analytical methodology and clinical applications. J Chromatogr 229:1–34
Krstulovic AM, Bertani-Dziedzic L, Bauttista-Cerqueira S, Gitlow SE (1982) Simultaneous determination of 4-hydroxy-3-methoxyphenylacetic (homovanillic) acid and other monoamine metabolites in human lumbar cerebrospinal fluid. An improved high-performance liquid chromatographic study with electrochemical detection. J Chromatogr 227:379–389
Krstulovic AM, Bertani-Dziedzic L, Dziedzic SW, Gitlow SE (1981) Quantitative determination of 3-methoxy-4-hydroxyphenylethyleneglycol and its sulfate conjugate in human lumbar cerebrospinal fluid using liquid chromatography with amperometric detection. J Chromatogr 223:305–314
Krstulovic AM, Colin H (1983) Clinical investigations of catecholaminergic systems using liquid chromatography with electrochemical detection. Trends Anal Chem 2:42–49
Krstulovic AM, Powell AM (1979) Use of native fluorescence measurements and stopped-flow scanning technique in the high-performance liquid chromatographic analysis of catecholamines and related compounds. J Chromatogr 171:345–356
Krstulovic AM, Zakaria M, Lohse K, Bertani-Dziedzic L (1979) Diagnosis of neural crest tumors by reversed-phase high-performance liquid chromatographic determination of urinary catecholamine metabolites. J Chromatogr 186:733–748
Lachâtre G, Nicot G, Magnet C, Rocca JL, Merle L, Valette JP (1982) Determination of plasma norepinephrine and epinephrine by high-performance liquid chromatography using a two-column system and an electrochemical detector. J Liquid Chromatogr 5:1947–1965

Magnusson O, Nilsson LB, Westerlund D (1980) Simultaneous determination of dopamine, DOPAC and homovanillic acid. Direct injection of supernatants from brain tissue homogenates in a liquid chromatography - electrochemical detection system. J Chromatogr 221:237–247

Mayer GS, Shoup RE (1983) Simultaneous multiple electrode liquid chromatographic - electrochemical assay for catecholamines, indoleamines and metabolites in brain tissue. J Chromatogr 255:533–544

Melander WR, Horvath C (1980) Mechanistic study on ion-pair reversed-phase chromatography. J Chromatogr 201:211–224

Mell Jr. LD, Dasler AR, Gustafson AB (1978) Pre-column fluorescent derivatization for high pressure liquid chromatography with o-phthalaldehyde: Separation of urinary catecholamines. J Liquid Chromatogr 1:261–277

Molnar I, Horvath CS (1976) Reverse-phase chromatography of polar biological substances. separation of catechol compounds by high-performance liquid chromatography. Clin Chem 22:1497–1502

Molnar I, Horvath CS (1976) Catecholamines and related compounds. Effect of substituents on retention in reversed-phase chromatography. J Chromatogr 145:371–381

Mori K (1981) Automated measurement of catecholamines in urine, plasma and tissue homogenates by high-performance liquid chromatography with fluorometric detection. J Chromatogr 218:631–637

Morrisey JL, Shihabi ZK (1979) Assay of 4-hydroxy-3-methoxyphenylacetic (homovanillic) acid by liquid chromatography with electrochemical detection. Clin Chem 25:2045–2047

Müller Th, Hofschuster E, Kuss HJ, Welter D (1979) A highly sensitive and precise radioenzymatic assay for plasma epinephrine and norepinephrine. J Neural Transmission 45:219–225

Nimura N, Ishida K, Kinoshita T (1980) Novel post-column derivatization method for the fluorimetric determination of norepinephrine and epinephrine. J Chromatogr 221:249–255

Okamoto KI, Ishida Y, Asai K (1978) Separation and detection of small amounts of catecholamines by high-performance liquid chromatography. J Chromatogr 167:205–217

Ong H, Capet-Antonini F, Yamaguchi N, Lamontagne D (1982) Simultaneous determination of free 3-methoxy-4-hydroxymandelic acid and free 3-methoxy-4-hydroxyphenylethyleneglycol in plasma by liquid chromatography with electrochemical detection. J Chromatogr 233:97–105

Peuler JD, Johnson GA (1977) Simultaneous single isotope radioenzymatic assay of plasma norepinephrine, epinephrine and dopamine. Life Sci 21:625–636

Refshauge C, Kissinger PT, Dreiling R, Blank L, Freeman R, Adams RN (1974) New high performance liquid chromatographic analysis of brain catecholamines. Life Sci 14:311–322

Riggin RM, Kissinger PT (1977) Determination of catecholamines in urine by reversed-phase liquid chromatography with electrochemical detection. Anal Chem 49:2109–2111

Santagostino G, Frattini P, Schinelli S, Cucci ML, Corona GL (1982) Urinary 3-methoxy-4-hydroxyphenylglycol determination using reversed-phase chromatography with amperometric detection. J Chromatogr 233:89–95

Scheinin M, Chang WH, Jimerson DC, Linnoila M (1983a) Measurement of 3-methoxy-4-hydroxy-phenylglycol in human plasma with high-performance liquid chromatography using electrochemical detection. Anal Biochem 132:165–170

Scheinin M, Chang WH, Kirk KL, Linnoila M (1983b) Simultaneous detection of 3-methoxy-4-hydroxyphenylglycol, 5-hydroxyindoleacetic acid, and homovanillic acid in cerebrospinal fluid with high-performance liquid chromatography using electrochemical detection. Anal Biochem 131:246–253

Schill G (1974) Isolation of drugs and related compounds by ion-pair extraction. In: Marinsky JA and Marcus Y (eds) Ion exchange and solvent extraction, vol 6. Marcel Dekker, New York, p 1

Schleicher ED, Kees FK, Wieland OH (1983) Analysis of total urinary catecholamines by liquid chromatography: Methodology, routine experience and clinical interpretations of results. Clin Chim Acta 129:295–302

Schüsler-van Hees MTIW, Beijersbergen van Henegouwen GMJ (1981) Determination of catecholamines and O-methylated metabolites by reversed-phase high-performance liquid chromatography with fluorimetric detection and its application to enzyme kinetics. J Chromatogr 196:101–108

Seki T (1978) Modified ethylenediamine condensation method and its application in the analysis of catecholamines by ion-exchange chromatography. J Chromatogr 155:415–420
Speek AJ, Odink J, Schrijver J, Schreurs WHP (1983) High-performance liquid chromatographic determination of urinary free catecholamines with electrochemical detection after prepurification on immobilized boric acid. Clin Chim Acta 128:103–113
Terweij-Groen CP, Heemstra S, Kraak J (1978) Distribution mechanism of ionizable substances in dynamic anion-exchange systems using cationic surfactants in high-performance liquid chromatography. J Chromatogr 161:69–82
Warsh JJ, Chiu AS, Godse DD (1982) Determination of biogenic amines and their metabolites by high-performance liquid chromatography. In: Techniques and instrumentation in analytical chemistry, vol 4: Evaluation of analytical methods in biological systems. de Zeeuw RA (ed) Part A: Analysis of biogenic amines . Baker GB, Coutts RT (eds). Elsevier, Amsterdam, p 203
Wong KP, Ruthven CRJ, Sandler M (1973) Gas chromatographic measurement of urinary catecholamines by an electron capture detection procedure. Clin Chem Acta 47:215–222
Yamaguchi M, Kuroda N, Ohkura Y (1982) Phosphorescence from 3,4-dihydroxyphenyl derivatives in basic solution and its use for the assay of epinephrine and norepinephrine in urine. Chem Pharm Bull 30:4557–4561
Yamatodani A, Wada H (1981) Automated analysis for plasma epinephrine and norepinephrine by liquid chromatography, including a sample cleanup procedure. Clin Chem 27:1983–1987
Yoshida JI, Yoshino K, Matsunaga T, Higa S, Suzuki T, Hayashi A, Yamamura Y (1980) An improved method for determination of plasma norepinephrine: Isolation by boric acid gel and assay by selected ion monitoring. Biomed Mass Spectrom 7:396–398
Yui Y, Fujita T, Yamamoto T, Itokawa Y, Kawai C (1980) Liquid chromatographic determination of norepinephrine and epinephrine in human plasma. Clin Chem 26:194–196
Yui Y, Itokawa Y, Kawai C (1980) A rapid and highly sensitive method for determination of picogram levels of norepinephrine and epinephrine in tissues by high-performance liquid chromatography. Anal Biochem 108:11–15
Yui Y, Kawai C (1981) Comparison of the sensitivity of various post-column methods for catecholamine analysis by high-performance liquid chromatography. J Chromatogr 206:586–588

Since this article was submitted to the Editors in October 1984, many new publications in the field have appeared. Of these, some important reviews are of particular interest. They contain references to recent original papers and are listed below.

1. A. M. Krstulovic, Quantitative Analysis of Catecholamines and Related Compounds. Ellis Horwood, Chichester and Wiley, New York, 1986.
2. P. Hjemdahl, Catecholamine Measurement in Plasma by High-Performance Liquid Chromatography with Electrochemical Detection. Methods in Enzymology, 142:521–534 (1987).
3. E. Morier-Teissier and R. Rips, Catecholamine Metabolite Measurements in Mouse Brain Using High-Performance Liquid Chromatography and Electrochemical Detection: Comparison of a One-Column Technique with a Two-Column Switching Technique. Methods in Enzymology, 142:535–549 (1987).

High-Performance Liquid Chromatography of the Neuropeptides: the Endogenous Opioid Peptides

A. T. McKnight, A. D. Corbett, S. J. Paterson, L. E. Robson, and H. W. Kosterlitz

Introduction

The first evidence for the existence of peptides in neurones of the mammalian nervous system was for the hypothalamic "neurohormones", oxytocin and vasopressin and thyrotropin releasing hormone, luteinising hormone releasing hormone and somatostatin (see Hughes, 1978; Burgen, Kosterlitz and Iversen, 1980; Gregory, 1982). In the last ten years, however, it has become apparent that peptide containing neurones are not confined to the magnocellular/tuberoinfundibular systems, and that these and other "neuropeptides" are widely distributed throughout the central and peripheral nervous systems (Table 1). The list of peptides in Table 1 contains 38 of the best known "neuropeptides", whose presence in neurones and/or nerve terminals is reasonably well established, but excludes the endogenous opioid peptides (Table 2); most of the peptides mentioned have likewise been found in the peripheral nervous system, or in the "gastroenteropancreatic and related endocrine system" (see Pearse, 1978; Hokfelt et al., 1980; Polak and Bloom, 1982; Gregory, 1982; Iversen, 1983; Iversen, 1984).

Perhaps the main factor that stimulated the explosion of interest in neuropeptides in the 1970s was the introduction of radioimmunoassay (Yalow, 1978) as a widely available technique for measurement of peptide levels in tissue extracts, and, more particularly, for visualisation by immunohistochemical and immunocytochemical methods of the presence and distribution of the various peptides in the brain and periphery. As is frequently pointed out, but just as frequently ignored, work carried out by such immunological methods must be interpreted with some reservation in view of the well know ability of antisera to cross-react with materials other than the immunogen. Apart from the likelihood of cross-reaction with known, structurally related peptides that may belong to the same family, or that may be precursors or intermediates (of the opioid peptides), there is always the possibility of cross-reaction with unknown peptides that are structurally related, or with peptides or proteins that have no obvious relation to the immunogen in terms of primary structure (e.g. Hokfelt et al., 1980). It seems to us that one of the particular strengths of HPLC is that it offers an extra dimension of specificity in conjunction to that available (or presumed to be available) with radioimmunoassay as an off-line assay.

Two important scientific landmarks heralded the "age of the neuropeptides". The first was the elucidation of the sequence of substance P (Chang and Leeman, 1970; Chang, Leeman and Niall, 1971), some forty years after its discovery

Table 1. Non-opioid neuropeptides. Peptides whose presence has been demonstrated, or inferred in neurones and nerve terminals in mammalian central nervous system. (After Iversen, 1984)

Pituitary peptides	Hypothamalic Releasing Hormones
corticotrophin (ACTH) growth hormone (GH) β-lipotropin (β-LPH) α-melanocyte stimulating hormone (α-MSH) β-melanocyte stimulating hormone (β-MSH) oxytocin vasopressin	corticotrophin releasing hormone (CRF) growth hormone releasing hormone luteinising hormone releasing hormone (LHRH) somatostatin thyrotropin releasing hormone (TSH)
Circulating Hormones	**Miscellaneous Peptides**
angiotensin atrial natriuretic peptide(s) calcitonin glucagon insulin	bombesin bradykinin calcitonin gene related peptide (CGRP) carnosine gastrin releasing peptide neurokinin A (substance K) neurokinin B (neuromedin K) neuropeptide K (NPK) neuropeptide Y (NPY) neurotensin proctolin Substance P
Gut Hormones	
avian pancreatic polypeptide cholecystokinin (CCK) galanin gastrin motilin pancreatic polypeptide PHI-27 (PHI) secretin vasoactive intestinal polypeptide (VIP)	

by von Euler and Gaddum (1931). Only when authentic substance P became available, was it possible to carry out the definitive work on this, the "doyen" of the neuropeptides, that confirmed the belief that this peptide was a likely neutrotransmitter or neuromodulator in the nervous system (see Pernow, 1983). The second landmark was the isolation and characterisation of the first of the endogenous opioid peptides, the enkephalins (Hughes, 1975; Hughes et al., 1975). The discovery of the enkephalins came as a result of a concerted effort to find the "endogenous ligand" for the opiate receptors that were known to exist throughout the nervous system, rather than by some serendipitous sequence of events. Such a claim cannot be made for all of the neuropeptides. The enkepha-

Table 2. Endogenous Opioid Peptides

Peptides Derived from Pro-opiomelanocortin (POMC):

β-endorphin [β-LPH(61-91)] β-END

Y G G F M T S E K S Q T P L V T L F K N A I I K N A Y K K G E *
1 5 16 26 31

The sequence for β-END is that of the human peptide. Camel and ovine β-END contain His in place of Tyr at position 27, and Gln and in place of Glu at position 31. The sequence of rat and porcine β-END are that for the camel except that Val replaces Ala in position 26 in the rat, and Val replaces Ile at position 23 in the pig. The sequences of β-END(1-26) and β-END(1-27) have been described in brain and pituitary, as have sequences 1-16 (α-endorphin) and 1-17 (γ-endorphin).

Peptides derived from pro-enkephalin A:

methionine-enkephalin	([Met]enkephalin)	ME	YGGFM*
leucine-enkephalin	([Leu]enkephalin)	LE	YGGFL
	[Leu]enkephalyl-Lys	LEK	YGGFLK
	[Met]enkephalyl-Arg	MER	YGGFMR
	[Met]enkephalyl-Lys	MEK	YGGFMK
	[Met]enkephalyl-Arg-Arg	MERR	YGGFMRR
	[Met]enkephalyl-Arg-Phe	MERF	YGGFMRF
	[Met]enkephalyl-Arg-Gly-Leu	MERGL	YGGFMRGL
metorphamide	[Met]enkephalyl-Arg-Arg-Val.NH_2	MERRV.NH	YGGFMRRV.NH_2

peptide E

Y G G F M R R V G R P E W W M D Y Q K R Y G G F L*
1 5 12 18 22 26

peptide F

Y G G F M K K M D E L Y P L E V E E E A N G G E V L
G K R Y G G F M*
27 34

The sequences of peptide E (1-12), (1-18) and (1-22) are reported to have been found in bovine adrenal medulla and are called BAM-12P, BAM-18P and BAM-22P respectively. The existence of peptide F (1-26).NH_2 (amidorphin) has been reported.

Peptides derived from pro-enkephalin B (Prodynorphin):

α-neo-endorphin		α-NEO	YGGFLRKYPK*
β-neo-endorphin		β-NEO	YGGFLRKYP
dynorphin A	[prodynorphin (209-225)]	DynA	YGGFLRRIRPKLKWDNQ
dynorphin A(1-8)		Dyn 1-8	YGGFLRRI
	prodynorphin (209-232)	1-24	YGGFLRRIRPKLKWD - - NQKRYGGFL
dynorphin B	[prodynorphin (228-240)]	DynB	YGGFLRRQFKVVWT
leumorphin	[prodynorphin (228-256)]	"Big B"	YGGFLRRQFKVVTRS - - QEDPNAYYEELFDV
	[Leu]enkephalyl-Arg	LER	YGGFLR
	[Leu]enkephalyl-Arg-Arg	LERR	YGGFLRR

* The alphabetical coding for the amino acid sequence is given in Table 2.1 in Stryer L (1981) Biochemistry. W H Freeman & Co, San Francisco. 2nd Edn., p 16

lins and the other opioid peptides (Table 2) are perhaps unique among the putative neuropeptides in that the pharmacology of the receptors that mediate their (presumed) physiological functions was well known before their discovery, largely due to the availability of numerous non-peptide opiate agonists and antagonists for the "morphine receptor". The existence of potent and selective antagonists was crucial to the discovery of the enkephalins by the Abderdeen group (Kosterlitz, 1978; 1979). As the result of the large body of information that is available on the effects of opioids in animals and man, the endogenous opioid peptides have been implicated in the regulation of sensory function, mood, respiration, the extrapyramidal motor system, the neuroendocrine regulatory system and the gastrointestinal system (see Kosterlitz and McKnight, 1981).

It is true, however, for the endogenous opioid peptides, as for all the other peptides that are found in the nervous system, that we are some way short of a full understanding of the functional importance of these molecules to the operation of the nervous system. Our understanding of the systems that may be presumed to make use of peptides as chemical messengers, is made even more difficult by the increasing number of instances where there is evidence for the coexistence in nerves of a (neuro)peptide and a "classical" transmitter (monoamine); this situation may be the rule, rather than the exception (see Hokfelt et al., 1986). It is clear that the power and flexibility of the several applications of HPLC methodologies will have an important role in unravelling these complexities.

Scope of the Chapter

The term "neuropeptide" is applied in general where the existence of the peptide in question in a population of neurones in the central or peripheral nervous systems has been demonstrated. As has been pointed out, however, the evidence for this may rely, initially at least, solely on immunohistochemical evidence, and the value of HPLC techniques in delineating the precise nature of the immunoreactive product cannot be overstated. Even before this verification has been obtained, however, it is often inferred from the demonstration of the "neuropeptide-like immunoreactivity" within the nervous system, that the peptide in question may have a "neurotransmitter role". In most instances of this kind the additional, or alternative role of "neuromodulator" may also be claimed for the peptide. The differentiation of "neurotransmitter" from "neuromodulator" may be difficult in a functional sense, since a precise definition of the latter has not been generally approved (see e.g. Desiderio, 1984), although it is clearly inherent in the concept of "neuromodulation" that this function is secondary in importance to that of "neurotransmission". It is axiomatic that the action of a neurotransmitter should be brief, and localised to a discrete locus, while the action of the (neuro)modulator may be found in a wider perisynaptic area, and be more prolonged, as in the case of the action of a hormone (see e.g. Kupferman, 1979). The idea that the actions of neuropeptides (and monoamines) in general, may represent forms of "slow chemical signalling" in the nervous system, as opposed to the "fast" means of signalling by the amino acids, has been the subject of much discussion (see Iversen, 1984; Iversen and Goodman, 1986).

Notwithstanding such considerations, we believe that, in general, the same basic criteria that must be fulfilled for any substance that is a putative neurotransmitter, before such a status should be conferred (Eccles, 1964), should apply to any type of chemical messenger in the nervous system. We shall, then, be limiting our discussion on the use of high performance liquid chromatography of neuropeptides to areas of research relevant to establishing the mechanisms of their biosynthesis, release and inactivation, and to investigating the nature of their interaction with specific receptors. Furthermore, in view of the large (and increasing) number of putative neuropeptides, and their heterogeneous nature, we shall confine ourselves to the class of peptides with which we have some practical experience: the endogenous opioid peptides.

Technical Considerations

It is important to realise that for the most part the application of HPLC methodologies with respect to separations of neuropeptides has been, strictly speaking, in the "preparative mode", albeit using analytical instruments and columns. For underivatised peptides, the level of sensitivity that can be obtained by UV detectors, even in the far ultra-violet, will not permit reliable measurement of amounts much less than 1 nmol. Although the levels of [Met]enkephalin, and more particularly cholecystokinin(s) in extracts of certain brain areas may contain such high levels (see Crawley, 1985), the contribution of the "biological matrix" will effectively reduce the sensitivity very markedly. A ten-fold, or greater, increase in sensitivity has been claimed for on-line analysis of the enkephalins, β-endorphin, and other neuropeptides (particularly those that contain tyrosine) by electrochemical detection (e.g. Feldman, Cohn and Blair, 1978; Mahy, Tusell and Gelpi, 1982; Mousa and Couri, 1983; Whitye, 1983) or by mass spectrometry (see Desiderio, 1984). However the use of isocratic conditions for elution, will, we feel, limit the usefulness of such applications.

In view of the multiple forms of structurally related peptides that may be present in the one extract (see below), we feel that it is imperative to optimise the resolving power of the HPLC system by the use of the appropriate choice of mobile phase, column and gradient for elution. We further believe that it will seldom be possible to obtain sufficient resolution by a single chromatographic step, for an unequivocal (off-line) measurement of the amounts of the various molecular species that may be present, and that re-chromatography of the relevant fractions from a first separation must be carried out, using a different set of chromatographic conditions.

Biosynthesis of the Enkephalins

It is now clear from the work of many groups that were involved in the isolation and characterisation of many peptides that contain the sequence of Tyr-Gly-Gly-

Phe at their N-terminus, and ultimately from the impressive work of Numa and his colleagues using recombinant gene technologies to characterise the propeptides (see e.g. Akil et al., 1984), that there exists three separate families of opioid peptides (Table 2).

Pro-opiomelanocortin contains only one copy of the sequence Tyr-Gly-Gly-Phe, pro-enkephalin has seven, and prodynorphin three. Processing of the precursor takes place largely at paired basic amino acids, and then by other enzymatic means of post-translational processing and modification (see e.g. Loh, Brownstein and Gainer, 1984; Akil et al., 1984), and the number or type of products will vary from region to region, dependent on the presence of the separate processing enzymes. The functional importance of the presence of so many different opioid peptides is not clear, especially since all of the possible end products, including those that are likely to be intermediates, are more or less able to activate opioid receptors (see below), so long as the N-terminal tyrosine is free.

At the time of our interest in the mechanisms of the biosynthesis of the enkephalins, only three endogenous opioid peptides were known to exist: [Met]enkephalin and [Leu]enkephalin themselves, and β-endorphin. It was thought possible at that time that the biosynthesis of the [Met]enkephalin might occur by way of β-endorphin, and its precursor β-lipotropin, and that similar molecules might exist that contained the sequence of [Leu]enkephalin (see McKnight et al., 1978). The anatomical evidence that was accumulated did not support the idea that peptides related to β-endorphin of local origin might be precursors of the enkephalins, and our experiments on the synthesis of the enkephalins in the corpus striatum of guinea-pig brain (McKnight, Hughes and Kosterlitz, 1979) or the myenteric plexus of guinea-pig intestine (Sosa et al., 1977) in vitro provided evidence that circulating peptides of pituitary origin were clearly not the source of the high levels of the enkephalins in these sites.

We determined the synthesis of the enkephalins as the incorporation of radioactive label that co-chromatographed with the synthetic, carrier peptides following a short period of incubation ("pulse") with [^{3}H]-labelled tyrosine or leucine, or [^{35}S]-methionine, and a further period of incubation without the label ("chase") for up to hours. At the end of the incubation period, the tissues were homogenised in 0.1 M HCl containing the marker enkephalins, and centrifuged. The enkephalins were isolated from the supernatant by successive chromatographic steps. First we used adsorption on Amberlite XAD-2 with elution in methanol to remove free amino acids, followed by cation exchange on HC-Pellionex-SCX with elution in ammonium acetate/methanol at pH 8.35. The fractions that corresponded to the retention time of the enkephalins on the SCX column, containing only 7% of the total radioactivity that was eluted from the column, were desalted by a second XAD-2 step and then applied to an anion-exchange column, AE-Pellionex-SAC (100 × 0.4 cm) with elution in 0.04 M acetic acid in 30% aqueous methanol. Only one major peak of radioactivity eluted from the SAX column, and this corresponded to the single peak of UV absorbance by the co-elution of [Met]- and [Leu]-enkephalin (Fig. 1). The two enkephalins were resolved by thinlayer chromatography (normal phase) and electrophoresis; the coeluting radioactivity was taken as an index of *de novo* biosynthesis.

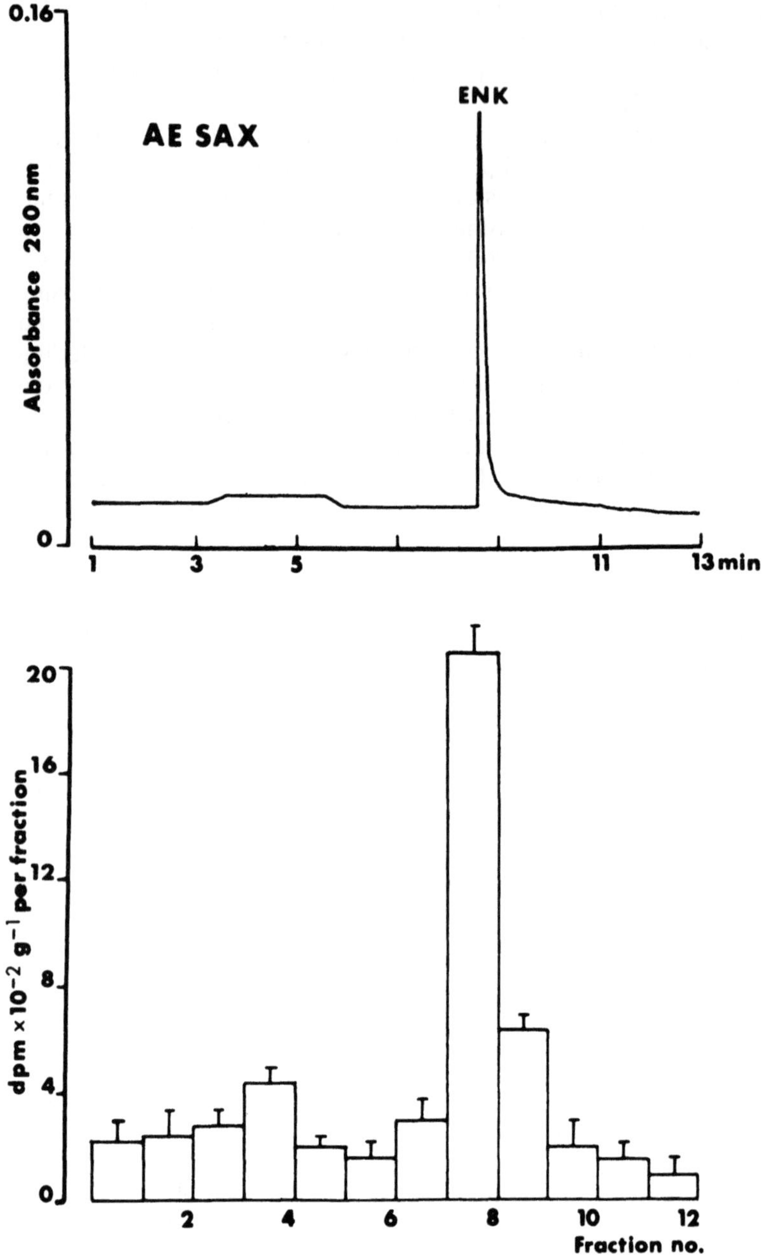

Fig. 1. Top: U.V. absorption profile of the eluate from an AE-Pellionex-SAX (Reeve Angel) anion-exchange column from a typical experiment in which slices of corpus striatum from guinea-pig brain were incubated for 5 h after a 30 min "pulse" with ^{3}H-Tyrosine. The single major peak is due to the co-eultion of synthetic [Met]enkephalin (70 nmol) and [Leuk]enkephalin (36 nmol) that were added as markers to the tissue extract. Bottom: elution profile of radioactivity in 1 ml fractions from four such experiments; the single major peak corresponds to the elution volume for the enkephalins. For further details see text.

By this means we were able to show that the incorporation of labelled amino acid into the enkephalins increased linearly with time after an initial lag period of one to two hours. When a blocker of ribosomal assembly, either puromycin or cycloheximide, was present during the labelling and lag periods then incorporation was blocked; when the blocker was present only after the lag period the incorporation was not reduced. We concluded that the enkephalins in the corpus striatum and myenteric plexus were synthesisised locally from a precursor, or precursors, also synthesised locally, and that the synthetic pathway involved ribosomal assembly of the precursor peptides (lag phase) followed by conversion to the enkephalins. It was inferred that since the rate of incorporation of labelled amino acid remained constant for some hours, that there was a large pool of rapidly synthesised precursor that was only comparatively slowly converted to enkephalin, and we were able to obtain some evidence for this for the myenteric plexus at least. We detected the presence of putative precursor as a species of molecular weight in the range 10–20K daltons in the eluate from Sephadex G50 or G75 columns; when fractions were digested with trypsin the presence of putative precursor was revealed as the appearance of material with the chromatographic and biological properties of enkephalin (McKnight et al., 1980).

A similar approach had been used by Udenfriend and his colleagues to identify precursors of the enkephalins from the striatum from several species, and in their case the isolation of the active opioid species was by reverse-phase chromatography using Lichrosorb RP-18 with off-line detection by radio-receptor assay (Lewis et al., 1978). This group and others were responsible for the isolation from brain and adrenal medulla of hexa- and heptapeptides that contained the sequence of [Met]enkephalin extended at the C-terminus by -Arg, or – Lys, or -Arg-Arg. These workers also were able to isolate and characterise the hepta- and octapeptides [Met]enkephalyl-Arg-Phe and [Met]enkephalyl-Arg-Gly-Leu, and several larger peptides that contained the enkephalin sequence (see Lewis and Stern, 1983). In these isolations the reverse-phase chromatography followed extensive separations e.g. on Sephadex columns, and entailed the use of propan-1-ol for gradient elution from 0.4/0.5 M formic or acetic acid solutions containing pyridine. On-line detection of peptides was by using fluorescamine in an automated fluorometric system; the detection of opioid activity was off-line by radio-receptor assay.

Release of Enkephalins

The content of enkephalins in the myenteric plexus of the small intestine of the guinea-pig is as high as that of the corpus striatum, and we believed that the use of this tissue in vitro would allow us to examine the release of the enkephalins by electrical stimulation of the nerve plexus, rather than by depolarisation by excess potassium or veratridine, as had been the approach for preparations of brain (see McKnight et al., 1978). After our initial failure to detect release of enkephalins into the bathing fluid following stimulation at various frequencies, we resorted to an indirect method that involved blocking *de novo* biosynthesis by cycloheximide, and measuring the loss of enkephalins from the tissue stores as an index

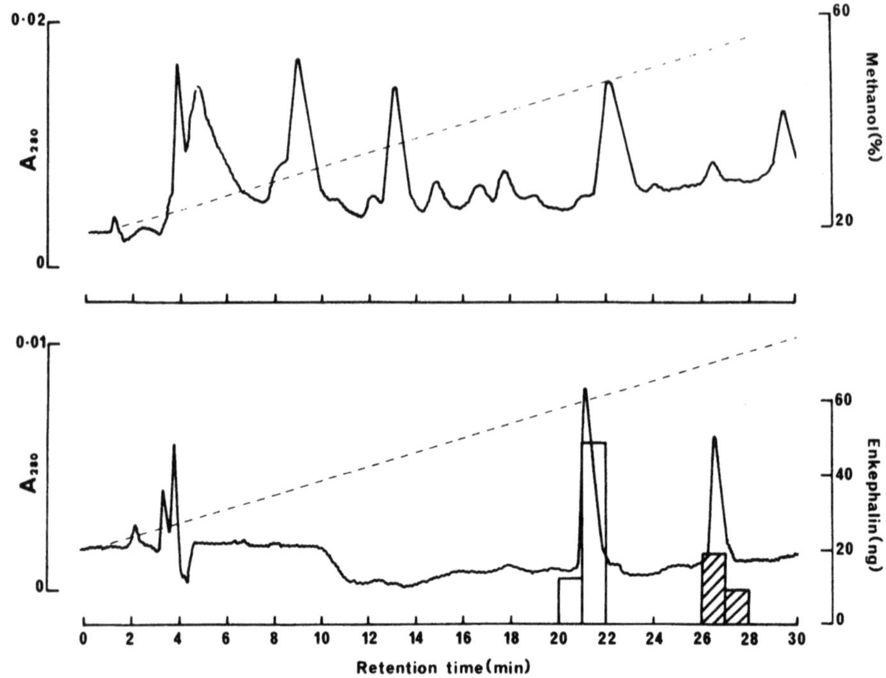

Fig. 2. HPLC separation of [Met]enkephalin and [Leu]enkephalin on a μBondapak C18 column (4 × 300 mm) with a linear gradient between 0.7 M acetic acid and methanol (80:20 to 40:60) in 30 min at 0.8 ml/min. Top: U.V. absorbance profile of a typical Amberlite XAD-2 extract of myenteric plexus - longitudinal muscle from the small intestine of the guinea-pig. Bottom: Elution of a mixture of synthetic [Met]enkephalin and [Leu]enkephalin (4 nmol). The histograms show [Met]enkephalin like acticity (open columns) and [Leu]enkephalinlike activity (hatched columns) in 0.8 ml (1 min) fractions of the eluate shown in the top panel. Waters chromatograph, absorbance at 280 nm

of the release. The enkephalins were isolated by extraction into 0.1 M HCl and adsorption onto, and elution from XAD-2, as before, and separation by reverse-phase HPLC on μBondapak C18 (Fig. 2) using a gradient to 60% methanol from equilibration in 0.5 M acetic acid in 20% methanol (pH 2.5); measurement was by bioassay on the mouse vas defernes. Thus we were able to show that the "release" was voltage-, frequency- and time-dependent, was blocked by tetrodotoxin, and increased by tetraethylammonium (McKnight et al., 1978; Corbett, McKnight and Kosterlitz, 1981).

When putative inhibitors of the breakdown of the enkephalins (see below), bacitracin and a mixture of dipeptides, were added to the bathing fluid we were above to recover 48 ± 13 pmol/g tissue and 16 ± 3 pmol/g tissue of [Met]- and [Leu]-enkephalin respectively after stimulation at 1 Hz for 30 min. These amounts represented 30–60% of those lost from the tissue stores.

We later refined the chromatographic separation by the use of an Ultrasphere ODS column (5 μm, 250 × 4.6 mm) in place of the μBondapak, and by using

1.1 M acetic acid/0.04 M ammonia (pH 3) as the aqueous phase. By this means we were able to increase the resolution between [Met]enkephalin and [Leu]enkephalin, with an improvement in peak shape, but we were also able to detect [Met]-enkephalyl-Arg/[Met]enkephalyl-Lys in the myenteric plexus as an early running peak of activity, and [Met]enkephalyl-Arg-Phe as a late running peak (Fig. 3b). Trace amounts of these species were detected as released material, but in this case it was necessary to introduce an additional chromatographic step between the XAD-2 and ODS separations. Here we used a high performance cation exchange column of Partisil SCX with a curvilinear programme (Waters) of elution from 0.01 M acetic acid/0.008 M ammonia in 20% methanol (ph 4) to 0.1 M acetic acid/0.11 m ammonia in 20% methanol (pH 6) (Fig. 3a); the "enkephalin fraction" was collected for re-chromatography on the ODS column, and the later "basic peptide fraction" was likewise collected for its separate rechromatrographic separation into the two activities (McKnight, Corbett and Kosterlitz, 1981). Later the replacement of methanol by acetonitrile allowed an even greater improvement in resolution and peak symmetry (Fig. 4). Furthermore with the former organic solvent it would be possible for us to test for the lipophylic peptides; β-endorphin, for example, that could not be eluted from a reverse-phase column using a methanol gradient.

Inactivation of Enkephalins

Our inability to measure directly the release of enkephalins from the myenteric plexus was attributed by us to the well known susceptibility of the enkephalins to rapid hydrolysis by a number of peptidases (see e.g. McKnight, Corbett and Kosterlitz, 1983). We therefore tested the possible value of various peptidase inhibitors for their ability to protect [Met]enkephalin, and thus increase its agonist potency at the opioid receptors in preparation of longitudinal muscle-myenteric plexus from guineas-pig ileum in vitro. The combination of the aminopeptidase inhibitor bestatin, the angiotensin converting enzyme inhibitor captopril, the "enkephalinase" inhibitor thiorphan and the dipeptide L-leucyl-L-leucine increased the potency of [Met]enkephalin in this assay for opioid activity 18-fold, without increasing the potency of β-endorphin, peptidase-resistant enkephalin analogues, or the non-peptide opioid normorphine. We also tested the effect of this combination of peptidase inhibitors on [Met]enkephalin activity in the mouse vas deferens, another standard assay for opioid acitivity (and our assay for HPLC fractions), and in the rat vas deferens where [Met]enkephalin was virtually inactive. The potency of [Met]enkephalin was increased 13-fold in the mouse vas, and 200-fold in the rat vas deferens. When we then went on to test systematically all of the available products of the processing of pro-enkephalin and pro-opiomelanocortin, we found all to be more or less active in all of the assays. With the pro-enkephalin peptides, all were increased in potency by the cocktail of peptidase inhibitors, but the potency of β-endorphin and the longer peptides that are derived from it, were little affected (McKnight et al., 1983).

We subsequently carried out similar experiments using the rabbit vas deferens in vitro as an assay for activity at the kappa-type of opioid receptor; only with the putative products of the processing of pro-dynorphin did we obtain significant agonist potency. We found that, as with the "parent peptide" dynorphin A, dynorphin A (1-8) and dynorphin A (1-9) were potent agonists at the kappa-receptor, both in the rabbit vas deferens, and in competitive binding assays in guinea-pig brain. Significantly, we found that with dynorphin A (1-8), a naturally occurring opioid peptide (Minamino et al., 1980), the potency was increased by the addition of peptidase inhibitors while with the heptadecapeptide dynorphin A the potency was not changed (McKnight et al., 1982, Corbett et al., 1982).

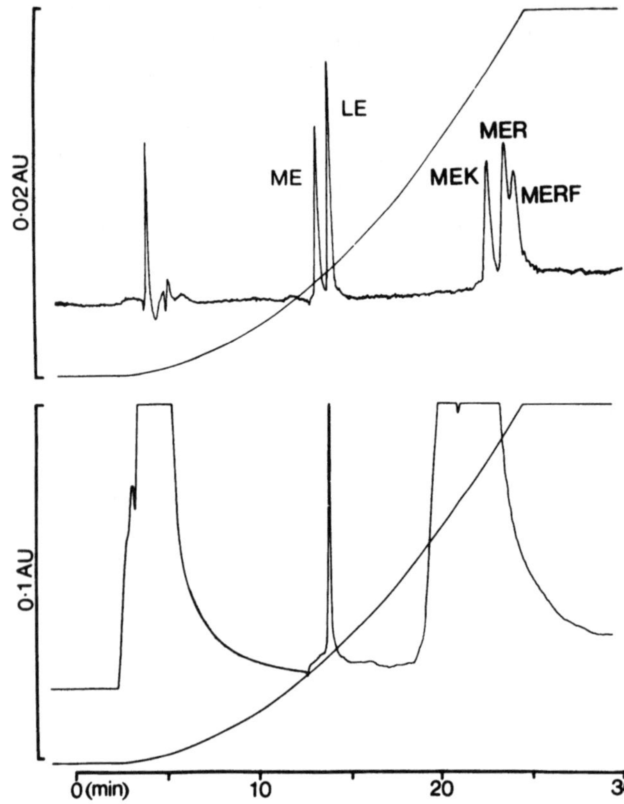

Fig. 3. HPLC separations of enkephalins. A: On a Whatman Partisil PXS 10/25 SCS (10 μm, 4.6 × 250 mm) cation-exchange column over a curvilinear (Waters) gradient between 0.01 M acetic acid/0.008 M ammonia in 20% methanol (pH 4) and 0.1 M acetic acid/0.11 M ammonia in 20% methanol (pH 6) in 25 min at 1 ml/min. The top panel shows the absorbance profile (280 nm) for the elution of synthetic standards (2.5–5 nmol) and the bottom panel shows the corresponding elution pattern of an Amberlite XAD-2 extract from the bath fluid in which preparations of myenteric plexus – longitudinal muscle from guinea-pig small intestine had been incubated. B: On an Altex Ultrasphere ODS (5 μm, 4.6 × 250 nm) column with a linear gradient between 1.1 m acetic acid/0.045 M ammonia (pH 3) and methanol (80:20 to 35:65) in 40 min at 1 ml/min. Top: elution of standards. Middle: elution of material from an XAD-2 extract of the preparation of guinea-pig intestine; the histograms represent the contents of the separate peptide species in unstimulated preparations (left), and after electrical stimulation (right), the scale on the rights is for peptide content in pmol/g tissue. Bottom: elution of the enkephalin-containing fractions from the SCX column with the estimation of the contents of separate species (pmol/g). Altex chromatograph, absorbance at 280 nm.

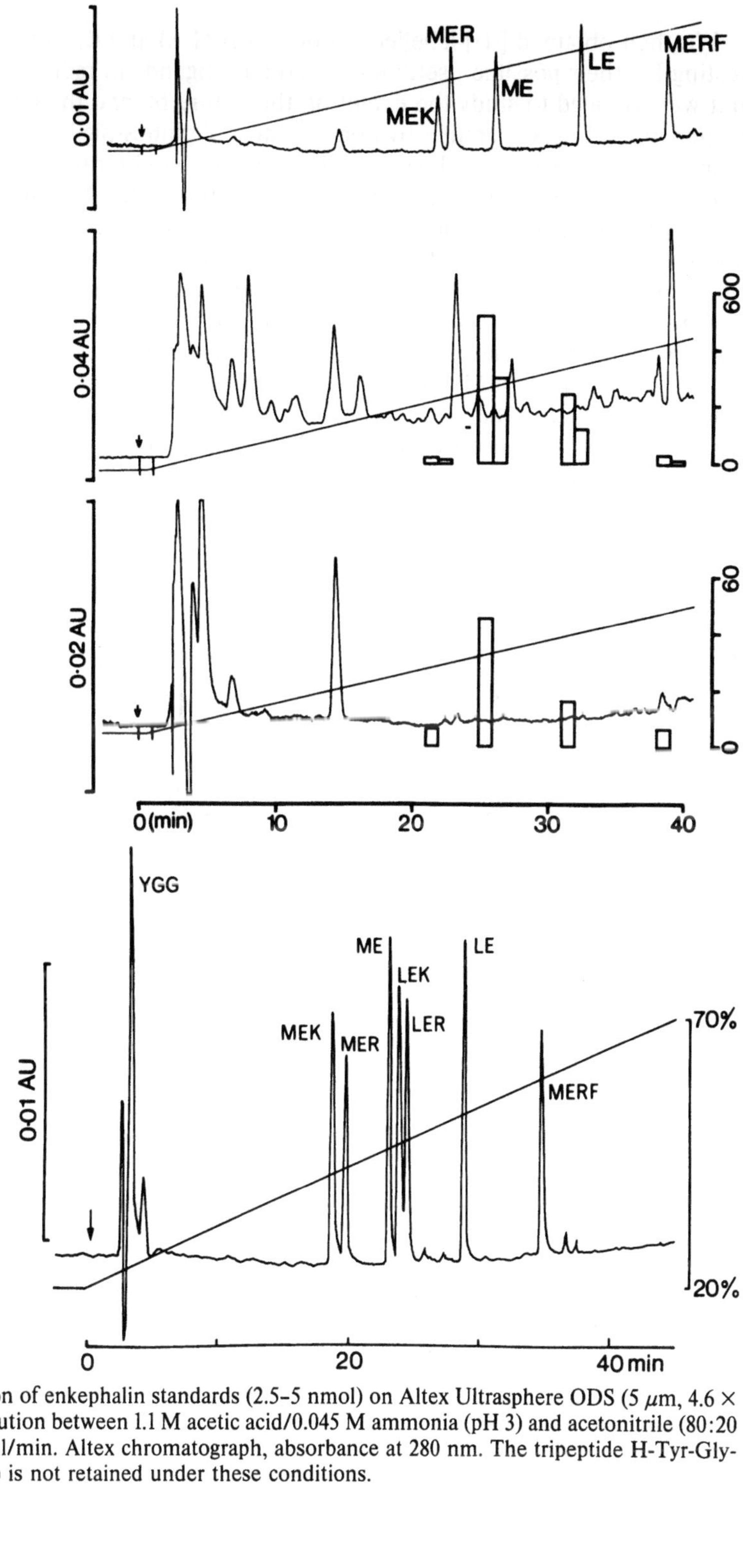

Fig. 4. Separation of enkephalin standards (2.5–5 nmol) on Altex Ultrasphere ODS (5 μm, 4.6 × 250 nm) with elution between 1.1 M acetic acid/0.045 M ammonia (pH 3) and acetonitrile (80:20 to 30:70) at 1 ml/min. Altex chromatograph, absorbance at 280 nm. The tripeptide H-Tyr-Gly-Gly-OH (YGG) is not retained under these conditions.

We then obtained [^{3}H]-labelled dynorphin A (1–8) and dynorphin A (1–9) for testing for their possible usefulness as primary ligands in radio receptor assays, but were obliged to study the extent of, the nature of, and the inhibition of the degradation of these peptides by homogenates of guines-pig brain using reverse-phase HPLC. Since the radioactive label was on the N-terminal tyrosine, it was necessary only to try to achieve base-line separation of the N-terminal fragments up to, and including the full octa- and nona-peptides (Fig. 5). We had been using Ultrasphere octyl columns (5 μm, 250 × 4.6 mm), and in attempting to use solutions containing trifluoroacetic acid (TFA) for elution (Dunlap, Gentleman and Lowney, 1978), we found that in order to avoid serious peak tailing with the more basic peptides it was necessary to use a much higher concentration of this putative counter-ion (26 mM) than those described for use with, for example, the μBondapak C18. In order then to avoid an aqueous phase whose pH was too low for safe use, it was necessary to back titrate the aqueous phase to pH 2.3 with triethylamine. The use of this base allowed us some stability, and maintained the

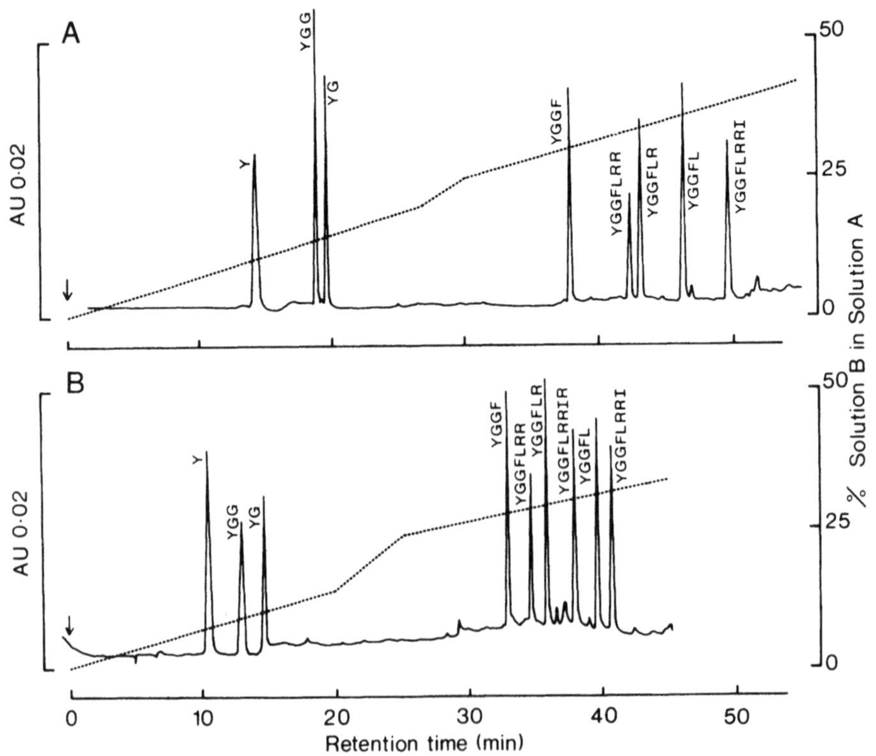

Fig. 5. Separation of peptide standards corresponding to the N-terminal fragments of dynorphin A (1–8) and dynorphin A (1–9) on Altex Ultrasphere Octyl (5 μm, 4.6 × 250 mm). A: system used for dynA (1–8). Solution A was 2% TFA in water with ammonia to pH 2.5, solution B was 1% TFA in acetonitrile; discontinuous linear gradient of 0–20% B in 27 min, 20–25% B in 3 min, 25–42% B in 25 min. B: system used for dynA (1–9). Solution A was 2% TFA in water with triethylamine to pH 2.3, solution B was as above; elution with a discontinuous gradient of 0–15% B in 20 min, 15–20% B in 5 min, 25–35% B in 20 min. Altex chromatograph, absorbance at 280 nm. Y = Tyr, G = Gly, F = Phe, L = Leu, R = Arg, I = Ile (Gillan et al., 1985).

UV transparency of the solution, although this was not a major consideration since all of the tyrosine-containing peptides absorbed strongly at 280 nm; the organic phase was 13 mM TFA in 98% acetonitrile. Separations similar to those shown in Figure 5 were subsequently obtained with solutions containing only 1% TFA (13 mM) using μBondapak C18 or Partisil C8 columns. In view of the differences in peak geometry that were obtained with the different (though ostensibly similar) stationary phases, we wonder if the usefulness of TFA does in fact reflect counter-ion properties.

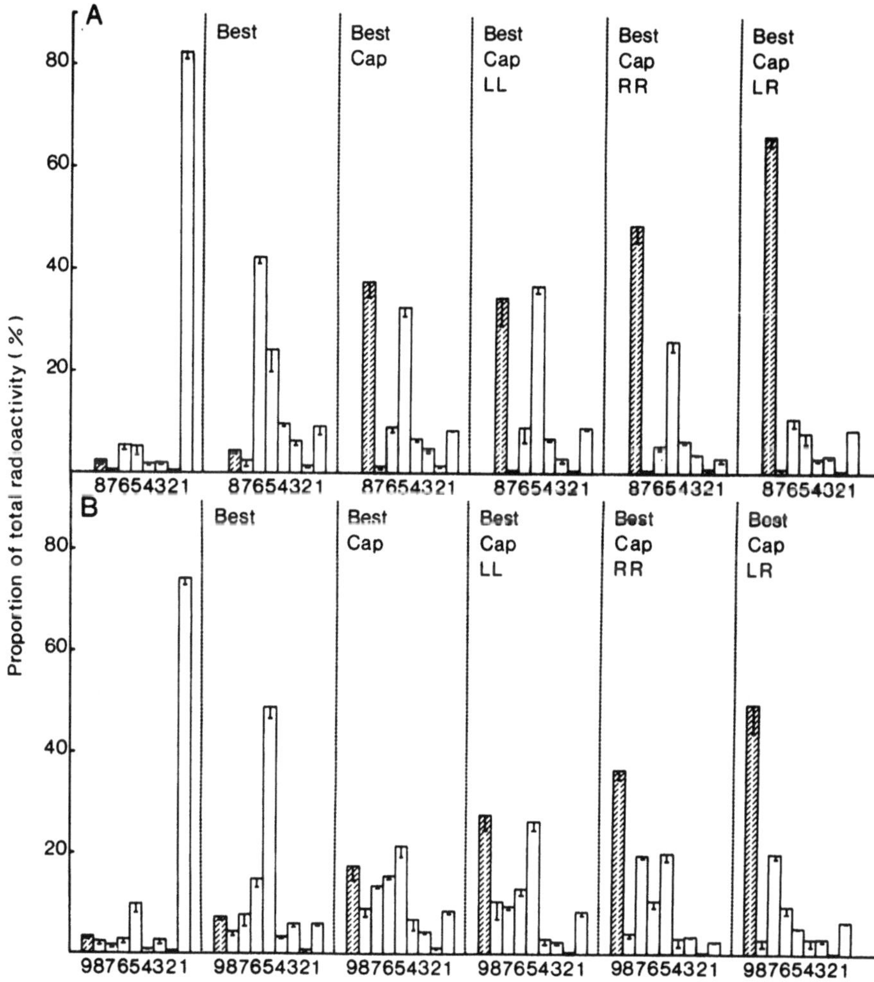

Fig. 6. The effects of inhibitors of peptidases on the degradation of [^{3}H]dynorphin A (1–9) (A) and [^{3}H]hynorphin A (1–8) (B) in suspensions of synaptic membranes from guinea-pig brain after incubation at 25° C for 30 min. The hatched columns indicate the proportion of radioactivity recovered from the supernatant as unchanged peptide; the open columns the radioactivity as metabolites. The numbers on the abscissae refer to the N-terminal fragments; 1 = Tyr, 8 = dynA (1–8), 9 = dynA (1–9). Best is bestatin (30 μM), Cap is captobril (300 μM), LL is L-Leucyl-L-Leucine (1 mM), RR is L-Arginyl-L-Arginine (0.5 mM), LR is L-Leucyl-L-Arginine (0.5 mM). (Gillan et al., 1985).

The results that we obtained (Fig. 6) suggested that degradation of both peptides in guinea-pig brain homogenate (in 50 mM Tris-HCL, pH 7.4) was extensive at 25° C, but was also found at 0° C. The inhibitors bestatin and captopril reduced degradation at 0° C, but for a similar degree of protection at 25° C, arginine-containing dipeptides were also required (Fig. 6). The main sites of degradation were at the Tyr^1 - Gly^2, Arg^6 - Arg^7 and Leu^5 - Arg^6 bonds, and with [^{3}H]dynorphin A (1-9) the Arg^7 - Ile^8 and Ile^8 - Arg^9 bonds were also liable to cleavage (Gillan et al., 1985). We believe, however, that homogenisation of the tissue may increase the capacity for degradation, perhaps by allowing to come into play enzymes that might not have a physiological role in the extracellular space. For example processing enzymes may then be allowed to act as degrading enzymes. Thus we have found by a similar approach to that illustrated for the brain, that in "intact" preparations of longitudinal muscle-myenteric plexus from guinea-pig ileum, the recovery of unchanged dynorphin A (1–8) is high if the cocktail of peptidase inhibitors is present, while in homogenates there is only very little recovery of the octapeptide under these conditions, but equivalent amounts of [Leu]enkephalin are found. This result suggests that in the process of homogenisation the capability for cleaving the Leu^5 - Arg^6 bond had been generated (McKnight, Corbett and Kosterlitz, unpublished).

Isolation of Endogenous Opioid Peptides

The original isolations of the enkephalins and β-endorphin did not rely on HPLC separations. It is probably true to say, however, that for all of the other natural opioid peptides that have been found, HPLC separation on a reverse-phase column was the final step in a lengthy isolation process. When the peptide in question was then characterised, an important test would be to confirm that the isolated material co-eluted from the HPLC column with synthetic material of the predicted amino acid sequence.

Other putative products of the processing of the pro-peptide pro-opiomelanocortin are β-endorphin (1–27), β-endorphin (1–26) and their α-N-acetylated derivatives (see Smyth, 1984). As already mentioned, the processing of pro-enkephalin at paired basic amino acid residues may generate [Met]enkephalyl-Arg-Phe and [Met]enkephalyl-Arg-Gly-Leu as well as [Met]- and [Leu]enkephalin, but that other putative products may be found. We were involved with the isolation and characterisation of another opioid peptide from bovine caudate, metorphamide, that seemed to require an unusual cleavage from pro-enkephalin. Between residues of 204 and 215 of pro-enkephalin is the sequence -Lys-Arg-Tyr-Gly-Gly-Phe-Met-Arg-Arg-Val-Gly-Arg-Pro-. From our experience with the fragments of dynorphin A, particularly dynorphin A (1–8), we believed that the paired basic amino acids -Lys-Arg- would be the signal for a cleavage of the "conventional" type, but that a cleavage at the single -Arg- at the C-terminus (by analogy with dynorphin A) would yield the peptide [Met]enkephalyl-Arg-Arg-Val-Gly. Oxidative processing would then produce the final product [Met]enkephalyl-Arg-Arg-Val-NH_2.

It is important to realise that this species was a postulated product of the processing of pro-enkephalin, and that the synthetic peptide was synthesised and antisera raised in advance of its isolation. The first step in the isolation from 200 g of bovine caudate was acid-acetone extraction followed by removal of lipid in heptane. The extract was neutralised with ammonia and centrifuged, and the supernatant was then acidified with acetic acid (to 5%) centrifuged, and adsorbed onto preparative C18 packings equilibrated with 5% acetic acid. The eluate in acid-acetone was evaporated, and the peptide-containing fractions were chromatographed on Sephadex G-50 in 50% acetic acid, with immunoreactive fractions being subjected then to successive reverse-phase separations on Ultrasphere ODS (5 μm, 250 × 4.6 mm) with elution in acetonitrile from equilibration in ammonium acetate solutions (50 mM, pH 2.7 and 100 mM, pH 7). The absorbance at 210 nm was measured on-line, and fractions were assayed for metrophamide-like-immunoreactivity off-line. Prior to amino acid analysis there was a desalting step on Ultrasphere ODS (3 μm, 75 × 4.6 mm in an acetonitrile gradient from 0.1% TFA, Weber et al., 1983).

Similar in their extensiveness have been the isolations and characterisations of the various opioid peptides that are derived from prodynorphin: α- and β-neoendorphin, and dynorphin A (1–8) by Matsuo's group, dynorphin A by Goldstein and Tachibana and their colleagues, and dynorphin B ("rimorphin") by Udenfriend's group (see Akil et al., 1984).

It is clear that if one seeks to investigate the nature of all of the constituent opioid peptides in a particular tissue, then extensive purification will be necessary prior to separation by HPLC and assay. From our experience of such situations, we feel that since one must now make provision for the possible presence of a large number of structurally related peptides, a resolution of the peptides by a single chromatographic run is not possible, and a subsequent separation of fractions from the first run must be carried out under entirely different conditions.

We are currently engaged in the characterisation of the full complement of endogenous opioid peptides in the myenteric plexus, and for this purpose we currently use a separation on Spherisorb S5 C8 columns (5 μm, 250 × 4.6 mm) with a 50 min gradient from 12.5 to 40% acetonitrile; the aqueous buffer is 1.1 M acetic acid/1.1 M ammonia in water (pH 5.75). We routinely collect ten fractions for further separation (Fig. 7). Our choice of ammonium acetate buffers was because they were volatile and simple to use, and with our ability to work at 280 nm, sufficiently UV-transparent. Originally, and with various columns, we had used buffers of much lower pH, but with the discovery of more and more opioid peptides of increasing basicity, it was necessary to resort to solutions of increasing pH to increase the effect of suppressing the positive charge on the peptides, and to reduce peak-tailing. This is illustrated for dynorphin A and its fragments in Figure 8. At pH 3 (1.1 M acetic acid/0.045 M ammonia) [Leu]enkephalin ("1–5") and [Leu]enkephalyl-Arg ("1–6") elute as sharp peaks, but the other three peptides elute in peaks whose broadness increases as the basic character of the peptide increased (Fig. 8a), while at pH 4.5 all five peptides elute as sharp peaks (Fig. 8b). Under these latter conditions the longer peptides, particularly dynorphin A (1–17) itself, elutes as a broad, skewed peak, but at pH 5.75 (Fig. 8c) acceptable peak geometry is obtained.

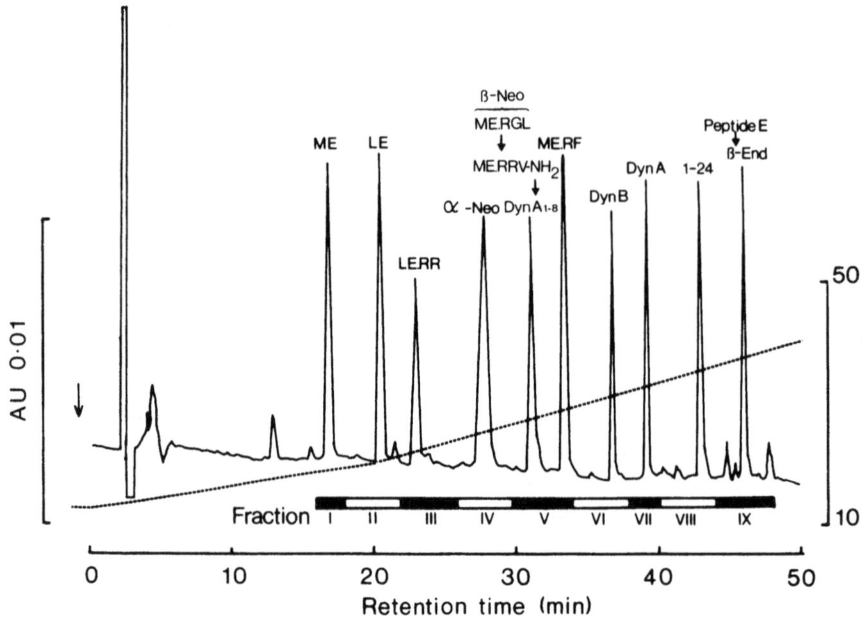

Fig. 7. Separation of endogenous opioid peptide standards (2.5–10 nmol) on Spherisorb S5 C8 (5 μm, 4.6 × 250 nm) with elution between 1.1 M acetic acid 1.1 M ammonia (pH 5.75) and acetonitrile (87.5:12.5 to 60:40) with a discontinuous gradient (12.5–20% acetonitrile in 20 min, 20–40% in 30 min) at 1 ml/min. Gilson chromatograph, absorbance at 280 nm. Note that the elution of β-neo-endorphin (not shown) is as complex, multicomponent pattern (see Figure 9). DynA coelutes with leumorphin ("Big B") (not shown)

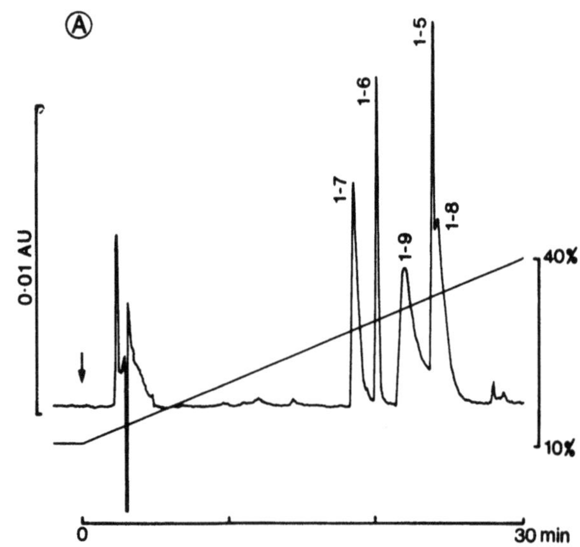

Fig. 8

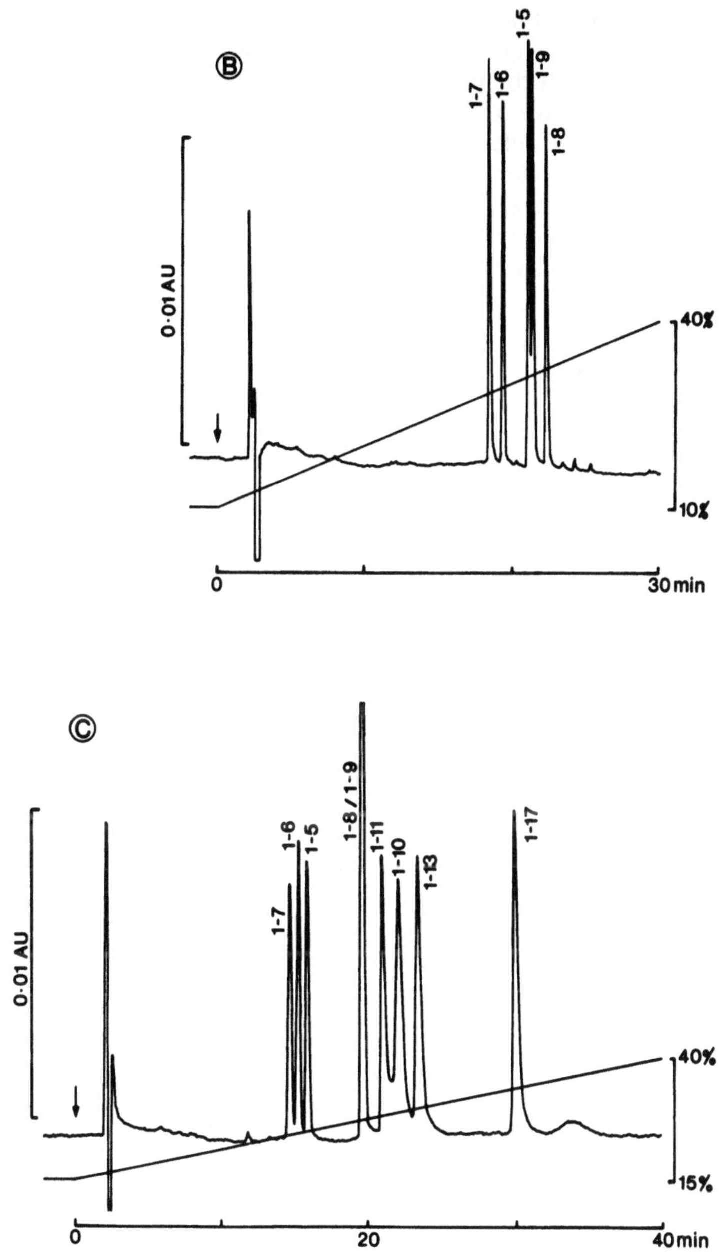

Fig. 8. HPLC separations of fragments of dynorphin A (2.5–10 nmol of each) on Altex Ultrasphere ODS (5 μm, 4.6 × 250 nm) between 1.1 M acetic acid/ammonia solutions and acetonitrile. A: 0.045 M ammonia, pH 2.5, 10–40% acetonitrile in 30 min at 1 ml/min. B: 0.45 M ammonia, pH 4.5, 10–40% acetonitrile in 30 min at 1 ml/min. C: 1.1 M ammonia, pH 5.75, 15–40% acetonitrile in 40 min at 1 ml/min. The numbering system for identifying the standards refers to the structure of the heptadecapeptide dynorphin A (YGGFLRRIRPKLKWDNQ), so that 1-5 is [Leu]enkephalin, 1-8 is dyn A (1-8) ([Leu]-enkephalyl-Arg-Arg-Ile). Altex chromatograph, absorbance at 280 nm

One problem that we have observed with the use of such mobile phases has been with β-neoendorphin. With this, but with no other opioid peptide, we have observed a multicomponent elution pattern (Fig. 9a, b). Such an effect is obtained irrespective of the pH or the stationary phase (and has been seen with a cation exchange column as in Fig. 10b). The effect is not peculiar to the use of ammonium acetate, since it was also observed in triethylamine-phosphate systems. This effect was not due to impure material, as we first thought, but may represent the effect obtained by the existence in solution and/or in combination with the stationary phase, of multiple ionisation states or conformations.

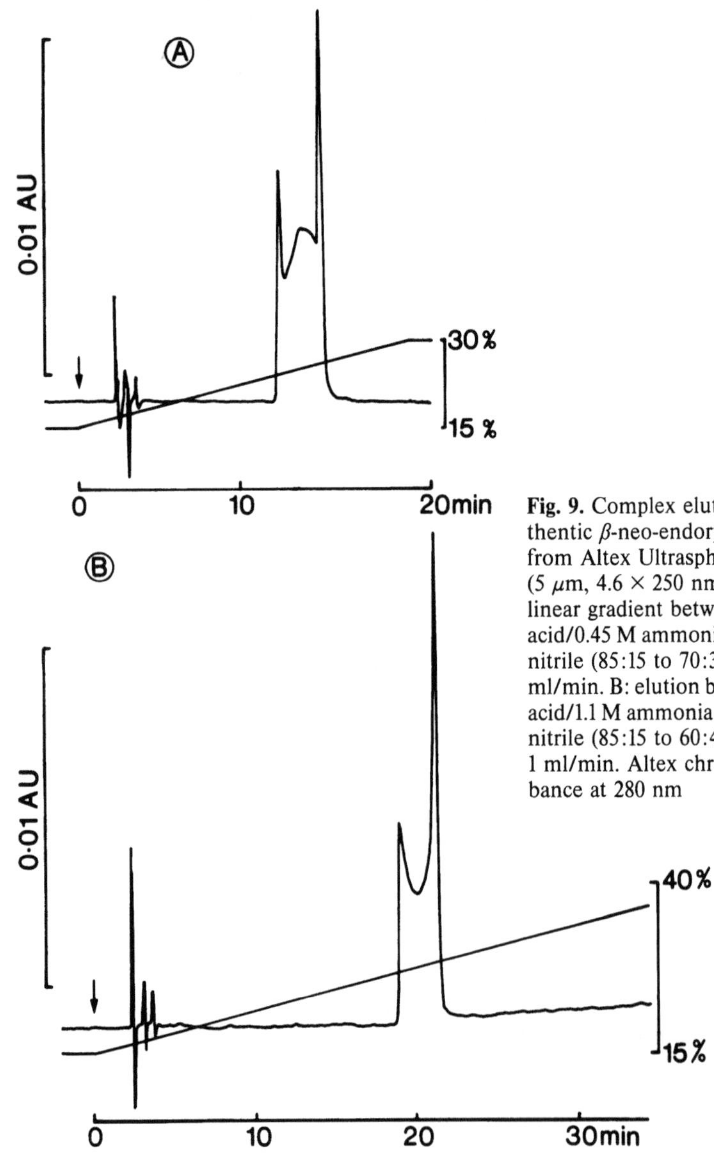

Fig. 9. Complex elution pattern of authentic β-neo-endorphin (13.5 nmol) from Altex Ultrasphere ODS columns (5 μm, 4.6 × 250 nm). A: elution with a linear gradient between 1.1 M acetic acid/0.45 M ammonia (pH 4) and acetonitrile (85:15 to 70:30) in 20 min at 1 ml/min. B: elution between 1.1 M acetic acid/1.1 M ammonia (pH 5.8) and acetonitrile (85:15 to 60:40) in 40 min at 1 ml/min. Altex chromatograph, absorbance at 280 nm

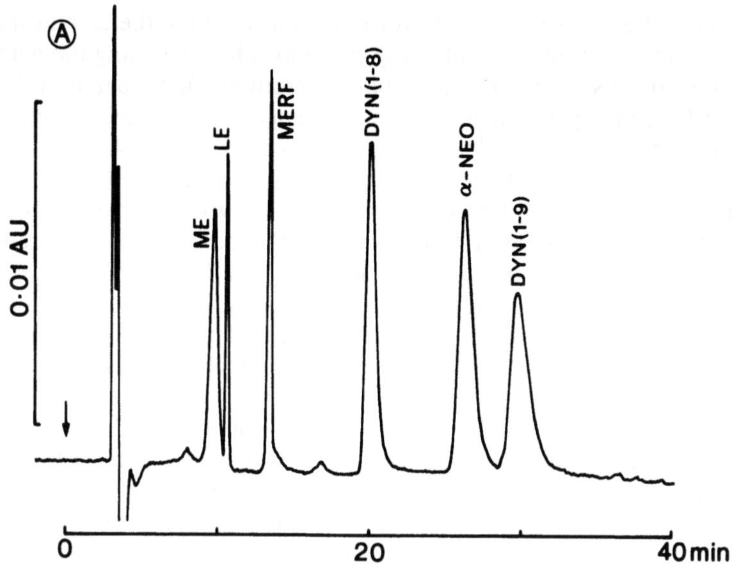

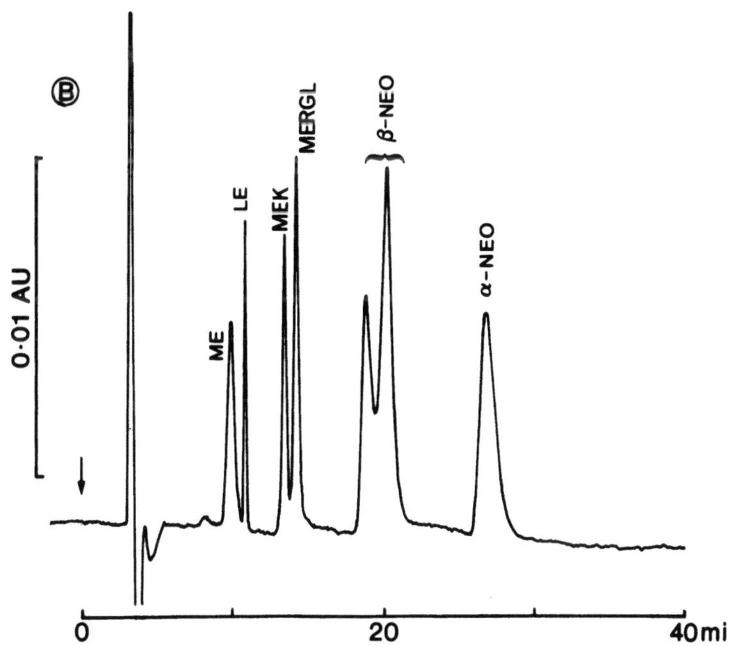

Fig. 10. HPLC separation of opioid peptide standards by cation-exchange on a Whatman Partisil PXS 10/25 SCX column (10 μm, 4.6 × 250 mm) with a linear gradient between 0.025 M acetic acid/0.002 M ammonia in 20% propan-2-ol (pH 3.7) and 0.45 M acetic acid/0.50 M ammonia in 20% propan-2-ol (pH 6.4) in 40 min at 1 ml/min. Note the biphasic elution pattern with authentic β-neo-endorphin (10 nmol). Dyn (1–9) is YGGFLRRIR; fragments of dynorphin A longer than the nonapeptide are retained by the column. Waters chromatograph, absorbance at 280 nm.

This effect was only one factor that contributed to the technical problem that we had to overcome, of obtaining conditions for separating the peptides that co-eluted, or closely eluted in fractions 4–6 from the first separation (Fig. 7). Separation by a high-performance cation exchanger was limited to peptides no longer than dynorphin A (1–9), but the co-elution of dynorphin A (1–8) and β-neoendorphin limited the usefulness of such an approach (Fig. 10). In a triethylamine-phosphate system under isocratic conditions it was possible to separate these two peptides from each other, but now α- and β-neoendorphin co-eluted (Fig. 11). We were able to effect a useful separation of α- and β-neoendorphins using gradients between 0.1% TFA in water and 0.1% TFA in 70% acetonitrile in water. Some important separations that we have obtained using a Partisil C8 column are illustrated in Figures 12–15. By re-chromatographing the separate fractions from the ODS column in such a way, we are able to obtain fractions for assay that allow us to determine the content of the separate opioid peptides.

By such a strategy we have been able to characterise the opioid peptides that were known from immunohistochemical studies to be present in heart and skin from the guinea-pig. In these cases the content of opioid peptides is comparatively low, and following the initial extraction of at acid pH, extraction in trichloroacetic acid (TCA) was also carried out. TCA extracts were extracted in ether and the aqueous phase was purified on a Sep-Pak C_{18} cartridge with application in 0.1% TFA in water and elution in 0.1% TFA in 80% propan-2-ol. The extracts from Sep-Pak were subjected to HPLC separations similar to that described above.

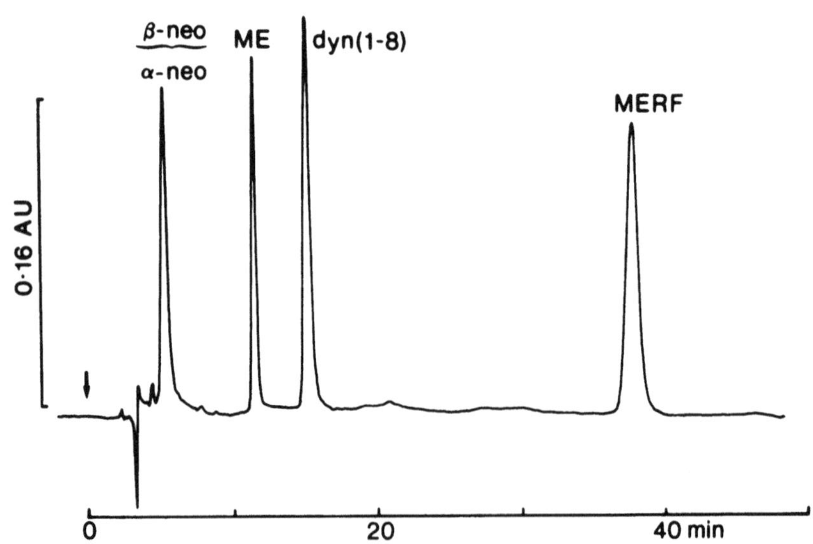

Fig. 11. Isocratic elution of endogenous opioid peptides (2.5-5 nmol of each) from an Altex Ultrasphere ODS column (5 μm, 4.6 × 250 mm) using 0.164 M ortho-phosphoric acid/0.189 M triethylamine in 18% aqueous acetonitrile (pH 4). Note that β-neo-endorphin elutes as a complex, multicomponent pattern under these conditions (not shown). Altex chromatograph, absorbance at 220 nm.

Fig. 12. Separation of *α*-neo-endorphin, *β*-neo-endorphin and [Met]enkephalyl-Arg-Gly-Leu (2.5–5 nmol of each) on Spherisorb S5 C8. Linear gradient between 0.1% in TFA in water and 0.1% TFA in 70% aqueous acetonitrile (75:25 to 55:45) in 45 min at 1 ml/min. Gilson chromatograph, absorbance at 280 nm. Note that the peptides that coelute in fraction IV in Figure 8 can thus be resolved.

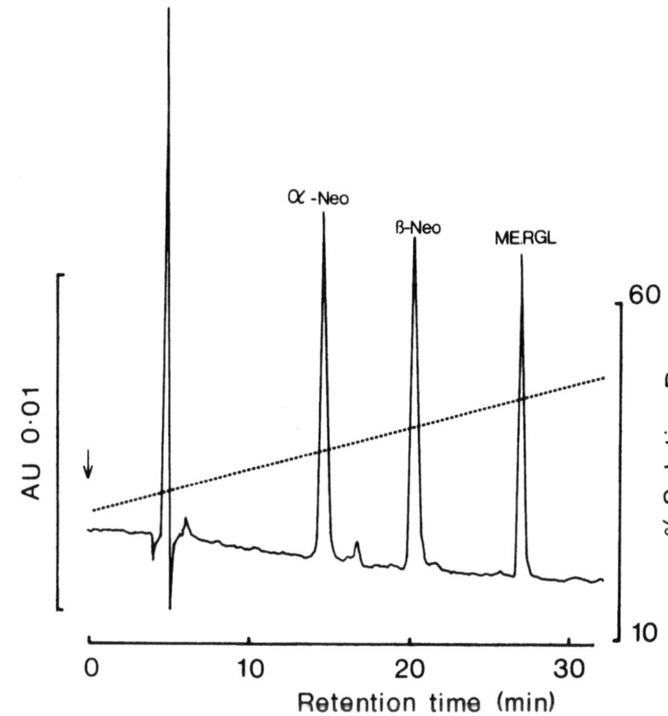

Fig. 13. Separation of [Met]enkephalyl-Arg-Arg-Val.NH_2, *β*-neo-endorphin, dnyorphin A (1–8) and [Met]enkephalyl-Arg-Gly-Leu (2.5–5 nmol) on Spherisorb S5 C8. Linear gradient between 0.1% TFA in water and 0.1% TFA in 70% acetonitrile (70:30 to 45:55) in 45 min at 1 ml/min. Note that the peptides co-eluting in fraction V in Figure 7 are thus resolved.

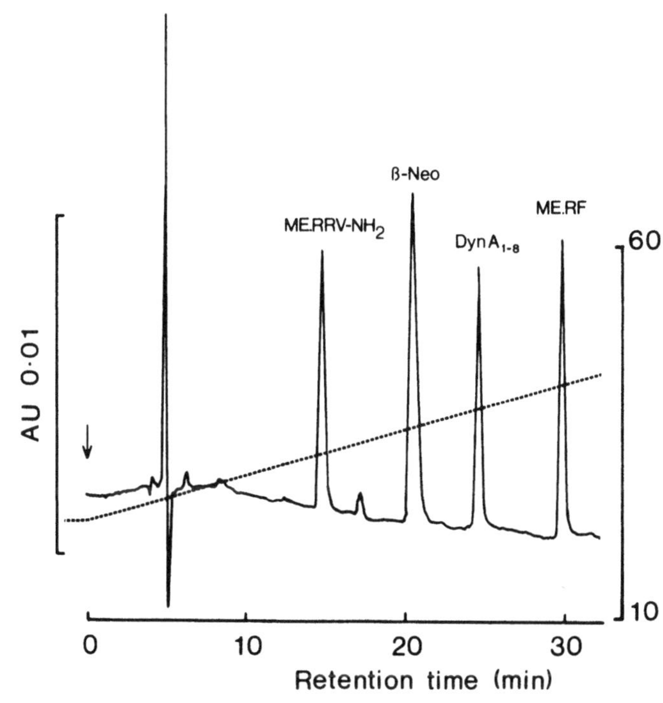

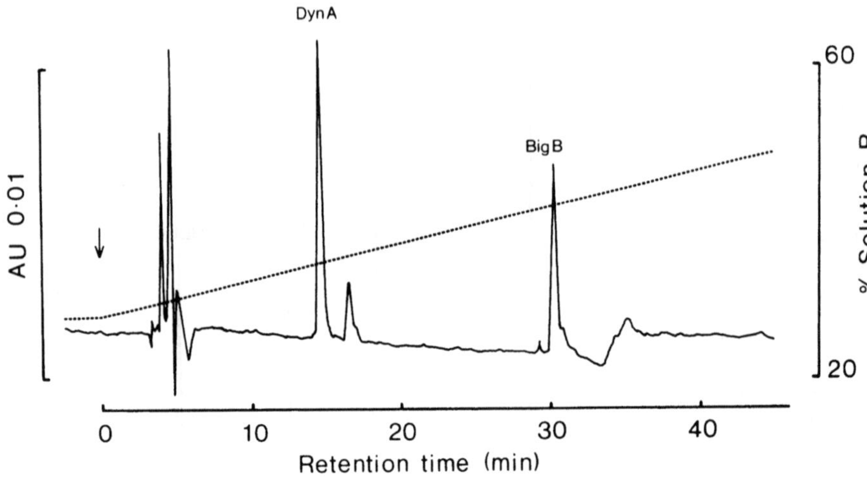

Fig. 14. Separation of dynorphin A and leumorphin ("Big B") (5 nmol) on Spherisorb S5 C8. Linear gradient between 0.1% TFA in water and 0.1% TFA in 70% acetonitrile (70:30 to 45:55) in 45 min at 1 ml/min. These two peptides coelute in fraction VII in Figure 7.

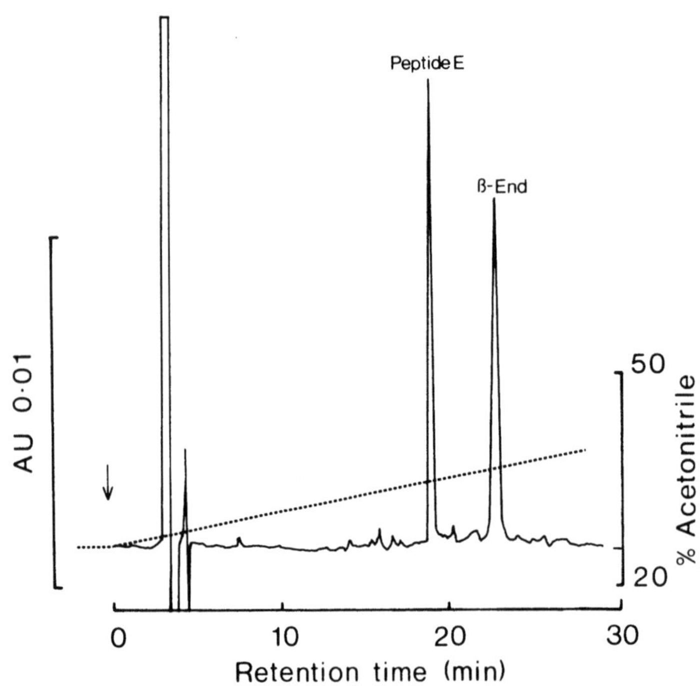

Fig. 15. Separation of peptide E and *β*-endorphin (human) (5 nmol) on Spherisorb S5 C8. Linear gradient between 0.1% TFA in water and 0.1% TFA in 70% acetonitrile (75:25 to 55:45) in 40 min at 1 ml/min. These two peptides coeluted in fraction IX in Figure 7.

Our findings indicated that opioid peptides were present in skin areas bearing the Merkel cell-rich sinus hair follicles and in atria, with low amounts, in the ventricles of the heart. The levels of [Met]enkephalin were approximately 3 pmol/g in the skin, 1 pmol/g in the atrium, with little or none in the ventricles. [Leu]enkephalin was detected in the ventricles, however, and the levels of this peptide in the atria were 5–6 times that for [Met]enkephalin; the presence of several products of the processing of prodynorphin suggests that in the heart the major source of opioid peptides is prodynorphin (Weihe et al., 1983; Weihe et al., 1985).

Opioid Peptides as Probes for Opioid Receptors

Our principle interest has been the characterisation of receptors and binding sites for opioids, and the analysis of the selectivity and potency or affinity of the various ligands that act at opioid receptors (see Kosterlitz et al., 1985). In this context the applications of HPLC are in the analytical mode, to check on the amount of opioid ligand in stock solutions, and to check on the stability of the ligand in solution, in storage or under experimental conditions. On several occasions it has unfortunately been necessary for us to extend this exercise in "quality control" to establishing the nature of the ligand in solution, where it may not be that which it purports to be.

In the case of the [^{3}H]-labelled ligands we routinely subject these to re-purification by elution from a reverse-phase column in methanol or acetonitrile gradients from ammonium acetate buffers, usually twice at different pH. By prior calibration with non-labelled standard we are able to confirm the identity of the material, and obtain our own estimation of specific activity. We use the synthetic peptide [D-Ala^2, Me-Phe^4, Gly-ol^5]enkephalin (RX 783006) as a selective probe for the μ-type of opioid receptor, this peptide is relatively hydrophillic species compared to the useful ligand for labelling the δ-receptor [D-Ala^2, D-Leu^5]enkephalin (BW 180 C) which elutes much later from for example an Ultrasphere ODS column (Fig. 16). It can be seen from Figure 16 that BW 180 C is clearly separated from the peptide BW 1064 C, the L-Leu^5-homologue [D-Ala^2, L-Leu^5]enkephalin. It was fortunate that we were aware of the relative retentions of these two peptides, since we were able to intercept a batch of [D-Ala^2, L-Leu^5]enkephalin that was being sold as the D-Leu^5-peptide.

It will be apparent that the usefulness of HPLC in this way depends upon the user having information with respect to the retention times under several chromatographic conditions of material whose authenticity has been verified by other means, such as FAB mass spectrometry and amino acid analysis. Several such examples of the need for stringent testing of commercially available peptides could be given, but the following one will serve to illustrate the power of HPLC in this regard.

We had obtained samples of two synthetic peptides that are highly selective for the δ-opioid receptor, [D-Pen^2, D-Pen^5]enkephalin and its L-Pen^5-homologue (see Corbett et al., 1984) in small amounts from an unimpeachable source and obtained the chromatographs that are shown in Figure 17. With 0.1% TFA as the

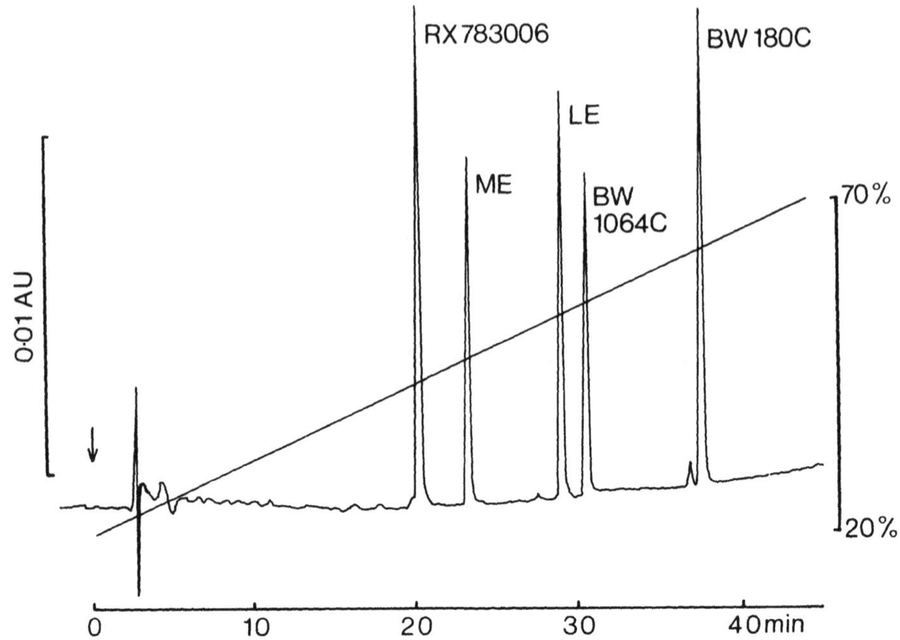

Fig. 16. Separation of [Met]enkephalin and [Leu]enkephalin from three synthetic enkephalin analogues (see text). Linear gradient between 1.1 M acetic acid/0.045 M ammonia (pH 3) and acetonitrile (80:20 to 30:70) in 45 min at 1 ml/min. Altex chromatograph, absorbance at 280 nm

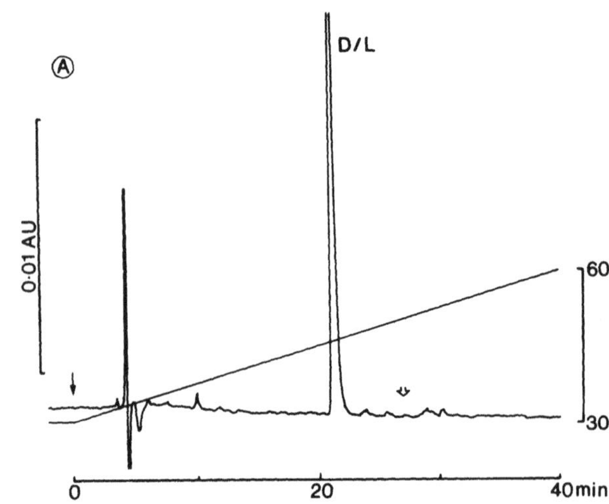

Fig. 17. HPLC separations of synthetic enkephalin analogues [D-$\overline{\text{Pen}^2, \text{D-Pen}}^5$]enkephalin (D) and [D-$\overline{\text{Pen}^2, \text{L-Pen}}^5$]enkephalin (L) under different conditions (Gilson chromatograph, absorbance at 280 nm). A: Partisil 5ODS3 (5 μm, 4.5 × 250 mm), linear gradient from 0.1% TFA in water to 0.1% TFA in aqueous acetonitrile (70:30 to 40:60) in 40 min at 1 ml/min. B-D: Spherisorb S5 C8 (5 μm, 4.5 × 250 mm), linear gradient between 1.1 M acetic acid/ammonia and acetonitrile (90:10 to 60:40) in 40 min at 1 ml/min. B: 0.045 M ammonia, pH 3. C: 0.45 M ammonia, pH 4.5. D: 1.1 M ammonia, pH 5.75. The arrows in A–C show the position for the elution of a commercially available product that purported to be the L-Pen^5-analogue

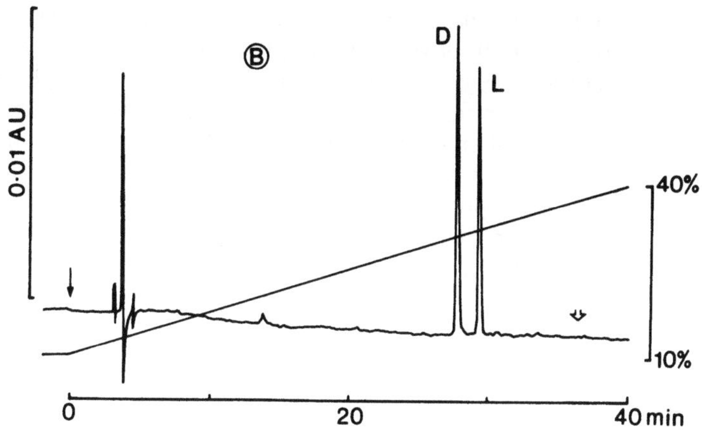

Fig. 17

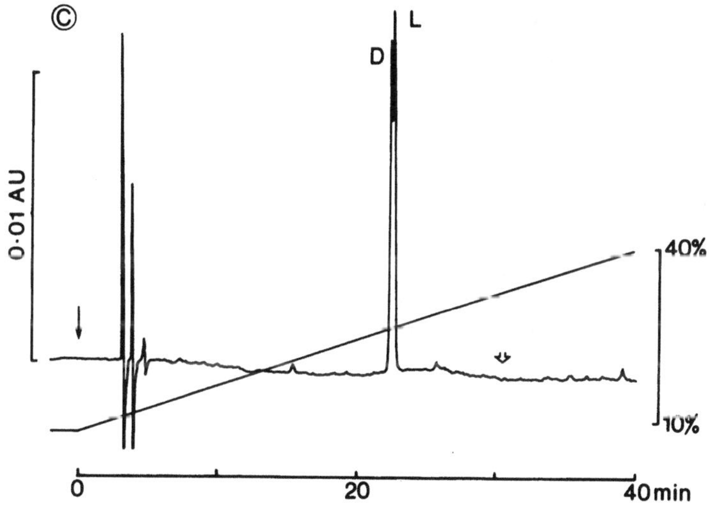

Fig. 17

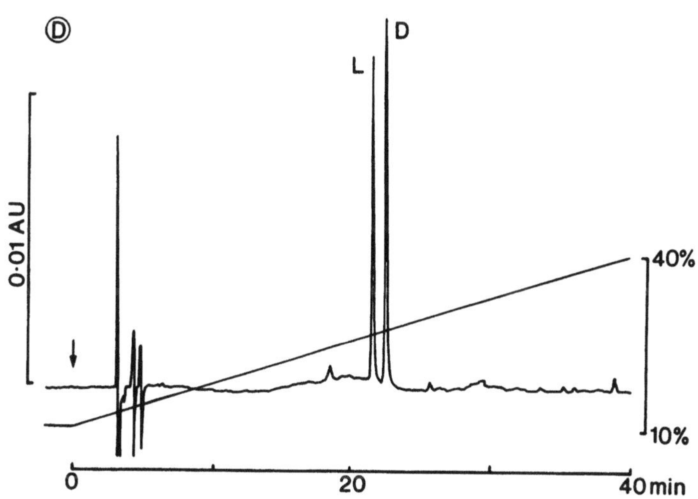

Fig. 17

aqueous buffer, or with an ammonium acetate system at pH 4, the two peptides co-elute (Fig. 17a, c), however if the pH of the ammonium acetate system is 3 (Fig. 17b), or 5.75 (Fig. 17d), the peptides are resolved, with their order of elution being different. By reference to these earlier results we were able to confirm that supplies of the D-Pen^5-compound that we had obtained from another source were reliable, but that the L-Pen^5-compound that we had purchased as the commercially available product (shown by the arrows in Fig. 17a–c) was not. The spurious material proved to be a major impurity in the synthesis of peptides of thes type that contain an intra-molecular disulphide bridge, the inter-molecular bridged dimer.

Conclusion

We have used several aspects of HPLC methodology to address various questions that are central to our interest in the functions of the endogenous opioid peptides. Such applications will be useful for neuropeptides, as a whole, however we believe that it is necessary to tailor the choice of column and mobile phases to achieve the precise set of conditions that is most appropriate to the desired separation. We further believe that the usefulness of on-line detection of neuropeptides may be limited, due to the low amounts of peptide that may have to be assayed, and to the likely presence of related molecular species, all of which may have to be resolved. Because of this last factor, the likely co-existence of various members of the one neuropeptide "family", we would suggest, in addition, that it will seldom be sufficient to effect only the one separation, rather that fractions from an initial spearation must be subjected to separations under different conditions from the first, for unequivocal results to be obtained.

References

Akil H, Watson SJ, Young E, Lewis ME (1984) Endogenous opioids: biology and function. Ann Rev Neurosci 7:223–255

Burgen A, Kosterlitz HW, Iversen L (eds) (1980) Neuroactive peptides. The Royal Society, London

Chang MM, Leeman SE (1970) Isolation of a sialogogic peptide from bovine hypothalamic tissue and its characterisation as substance P. J Biol Chem 245:4784–4790

Chang MM, Leeman SE, Niall HD (1971) Amino acid sequence of substance P. Nature [New Biol] 232:86–87

Corbett AD, Gillan MGC, Kosterlitz HW, McKnight AT, Paterson SJ, Robson LE (1984) Selectivities of opioid peptide analogues as agonists and antagonists at the δ-receptor. Brit J Pharmacol 83:271–279

Corbett AD, McKnight AT, Kosterlitz HW (1981) Tetraethylammonium facilitates the stimulation-evoked loss of the enkephalins from the myenteric plexus of the guinea-pig ileum. Proc Roy Soc Series B 213:171–182

Corbett AD, Paterson SJ, McKnight AT, Magnan J, Kosterlitz HW (1982) Dynorphin$_{1-8}$ and dynorphin$_{1-9}$ are ligands for the K-subtype of opiate receptor. Nature 299:79–81

Crawley JN (1985) Comparative distribution of cholecystokinin and other neuropeptides. Ann NY Acad Sci 448:1–8

Desiderio DM (1984) Analysis of neuropeptides by liquid chromatography and mass spectrometry. Techniques and Instrumentation in Analytical Chemistry, Vol 6, Elsevier, Amsterdam
Dunlapp III CE, Gentleman S, Lowney LI (1978) Use of trifluoroacetic acid in the separation of opiates and opioid peptides by reversed-phase high-performance liquid chromatography. J Chromatogr 160:191–198
Eccles JC (1964) The Physiology of Synapses. Springer-Verlag, Berlin
Euler US v, Gadum JH (1931) An unidentified depressor substance in certain tissue extracts. J Physiol 72:74–87
Feldman JA, Cohn ML, Blair D (1978) Neuroendocrine peptides - analysis by reversed phase high performance liquid chromatography. J Liq Chromatogr 1:833–848
Gillan MGC,Robson LE, McKnight AT, Kosterlitz HW (1985) K-binding and degradation of [^{3}H]dynorphin A(1–8) and]^{3}H]dynorphin A(1–9) in suspensions of guinea-pig brain membranes. J Neurochem 45:1034–1042
Gregory RA (ed) (1982) Regulatory Peptides of Gut and Brain. Churchill Livingstone, London
Hokfelt T, Everitt B, Holets VR, Meister B, Melander T, Schalling M, Staines W, Lundberg JM (1986) Coexistence of peptides and other active molecules in neurones: diversity of chemical signalling potential. In: Iversen LL, Goodman E (eds) Fast and Slow Chemical Signalling in the Nervous System, Oxford University Press, Oxford pp 205–231
Hokfelt T, Lundberg JM, Schultzberg M, Johansson O, Skirboll L, Anggard A, Fredholm B, Hamberger B, Pernow B, Rehfeld J, Goldstein M (1980) Cellular localization of peptides in neural structures. Proc R Soc Lond (Biol) 210:63–77
Hughes J (1975) Isolation of an endogenous compound from the brain with pharmacological properties similar to morphine. Brain Res 88:295–308
Hughes J (ed) (1978) Centrally Acting Peptides. Macmillan, London
Hughes J, Smith TW, Kosterlitz HW, Fothergill LA, Morgan BA, Morris HR (1975) Identification of two related pentapeptides from the brain with potent opiate agonist activity. Nature 258:577–579
Iversen LL (1983) Nonopioid neuropeptides in mammalian CNS. Ann Rev Pharmacol Toxicol 23:1–27
Iversen LL (1984) Amino acids and peptides: fast and slow chemical signals in the nervous system? The Ferrier Lecture 1983. Proc R Soc London (Biol) 221:245–260
Iversen LL, Goodman E (eds) (1986) Fast and Slow Chemical Signalling in the Nervous System, Oxford University Press, Oxford
Kosterlitz HW (1978) Endogenous opioid peptides: historical aspects. In: Hughes J (ed) Centrally Acting Peptides. Macmillan, London, pp 157–159
Kosterlitz HW (1979) The best laid schemes o' mice an' men gang aft agley. Ann Rev Pharmacol Toxicol 19:1–12
Kosterlitz HW, Corbett AD, Gillan MGC, McKnight AT, Paterson SJ, Robson LE (1985) Recent developments in the bioassay of opioids. Regulatory Peptides 11: suppl 4, 1–7
Kosterlitz HW, McKnight AT (1981) Opioid peptides and sensory function. Progr in Sensory Physiol 1:31–95
Kupfermann I (1979) Modulatory actions of neurotransmitters. Ann Rev Neurosci 2:447–465
Lewis RV, Stern AS (1983) Biosynthesis of the enkephalins and enkephalin-containing polypeptides. Ann Rev Pharmacol Toxicol 23:353–372
Lewis RV, Stern S, Gerber LD, Rubinstein M, Udenfriend S (1978) High molecular weight opioid-containing proteins in striatum. Proc Natl Acad Sci USA 75:4021–4023
Loh YP, Brownstein MJ, Gainer H (1984) Proteolysis in neuropeptide processing and other neural functions. Ann Rev Neurosci 7:189–222
McKnight AT, Corbett AD, Kosterlitz HW (1981) Inhibition of their breakdown and release of endogenous opioid peptides from the myenteric plexus-longitudinal muscle of the guinea-pig ileum. In: Tagaki H (ed) Advances in Endogenous and Exogenous Opioids. Kodansha, Tokyo, pp 188–190
McKnight AT, Corbett AD, Kosterlitz HW (1983) Increase in potencies of opioid peptides after peptidase inhibition. Eur J Pharmacol 86:393–402
McKnight AT, Corbett AD, Paterson SJ, Magnan J, Kosterlitz HW (1982) Comparison of in vitro potencies in pharmacological assays and binding assays after inhibition of peptidases reveals that dynorphin (1–9) is a potent κ-agonist. Life Sci 31:1725–1728

McKnight AT, Hughes J, Kosterlitz HW (1979) Synthesis of enkephalins by guinea-pig striatum *in vitro*. Proc R Soc London (Biol) 205:199–207
McKnight AT, Sosa RP, Corbett AD, Kosterlitz HW (1980) Enkephalin precursors from guinea-pig myenteric plexus. In: Way EL (ed) Endogenous and Exogenous Opiate Agonists and Antagonists. Pergamon Press, New York, pp 213–216
McKnight AT, Sosa RP, Hughes J, Kosterlitz HW (1978) Biosynthesis and release of enkephalins. In: van Ree JM; Terenius L (eds) Characteristics and Function of Opioids. Elsevier/North-Holland Biomedical Press, Amsterdam, pp 259–269
Mahy N, Tuselll JM, Gelpi E (1982) Study of leucine enkephalin in rat brain by a new high performance liquid chromatographic method. J Chromatogr 233:115–122
Minamino N, Kangawa K, Fukuda A, Matsuo H, Igarashi M (1980) A new opioid octapeptide related to dynorphin from porcine hypothalamus. Biochem Biophys Res Commun 95:1475–1481
Mousa S, Couri D (1983) Analysis of enkephalins, β-endorphin and small peptides in their sequences by highly sensitive high-performance liquid chromatography with electrochemical detection: implications in opioid peptide metabolism. J Chromatogr 267:191– 198
Pearse AGE (1978) Diffuse neuroendocrine system: peptides common to brain and intestine and their relationship to the APUD concept. In: Hughes J (ed) Centrally Acting Peptides. Macmillan, London, pp 49–57
Pernow B (1983) Substance P. Pharmacol Rev 35:85–141
Polak JM, Bloom SR (1982) Peripheral regulatory peptides: a newly discovered control system. In: Fink G, Whalley LJ (eds) Neuropeptides: Basic and Clinical Aspects. Churchill Livingstone, Edinburgh, pp 118–147
Smyth DG (1984) Chromatography of peptides related to β-endorphin. Analyt Biochem 136:127–135
Sosa RP, McKnight AT, Hughes J, Kosterlitz HW (1977) Incorporation of labelled amino acids into the enkephalins. FEBS Lett 84:195–198
Weber E, Esch FS, Bohlen P, Paterson SJ, Corbett AD, McKnight AT, Kosterlitz HW, Barchas JD, Evans CJ (1983) Metorphamide: isolation, structure, and biological activity of an amidated opioid octapeptide from bovine brain. Proc Natl Acad Sci USA 80:7362–7366
Weihe E, McKnight AT, Corbett AD, Hartschuh W, Reinecke M, Kosterlitz HW (1983) Characterization of opioid peptides in guinea-pig heart and skin. Life Sci 33:suppl I, 711–714
Weihe E, McKnight AT, Corbett AD, Kosterlitz HW (1983) Proeenkephalin- and prodynorphin-derived opioid peptides in guinea-pig heart. Neuropeptides 5:453–456
White MW (1983) High-performance liquid chromatography of tyrosinerelated peptides with electrochemical detection. J Chromatogr 262:420–425
Yalow RS (1978) Radioimmunoassay: a probe for the fine structure of biologic systems. Science 200:1236–1245

High-Performance Liquid Chromatography of Vitamin D and its Application to Endocrinology

G. JONES, and H. F. DELUCA

HPLC of Vitamin D and its Metabolites

Introduction

The newest arrival among the steroid hormones is 1,25-dihydroxyvitamin D_3 [1,25-$(OH)_2D_3$], which is derived from the vitamin D molecule. Although vitamin D has been known since 1919–1922 (McCollum, 1922; Mellanby, 1919), an appreciation that it is a precursor of a steroid hormone did not exist until the discovery that the vitamin D molecule must be altered metabolically in the liver and kidney before function (DeLuca, 1972, 1974). This led to a delineation of the vitamin D endocrine system based in the kidney which gave new understanding of metabolic bone diseases as well as provided new tools for diagnosis and treatment. This exciting development has led to a great deal of effort in the measurement of the vitamin D compounds, not only for nutritional studies but more importantly for clinical investigation of disease in preparation for treatment. Since the vitamin D hormone and some of its important anlogs are currently used for treatment of metabolic bone disease, a large number of analogs have now been prepared, some of which will find their way into human as well as agricultural use. Thus, analysis of the metabolites derived from vitamin D and their analogs has become an exceedingly important field. This chapter is devoted to those measurements and their meaning.

Figure 1 provides an illustration of the principal nutritional forms of vitamin D, one of which is synthesized in skin as a result of ultraviolet light. Although there are potentially five nutritional forms of vitamin D, there are only two of any significance. Vitamin D_3 or cholecalciferol or in the new terminology "calciol" is prepared commercially from chemically synthesized or isolated 7-dehydrocholesterol. It is also this compound that is synthesized in skin as a result of the photochemical process induced by ultraviolet light of 282 nm wavelength. It represents the most significant nutritional form of vitamin D since it is universally applied to all species including man.

The next most important form of vitamin D is vitamin D_2, ergocalciferol, or "ercalciol" by modern nomenclature. This compound is prepared commercially by ultraviolet irradiation of the plant and yeast sterol, ergosterol. Historically, this was the major form of vitamin D produced after discovery of vitamin D at the early part of this century. More recently with the improved methods of chemical synthesis of 7-dehydrocholesterol, ergocalciferol or vitamin D_2 has

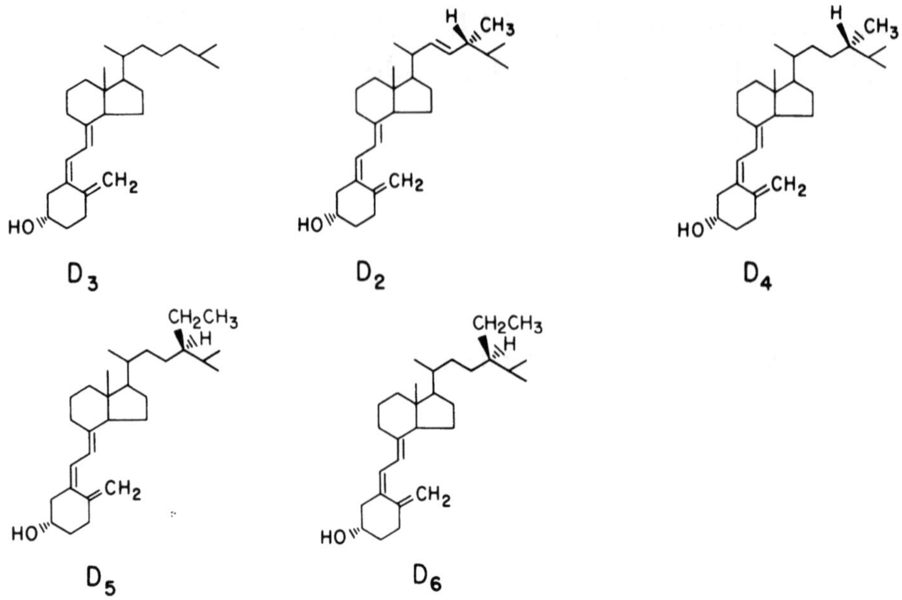

Fig. 1. The nutritional forms of vitamin D. Only vitamins D_2 and D_3 are of practical significance; the other nutritional forms are chemical or biochemical curiosities since they do not occur to any degree in nature nor are they of practical significance

diminished both in commercial and medical importance. It is nevertheless widely used as a constituent of human vitamin supplements and in feeding domestic animals other than birds. Vitamin D_4 which is the reduction product of ergocalciferol is only of biochemical interest as is vitamin D_5. Vitamin D_1 does not exist, inasmuch as it was misidentified as a separate form of vitamin D when in actual fact it was a mixture of vitamin D_2 and tachysterol$_2$ (Bills, 1954).

The principal physical constants for the vitamin D compounds are provided in Table 1 (DeLuca, 1984). Vitamin D is a seco sterol in which the B ring is replaced by a 5,7-diene bridge. The 5,7-diene bridge, together with the 10,19-double bond, constitutes a cis-triene structure that provides the chromophore for this molecule. It has a maximum ultraviolet absorption at 265 with a minimum at 228. The molar extinction coefficient is 18 200 for all forms of vitamin D. The mass spectral fragments and NMR absorption bands are also illustrated in Table 1 for vitamins D_2 and D_3. From these two compounds, therefore, are derived the metabolites and analogs to be discussed below.

Biochemistry of Vitamin D

Vitamin D_2 and vitamin D_3 are not found in abundance in nature (DeLuca, 1984). In fact, they are virtually absent from the food supply with the notable exception

Table 1. Physical Properties of Vitamin D Compounds

Parameter	Vitamin D_2	Vitamin D_3
u. v. absorption		
λ_{max} (nm)	265	265
λ_{max} (nm)	228	228
Extinction coefficient	18 200	18 200
Mass spectra (m/z)		
M*	396	384
M-side chain	271	271
M-side chain-H_2O	253	253
A-ring + C_6 + C_7	136	136
A-ring + C_6 + C_7-H_2O	118	118
NMR (PPM)		
(J = CPS)		
C_{18}-H_3	s 0.54	s 0.54
C_{21}-H_3	d 0.93 (J = 5)	d 0.93 (J = 5)
$C_{26,27}$-H_6	d 0.87 (J = 7)	d 0.87 (J = 6.5)
	d 0.98 (J = 7)	
$C_{6,7}$-H_5	d 6.02 (J = 11)	d 6.02 (J = 11)
	d 6.24 (J = 10.5)	d 6.24 (J = 10.5)
C_{19}-H_2	s 4.81	s 4.81
	s 5.03	s 5.03

* Molecular ion

of fish liver oils and other such unusual food sources. Thus, an individual not consuming fish liver oils would likely become deficient in vitamin D were it not for the fact that vitamin D is normally produced in skin as a result of ultraviolet irradiation. Therefore, the first step in vitamin D biochemistry is its photochemical production in skin from the sterol, 7-dehydrocholesterol (Holick, 1981). In the malpighian layer of skin, as in the preputial glands, is found abundant quantities of 7-dehydrocholesterol (Bloch, 1965; Kandutsch, 1964). This compound is a possible intermediate in the biosynthesis of cholesterol, needing only a Δ^7 reduction to become the basic biological constituent, cholesterol. 7-Dehydrocholesterol in skin serves not only as a possible precursor for cholesterol but is obviously a precursor of vitamin D. This compound absorbs 282 nm ultraviolet light bringing about a photolysis reaction involving rupture of the 9,10-bond and a sigmatropic shift to produce previtamin D as shown in Figure 2 (Velluz et al., 1949). Previtamin D is a biologically inert form of the molecule and in fact remains in skin until it isomerizes to the vitamin D structure (Holick & Clark, 1978). At body temperatures, approximately 36 hours are required for the equilibration of previtamin D to vitamin D. A protein in the plasma called vitamin D transport protein or DBP does not recognize previtamin D, whereas it binds vitamin D_3 itself (Holick & Clark, 1978). This plasma transport protein therefore transports vitamin D to the liver where it begins its initial metabolic alteration required for function (Olson et al., 1976; Ponchon & DeLuca, 1969). It is, therefore, important to realize that the natural source of vitamin D is the photochemi-

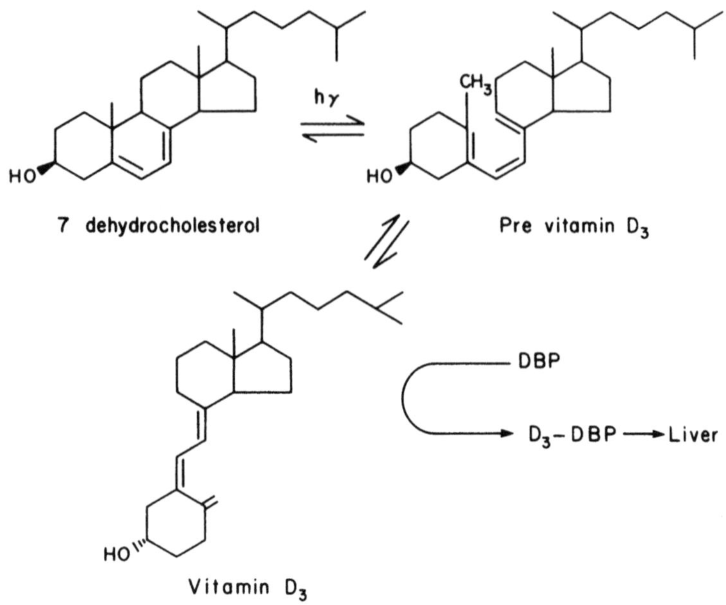

Fig. 2. Conversion of 7-dehydrocholesterol to vitamin D_3 by a chemical photolysis-equilibration reaction. The vitamin D transport protein (DBP) binds vitamin D_3 but does not bind previtamin D_3 in order to transport it to the liver for the beginning of its metabolic activation

cal process occuring in skin. This process does not require enzymes or proteins. It is quite analogous to the chemical photolysis reaction well known for the production of vitamin D since the discovery of Steenbock and his associates in 1924 (Steenbock). The details of the photochemical process have been carefully worked out by Havinga, Velluz, Amiard, and others (Havinga, 1973; Velluz & Amiard, 1949; Velluz et al., 1949).

As discovered in 1968, vitamin D_3 is converted in the liver to the circulating form of the vitamin, 25-hydroxyvitamin D_3 (25-OH-D_3) (Blunt et al., 1968; Ponchon et al., 1969) as shown in Figure 3. This reaction is carried out by a micro-

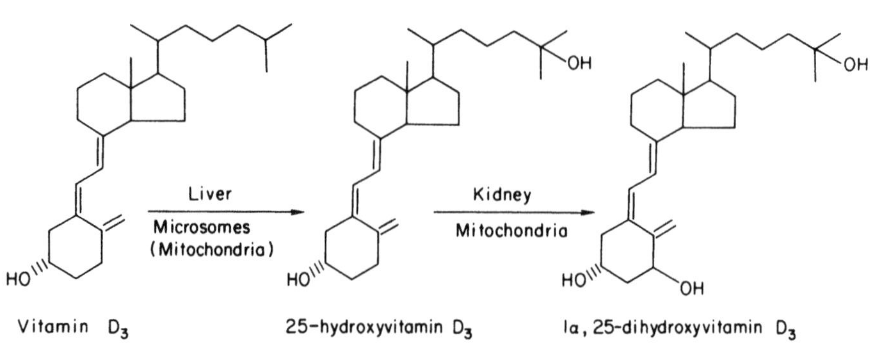

Fig. 3. Metabolism of vitamin D required for function

somal system requiring NADPH, molecular oxygen, a flavoprotein and a cytochrome P-450 (Madhok & DeLuca, 1979). Additional 25-hydroxylation is carried out by a mitochondrial cholesterol 25-hydroxylase (Björkhem et al., 1980) as well as an ill-defined system in intestine and kidney (Tucker et al., 1973). However from a physiologic point of view, virtually all of the 25-hydroxylation occurs in the liver (Olson et al., 1976). 25-Hydroxylation of vitamin D takes place on both vitamins D_2 and D_3 (Jones et al., 1976). 25-OH-D_3 is the common intermediate for all known metabolism of vitamin D (DeLuca & Schnoes, 1983). It is the major circulating form of the vitamin, being present at between 10–40 ng/ml. It is this form of vitamin D that is commonly measured to assess the vitamin D status of patients. This compound, however, does not function physiologically directly in mammals and higher animals but must be metabolized further (Boyle et al., 1972; Holick et al., 1972a). In the non-pregnant mammal, this metabolism occurs exclusively in the kidney and exclusively in the mitochondrial fraction (Fraser & Kodicek, 1970; Reeve et al., 1983; Shultz et al., 1983). This system converts 25-OH-D_3 to 1α,25-$(OH)_2D_3$ (Gray et al., 1972; Holick et al., 1971). This 1α-hydroxylation is carried out by a system located in the mitochondrial membrane involving NADPH, a flavoprotein, and iron sulfur protein, and a cytochrome P-450 (Ghazarian et al., 1974; Pedersen et al., 1976). The P-450 portion of this system is highly regulated as will be discussed in the endocrine section of this chapter. 1,25-$(OH)_2D_3$ is then transported from the proximal convoluted tubule cells where 1α-hydroxylation takes place (Brunette et al., 1978), to the enterocytes of the small intestine (Stumpf et al., 1979), to the distal renal tubule sites of the kidney (Stumpf et al., 1980), and to the bone lining cells, osteoblasts, and promyelocyte stem cells of the bone (Stumpf et al., 1981). In these cells, 1,25-$(OH_2D_3$ interacts with a specific receptor for 1,25-$(OH)_2D_3$ found in the nucleus (Brumbaugh & Haussler, 1975; Kream et al., 1976). As a result of the interaction between the ligand and the receptor, transcription of specific structural genes takes place resulting in messenger RNA that code for calcium and phosphorus transport proteins (DeLuca, 1988). These calcium and phosphorus transport proteins then stimulate the transfer of calcium transcellularly from the lumen of intestine to the plasma compartment, from the distal renal tubule fluid to plasma, and from the bone fluid compartment to the plasma compartment (DeLuca et al., 1982). The molecular mechanism whereby 1,25-$(OH)_2D_3$ with its receptor initiates the expression of these specific genes and the nature of the gene products as well as how they bring about the transfer of calcium and phosphorus remains largely unknown and represents a major area of current investigation. Recent work by Suda's group (Miyaura et al., 1981; Abe et al., 1981; Abe et al., 1983) suggests that 1,25-$(OH)_2D_3$ may play an additional receptor-mediated role in bone cell differentiation by inducing maturation of monocytes into macrophages and stimulating fusion of macrophages to produce multinucleated cells resembling osteoclasts.

1,25-$(OH)_2D_3$, once it functions in the target cells, is metabolized. Little is known concerning the physiologic metabolism of 1,25-$(OH)_2D_3$. However as is shown in Figure 4, 1,25-$(OH)_2D_3$ can be hydroxylated in the 24R-position to form 1,24,25-dihydroxyvitamin D_3. It can also be oxidized to the C-23 carboxylic acid, called calcitroic acid. The pathway of this metabolism remains largely unknown

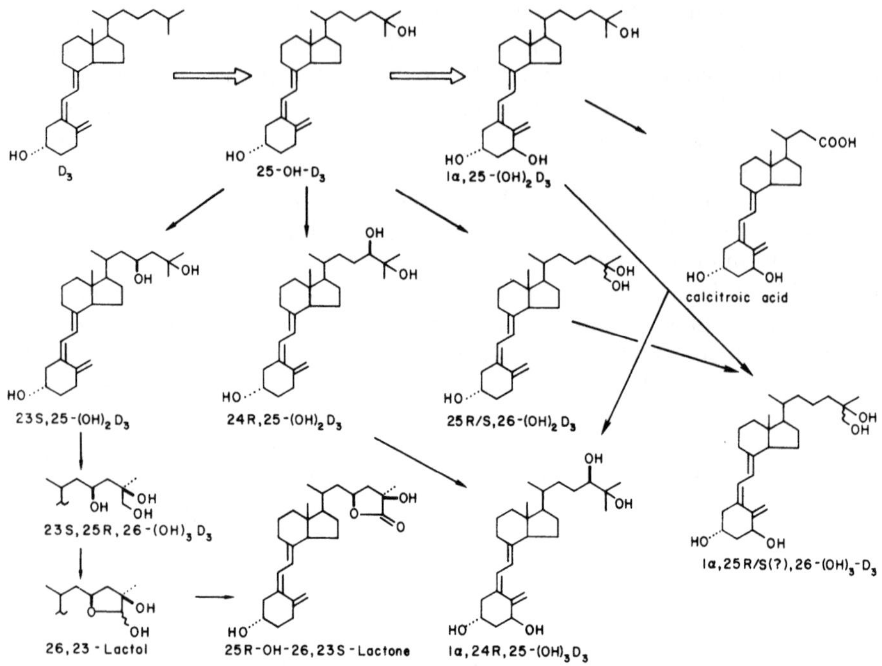

Fig. 4. Physiologically significant metabolism of vitamin D. Large arrows represent activation of vitamin D to its hormonal form. Other arrows represent reactions considered to be either degradation reactions or reactions whose significance remains inknown. These metabolites are those that have been identified to occur under physiologic circumstances *in vivo*

and many postulated intermediates have been reported (DeLuca & Schnoes, 1984). For example, one incorrect intermediate has been postulated to be 23-oxo-1,25-$(OH)_2D_3$ (Ohnuma & Norman, 1982) that was then later identified as 24-oxo-1,25-$(OH)_2D_3$ (Mayer et al., 1983). Although the 24-oxo compound probably does exist in *in vitro* incubations and in animals given large doses of vitamin D, whether it occurs as the correct pathway leading to calcitroic acid remain ed unclear until recently.

Recent studies using the perfused rat kidney and the rat osteosarcoma UMR-106 (Lohnes & Tones, 1987; Lohnes et al., 1988) now suggest-that 1,25-$(OH)_2D_3$ is metabolised first to 1,24,25-$(OH)_3D_3$, then to 24-oxo-1,25-$(OH)_2D_3$ and 24-oxo-1,23,25-$(OH)_3D_3$ before side-chain clearage to tetranor-1,23-$(OH)_2D_3$. This C-23 alcohol is ultinately converted by a 1,25-$(OH)_2D_3$-inducible enzyme to the C-23 carboxylic acid, calcitroic acid inside target cells (Lohnes et al., 1988).

Besides formation of the hormonal form of vitamin D, namely 1,25-$(OH)_2D_3$, 25-OH-D_3 can be converted to a variety of other metabolites. Many new metabolites have been reported from *in vitro* incubations, from perfusion of large amounts of compound into isolated organs, or from incubation of tissue with large amounts of substrate (DeLuca & Schnoes, 1984; Horst et al., 1983; Horst et al., 1984; Jones et al., 1983; Jones et al., 1984; Mayer et al., 1983; Napoli & Horst, 1983a, b; Ohnuma & Norman, 1982; Reinhardt et al., 1981; Simpson et al., 1984; Takasaki et al., 1981; Yamada et al., 1983). Whether these metabolites exist *in vivo* under physiological circumstances or not remains to be determined. However, the major pathways of vitamin D metabolism that can be documented as being present *in vivo* are shown in Figure 4. The most important reactions are 26-hydroxylation to form the 25,26-$(OH)_2D_3$, as an epimeric mixture, 24R-hydroxylation to form 24R,25-dihydroxyvitamin D_3 (Holick et al., 1972b) and 23-oxidation followed by oxidation of the 26-carbon resulting in the 26,23-lactol, ultimately becoming 26,23-lactone (Tanaka et al., 1981a, b). 26-Hydroxylation and lactone formation result in largely much less active, if not biologically inactive, metabolites as far as calcium and phosphorus homeostasis is concerned. 24R-Hydroxylation results in a compound that retains biological activity when administered to vitamin D-deficient animals presumably because of further in vivo 1α-hydroxylation (Tanaka et al., 1975). Furthermore, this compound is formed when 1α-hydroxylation is shut down under physiological conditions (DeLuca, 1984).

24,25-Dihydroxyvitamin D_3 has been postulated as being required for a number of functions of vitamin D for which there is inadequate evidence (Henry & Norman, 1978; Henry et al., 1977; Norman et al., 1980; Ornoy et al., 1978; Rasmussen & Bordier, 1978). Chemical synthesis of the 24,24-difluoro-25-OH-D_3, 23,23-difluoro-25-OH-D_3, and 26,26,26,27,27,27-hexafluoro-25-OH-D_3 (Kobayashi et al., 1980, 1982; Taguchi et al., 1984; Yamada et al., 1979) have led to experiments in which it can be clearly demonstrated that the side-chain modifications play no role in the physiologic funtions of vitamin D. Thus, all of the known and proven functions of vitamin D are completely satisfied by these fluorinated derivatives for which modification on the fluorinated carbons cannot take place (Brommage et al., 1983; Jarnagin et al., 1983; Nakada et al., 1985; Okamoto et al., 1981, 1983; Parfitt et al., 1984; Tanaka et al., 1979a, 1982). On the other hand, substitution of the fluoro groups on the 25-position or the 1-position virtually eliminates biological activity of the vitamin D molecule (Napoli et al., 1979; Onisko et al., 1979). Thus, 1α-hydroxylation and 25-hydroxylation are the only known modifications of the vitamin D molecule required for function.

Placenta, fetal renal tissue, and sarcoid tissue (see below) are known in vivo sites of 1α-hydroxylation of 25-OH-D_3 (Lester et al., 1981; Tanaka et al., 1979b). The physiologic significance of these sites of 1α-hydroxylation remains unknown except they may provide 1,25-$(OH)_2D_3$ in large quantities when required for physiologic functions requiring large amounts of calcium and/or phosphorus. The nature of the hydroxylases found in these two organs has not yet been determined. Furthermore the dramatic reports of extra-renal synthesis of 1α,25-$(OH)_2D_3$ by bone cells, intestinal cells, and other tissues must be regarded as tissue culture curiosities since nephrectomy of animals or man eliminates the appearance of 1α,25-$(OH)_2D_3$ in blood and tissues (Reeve et al., 1983; Shultz et al., 1983). Thus for all practical purposes, the kidney and placenta represent the only significant sites of 1α-hydroxylation of vitamin D to form the final vitamin D hormone.

The Vitamin D Endocrine System

The production of 1α,25-$(OH)_2D_3$ by the kidney is very tightly regulated. Under conditions of low blood calcium (Boyle et al., 1971), low dietary calcium (Boyle et

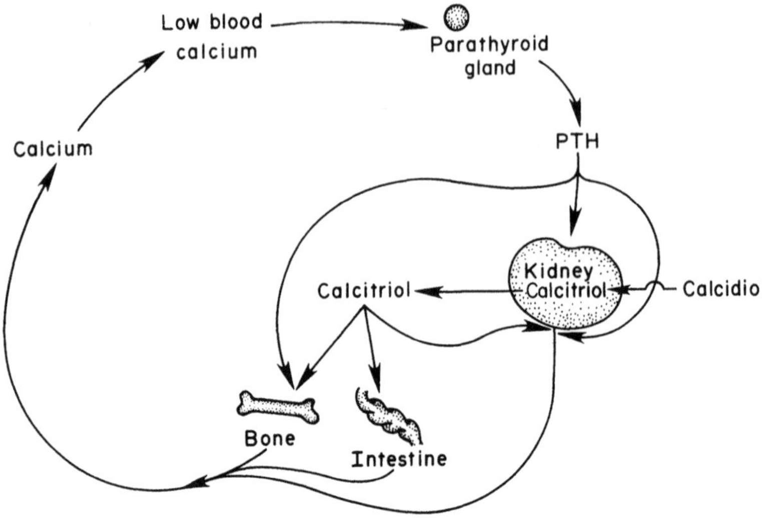

Fig. 5. Diagrammatic representation of the vitamin D endocrine system. Calcidiol represents 25-OH-D; calcitriol represents $1\alpha,25\text{-}(OH)_2D_3$. Low blood calcium stimulates parathyroid glands to secrete parathyroid hormone. Parathyroid hormone triggers conversion of $25\text{-}OH\text{-}D_3$ or calcidiol to calcitriol. Calcitriol then functions in bone, intestine, and elsewhere in the kidney with or without parathyroid hormone to bring about the mobilization of calcium. This causes a rise in blood calcium that then suppresses the parathyroid glands

al., 1971), or low blood phosphorus (Tanaka & DeLuca, 1973), $1,25\text{-}(OH)_2D_3$ production is markedly increased, $1,25\text{-}(OH)_2D_3$ then stimulates intestine, bone, and kidney to provide calcium and phosphorus as required (DeLuca, 1984). The discovery of this regulation led to an elucidation of the vitamin D endocrine system as illustrated in Figure 5. Low blood calcium stimulates the parathyroid glands to secrete parathyroid hormone. Parathyroid hormone stimulates the kidney by blocking proximal convoluted tubule reabsorption of phosphorus and in the same cell, stimulates $25\text{-}OH\text{-}D\text{-}1\alpha$-hydroxylase (Brunette et al., 1978; Garabedian et al., 1972). The $1,25\text{-}(OH)_2D_3$ thus produced stimulates the intestine to absorb calcium and phosphorus (DeLuca, 1974). It stimulates bone to mobilize calcium from the bone fluid compartment when directed by parathyroid hormone (Garabedian et al., 1974) and, furthermore, it stimulates renal reabsorption of calcium when directed by parathyroid hormone (Sutton & Dirks, 1978). In the case of the kidney and bone, both parathyroid hormone and $1,25\text{-}(OH)_2D_3$ are involved in calcium mobilization. In the case of the small intestine, parathyroid hormone is not further involved once $1,25\text{-}(OH)_2D_3$ has been secreted (Garabedian et al., 1974). These sources of calcium raise the blood calcium concentration, suppressing parathyroid secretion thereby reducing 1α-hydroxylation. By this mechanism, therefore, calcium mobilization is very tightly regulated.

Under conditions of great calcium need, the parathyroids stimulate $1,25\text{-}(OH)_2D_3$ production. Thus, high concentrations of this hormone are found under physiologic circumstances, as for example when animals are put on a low calcium diet (Tanaka & DeLuca, 1983) when excessive calcification of organic matrix of bone is required (Papapoulos et al., 1980), during last stages of pregnancy when bone mineral is being laid down (Halloran et al., 1979), and during lactation when large amounts of calcium are required for milk formation (Halloran et al., 1979). On the other hand, high calcium diets suppress 1,25-$(OH)_2D_3$ production (Tanaka & DeLuca, 1983) as does removal of the parathyroid glands (Garabedian et al., 1972). Thus, the vitamin D system is eminently involved in the regulation of plasma calcium concentration. Furthermore, the vitamin D hormone is the only hormone involved in utilization of calcium from the environment, namely by stimulating the intestinal calcium absorption (DeLuca et al., 1982). This mechanism is extremely important in bringing in environmental calcium thereby protecting against loss of calcium by the skeleton (DeLuca, 1985). A condition of osteporosis may well result from failure to synthesize adequate amounts of the vitamin D hormone when required by the organism. This can result in a continued reliance upon bone calcium for soft tissue needs bringing about a bone loss (DeLuca, 1985). Thus, the vitamin D endocrine system is of central importance in protecting the skeleton against physiological loss of calcium.

It is obvious that measurement of vitamin D hormone in a variety of circumstances therefore is of great importance from a diagnostic point of view. It is also obvious that a great deal remains to be learned concerning metabolism and function of the vitamin D substances, necessitating therefore special use of extraction, separation, and detection methods of the vitamin D compounds. The sections below are devoted to a presentation of the current methods in use.

Separation Techniques

General Remarks

Vitamin D and its metabolites represent a family of lipid-soluble, non-polar, multihydroxylated sterols. The fact that they have been termed "vitamins" implies that they are trace materials present in biological fluids at a nanomolar concentration or less. Since vitamin D metabolites are usually present in an environment richer in proteins and lipids the usual strategy, prior to attempting separation techniques, is to remove the bulk of the proteins and some charged lipids. The extraction and partial purification of vitamin D metabolites prior to HPLC have the combined effects of greatly reducing the possibility of overloading the HPLC column with solute and also minimising contamination with UV interfering materials thereby making possible UV detection techniques during HPLC.

Vitamin D and its metabolites have been traditionally separated by adsorption and liquid-liquid partition chromatography. Thus HPLC on microparticulate silica and reverse phase bonded packings are still favoured. Since these molecules cover a narrow polarity range they can be separated easily by isocratic or shallow gradient solvent systems based upon hexane or methylene chloride for adsorption chromatography; or based upon methanol or acetonitrile for reverse-phase HPLC. Currently, resolution of isomers or metabolites, once considered difficult or impossible by open column techniques, is being routinely achieved by HPLC in most vitamin D laboratories.

Extraction and Purification Techniques

Several approaches exist for extraction of vitamin D and its metabolites from biological fluids. These can be listed under the following headings:

Total lipid Extraction

This was pioneered by Bligh & Dyer (1959) and provides, as the name implies, quantitative extraction of most neutral and some polar lipid after monophasic dispersion of the aqueous biological tissue or fluid in an organic environment consisting of methanol and methylene chloride. Since lipid contents of extracts from this technique are high; there is an increased burden placed upon the purification techniques combined with total lipid extraction. Several purification techniques have been used to reduce the lipid content of extracts prior to HPLC including miniature Sephadex LH-20 columns (0.9 cm × 10 cm) (Eisman et al., 1976), small silica columns (Belsey et al., 1971) or commercially available cartridges (e.g. Sep-Pak SIL, Waters, Milford, Mass.) (Adams et al., 1981). Such a combination of total lipid extraction and a mini open column offers the advantage of a reliable quantitative yield of most D metabolites from protein-lipid complexes and in addition some subdivision of the vitamin D metabolites into general classes on the basis of their hydroxyl content. The main disadvantage of this method is that it is time consuming.

Selective Extraction

Uses a combination of organic solvent(s) which selectively partitions out the vitamin D or its metabolite into the organic phase without the bulk of the other lipids. Such methods are usually designed to extract only one or perhaps a class of hydroxylated metabolites and rarely can be applied to all vitamin D metabolites. Nevertheless, these are extremely useful techniques since the extracts one obtains are generally less contaminated with lipids that those extracts derived from total lipid extraction. Solvents used for this type include: methylene chloride (Eisman et al., 1976), ether (Bishop et al., 1980), cyclohexane-ethyl acetate (1:1) (Bouillon et al., 1980) hexane-isopropanol (1:2) (Parvainen et al., 1981) or hexane-isopropanol-n-butanol (93:3:4) (Jongen et al., 1981). Such an approach using selective extraction of a vitamin D metabolite does not preclude use of a purification technique such as a mini-column or cartridge prior to HPLC. In fact the successful assay procedure developed by Eisman et al., 1976 for 1,25-$(OH)_2$D combines a selective extraction procedure with a mini-Sephadex column. Advantages of this method are that it produces an extract essentially devoid of lipid thereby permitting use of narrower diameter HPLC columns without fear of over-loading. Disadvantages include the time-taken to develop a technique that provides quantitative yields of a vitamin D metabolite and the unsuitability of selective extraction solvents for a broad range of metabolites.

Direct Extraction and Purification on Cartridges

Of late there has been considerable interest in a variant of the selective extraction technique described above which combines extraction and purification steps into one by use of a commercially available cartridge. Deproteinated plasma or serum containing the vitamin D and its metabolites is immobilised on a C_{18}-bonded cartridge (e.g. Sep-Pak C_{18}, Waters Associates, Milford, Mass.; or Bond-Elut C_{18}, Analytichem International Harbor City, Cal.) and the bulk of the contaminating polar lipids are eluted with aqueous rich solvent mixtures. Subsequently vitamin D metabolites are eluted in general classes with suitable polarity solvents (e.g. 80% methanol in water elutes 25-OH-D) (Dabek et al., 1981; Turnbull et al., 1982; Kao and Heser, 1984; Reinhardt et al., 1984). Neutral lipids are left on the C_{18}-packing material. A method involving both C_{18}-cartridges and SIL-cartridges used previously with the total lipid extraction or selective extraction techniques described above, thereby provides a purification approach which combines the virtues of two different chromatographic techniques used for vitamin D: reverse phase liquid-bonded chromatography and straight phase adsorption chromatography. In clinical assays, some (Reinhardt et al., 1984; Manolagas, 1983) claim that a combination of cartridges provides such excellent purification of 1,25-$(OH)_2D$, that need for HPLC as a separation tool can be avoided. In such cases detection of vitamin D is achieved by radioreceptor or radioimmunoassay. Use of cartridges offers the advantages of consistent, convenient chromatography and a reduction in the number of purification steps. Major disadvantages are cost and the tendency of such cartridges to bleed UV-contaminants into chromatographic effluents.

HPLC Techniques

HPLC can be performed on any commerically available apparatus capable of withstanding pressures of 2000 psi (14 300 kpascals) and producing pulse-free flow rates of 0.1 to 10 ml/min. Care must be taken to avoid particulate matter in extracts and we have used either filtration of samples using an organic filtration kit (Waters, Milford, Mass.) or centrifugation in conical tubes (e.g. Reactivials). Use of a guard column or in-line filter between sample injector and column also increases column lifetime and minimises the effect of extraction impurities on column performance. Standard 4.6 mm × 25 cm HPLC columns are generally adequate for vitamin D separations. In some unusual cases, the need for sample purification prior to HPLC is less critical because of high vitamin D content or because UV contaminants or lipids are dilute. In these samples (e.g. kidney perfusates), extracts generated using the total lipid extraction technique described above involving extraction but no minicolumn or cartridge can be injected directly onto HPLC columns without sacrificing column performance. In such cases, however, we have found that the "standard" size 4.6 mm × 25 cm HPLC columns are best replaced by larger "semi-preparative" 6.2 mm × 25 cm columns to avoid over-loading problems (Jones 1978).

Separation of Vitamin D and Irradiation Precursors

The synthesis of vitamin D from its provitamin D precursor (7-dehydrocholesterol or ergosterol) is a commercially important process. Irradiation of provitamin D with UV light at 0–4° C produces a mixture of photoisomers, the principal ones being previtamin D, tachysterol and lumisterol. This mixture is easily resolved using HPLC with microparticulate silica and the solvent, hexane-isopropanol, 99:1 (Fig. 6). The different spectral properties of the photoirradiation products makes identification of the separated components quite straightforward. Separation of the previtamin D at this stage can be followed by thermal isomerisation to vitamin D. Tachysterol is poorly resolved from vitamin D using this solvent (hexane-isopropanol, 99:1) and heating of the irradiation mixture prior to separation of previtamin results in a mixture of tachysterol and vitamin D which is a difficult separation. Better resolution of tachysterol and vitamin D can be achieved with other solvent systems such as methylene chloride-methanol, 99.75:0.25 (Fig. 7). Note that the order of elution of tachysterol and vitamin D_2 is reversed with the methylene chloride:methanol (99.75:0.25) solvent. Burns et al. (1980) have suggested that, generally speaking, the elution order of the photoisomers of vitamin D correlates inversely with their molecular planarity. Provitamin D_2 (ergosterol) is the most planar molecule, interacts with the silica surface most readily and elutes last whereas previtamin D_2 is essentially a non-planar molecule and elutes first. Burns et al., ascribe the reverse elution order of $tachysterol_2$ and vitamin D_2 observed in methylene chloride:methanol (99.75:0.25) as being due to a slight conformational change induced by interaction with the solvent; any acidic species in the solvent could effect a change in molecular conformation.

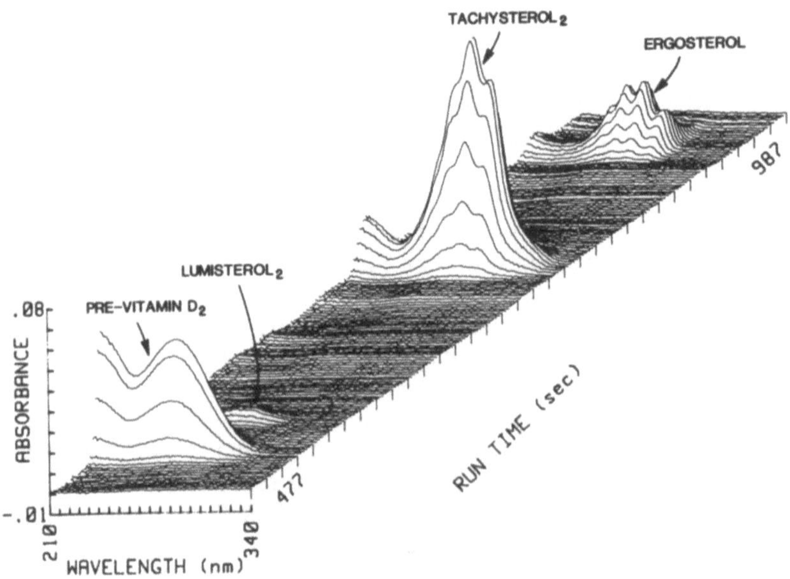

Fig. 6. HPLC of irradiation mixture of ergosterol. 3-D Plot of Absorbance v Wavelength v Run Time Zorbax-SIL (6.2 mm × 25 cm); hexane-isopropanol, (99/1) 2.0 ml/min

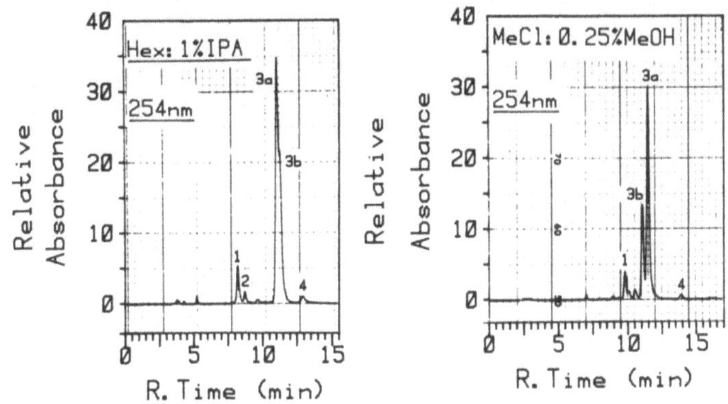

Fig. 7. Separation of heated irradiation mixture of ergosterol. (a) Using hexane-isopropanol, (99/1); 2.0 ml/min (b) Using methylene chloride-methanol, (99.75/0.25); 1.0 ml/min. Metabolites: 1 = previtamin D_2; 2 = lumisterol; 3a = vitamin D_2; 3b = tachysterol$_2$; 4 = ergosterol. Both systems using Zorbax-SIL (6.2 mm × 25 cm); 254 nm

Separation of Vitamins D_2 and D_3, 25-OH-D_2 and 25-OH-D_3

Zorbax-SIL eluted with the solvent hexane-isopropanol (97.5:2.5) will resolve vitamin D_3 from its principal blood metabolite, 25-OH-D_3 (Jones and DeLuca, 1975). This HPLC system also provides baseline resolution of 25-OH-D_2 from 25-OH-D_3 but not of vitamin D_2 from vitamin D_3 (Fig. 8). These data suggest a masking influence of the 24-methyl group on the 25-hydroxyl group in the side chain of 25-OH-D_2. Other solvents used with microparticulate silica for the separation of 25-OH-D_2 and 25-OH-D_3 include hexane-ethanol (95:5) (Gilbertson and Stryd, 1977) and hexane-isopropanol-methanol (96:3:1) (Jones, 1980).

Vitamins D_2 and D_3 can be separated by reverse phase chromatography on Zorbax-ODS or other octadecasilane packings. One solvent system used for this separation consists of methanol-water (98.5:1.5) (Jones, 1978) (Fig. 9). 25-OH-D_2 and 25-OH-D_3 can also be resolved with a Zorbax-ODS column but solvent strength must be increased to methanol-water (91:9) (Jones, 1978), (Fig. 10). The solvent acetonitrile-methanol-water (90:5:5) with Zorbax-ODS provides equivalent resolution of 25-OH-D_2 and 25-OH-D_3 with much reduced back pressure (Koshy and Vanderslik, 1976).

Separation of Dihydroxylated Vitamin D Metabolites

Reverse phase HPLC using Permaphase-ODS was the first HPLC technique to be used for the resolution of dihydroxylated metabolites of vitamin D (Matthews et al., 1974). Unfortunately, it provides poor resolution of 1,25-$(OH)_2D_3$ and 24,25-$(OH)_2D_3$ and is rarely used. Instead most separations are carried out on microparticulate silica using hexane or methylene chloride based solvent systems. Jones and DeLuca (1975) pioneered the use of microparticulate silica with the solvent

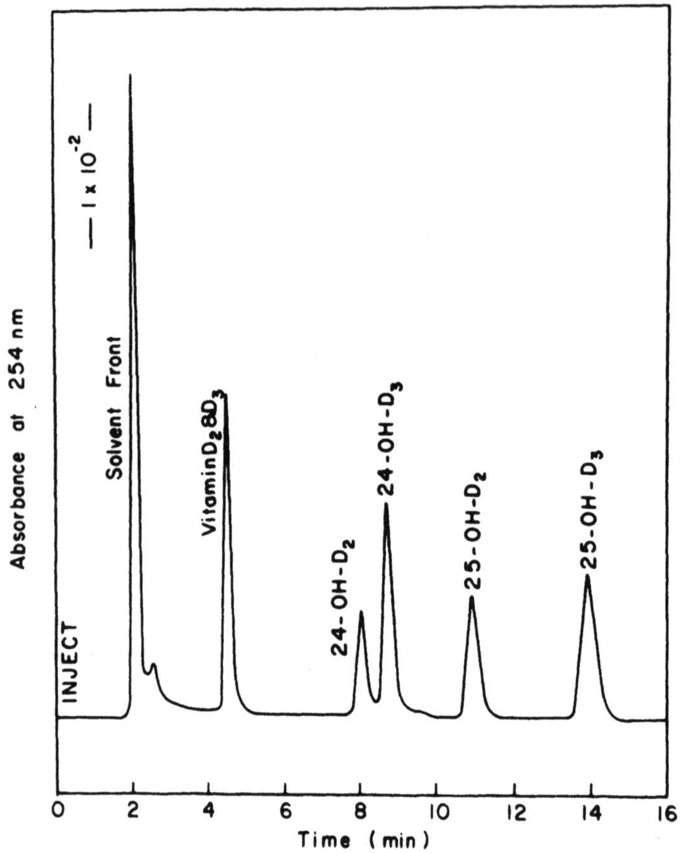

Fig. 8. HPLC of vitamins D_3 & D_2 and their mono-hydroxylated metabolites on Zorbax-SIL. 10 ng vitamin D_3, 6 ng vitamin D_2, 19 ng 25-OH-D_3, 13 ng 25-OH-D_2, 16 ng 24-OH-D_3, 7 ng 24-OH-D_2 were applied to the column in 10 μl eluting solvent. Conditions: two Zorbax-SIL (2.1 mm × 25 cm) columns in series; Skellysolve B-isopropanol, (97.5/2.5); flow rate 0.70 ml/min; pressure 4000 psi; 254 nm. From Jones and DeLuca (1975)

system hexane-isopropanol (90:10) to separate 24,25-$(OH)_2D_3$ 25,26-$(OH)_2D_3$ and 1,25-$(OH)_2D_3$ (Fig. 11). This HPLC system has been widely used over the past decade for the separation of dihydroxylated vitamin D metabolites in a variety of biochemical and clinical studies. 1,25-$(OH)_2D_3$, in particular, is retarded by its interaction with Zorbax-SIL in the hexane environment, suggesting that this vitamin D molecule has its 1α-OH group more exposed in this solvent. A refinement suggested by Jones (1980) involves the addition of small amounts of methanol to hexane-isopropanol mixtures to ensure improved plate counts and decrease peak skewing for dihydroxylated vitamin D metabolites, especially 24,25-$(OH)_2D_3$ (Fig. 12).

Ikekawa and Koizumi (1976) reported the use of a gradient of 0.02 to 6% methanol in methylene chloride with Zorbax-SIL for the separation of mixtures

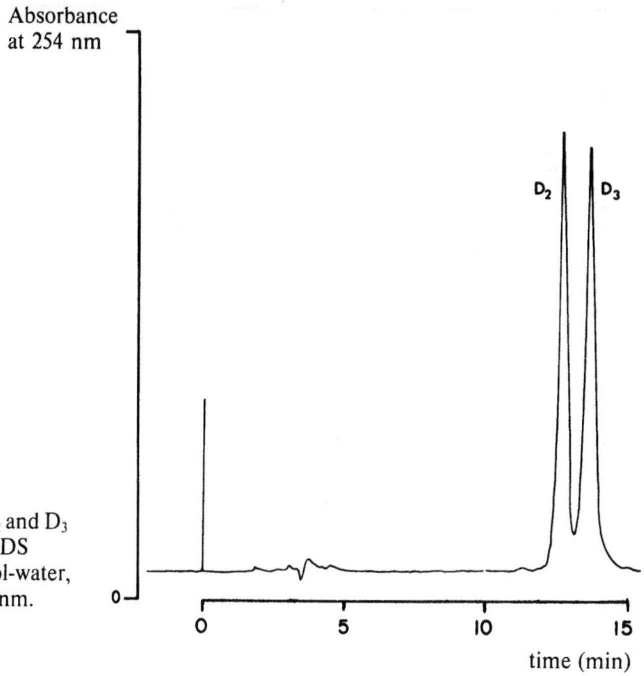

Fig. 9. HPLC of vitamins D_2 and D_3 on Zorbax-ODS. Zorbax-ODS (6.2 mm × 25 cm); methanol-water, (98.5/1.5); 1.5 ml/min; 254 nm. From Jones (1977)

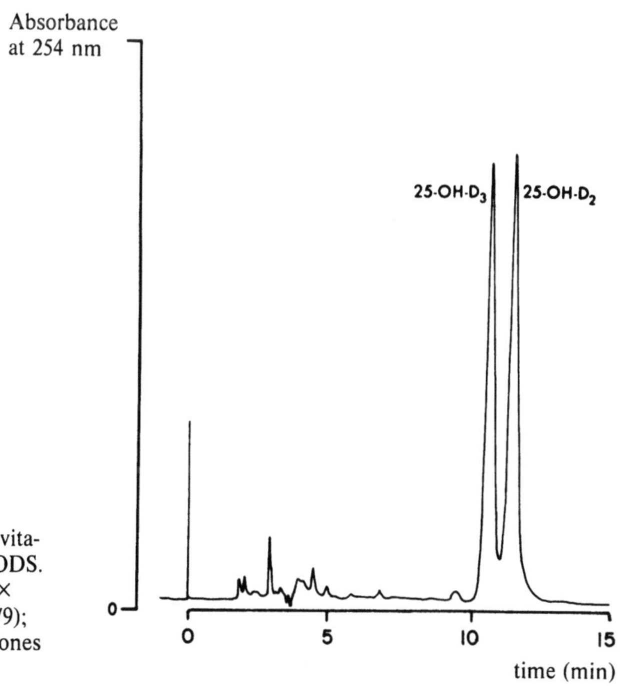

Fig. 10. HPLC of 25-hydroxyvitamins D_2 and D_3 on Zorbax-ODS. Zorbax ODS-ODS (6.2 mm × 25 cm); methanol-water, (91/9); 1.5 ml/min; 254 nm. From Jones (1977)

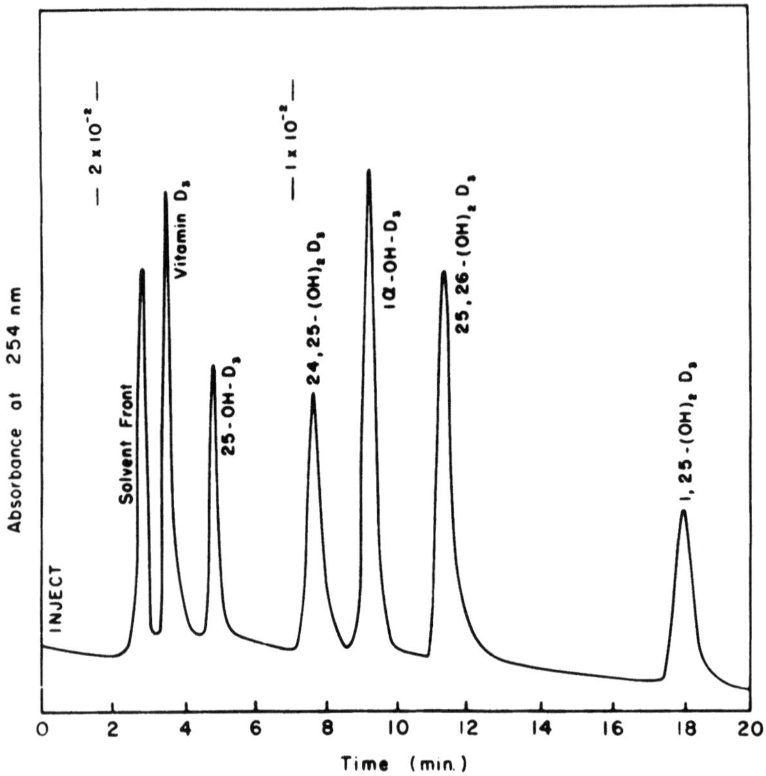

Fig. 11. HPLC of vitamin D_3 & its metabolites on Zorbax-SIL. 40 ng vitamin D_3, 30 ng 25-OH-D_3, 25 ng 24,25-$(OH)_2D_3$ 40 ng 1-alpha-OH-D_3, 40 ng 25,26-$(OH)_2D_3$, 25 ng 1,25-$(OH)_2D_3$ injected in 10 microliters eluting solvent. Conditions: two Zorbax-SIL (2.1 mm × 25 cm) columns in series; Skellysolve B-isopropanol, (90/10); flow rate 0.5 ml/min; pressure 3000 psi; 254 nm. From Jones and DeLuca (1975)

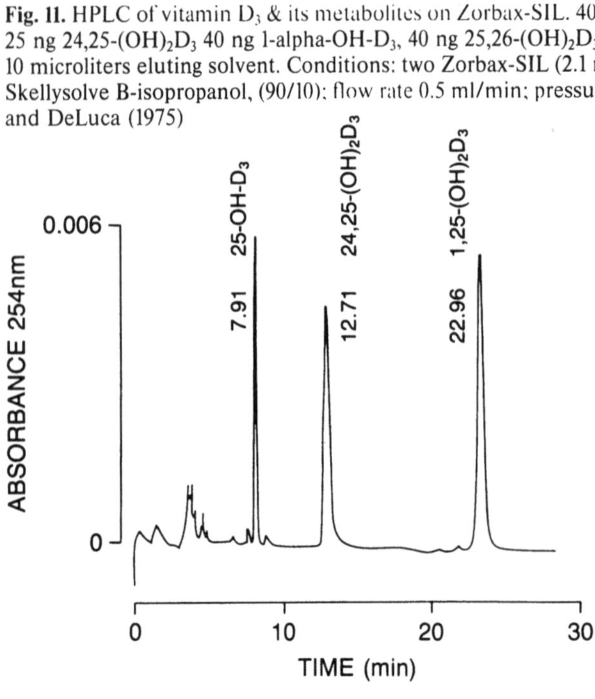

Fig. 12. HPLC of 25-OH-D_3, 24,25-$(OH)_2D_3$ and 1,25-$(OH)_2D_3$ on Zorbax-SIL. Zorbax-Sil (6.2 mm × 25 cm); hexane/isopropanol/methanol, (87/10/3); flow rate 1.5 ml/min; pressure 600 psi; 254 nm. From Jones (1980)

of vitamin D metabolites ranging from vitamin D through to 1,24,25-$(OH)_3D_3$. In general, mixtures of methylene chloride and methanol often provide higher plate counts with a given column than similar strength mixtures of isopropanol and hexane (Jones, 1980). It should be noted, however, that methylene chloride-based solvent mixtures also have disadvantages including: a poor solubility for lipid extracts, a tendancy to produce bubbles if not degassed, a higher UV cutoff (~ 240 nm) than hexane, and the possibility that they contain small amounts of hydrochloric acid which can attack the stainless steel used in HPLC construction. Nevertheless, methylene chloride-based solvent systems have been used effectively in purification schemes for vitamin D metabolites and are capable of some resolutions impossible with hexane-isopropanol mixtures. For example, 24,25-$(OH)_2D_3$ and 25-OH-D_3-26-23-lactone are widely separated using methylene chloride-isopropanol (96.5:3.5) (Horst, 1979) but virtually unresolved using hexane-isopropanol (Tab. 2). Dreyer and Goodmann, (1981) have exploited this difference in solvent properties in designing a spectrophotometric assay for 24,25-$(OH)_2D_3$ based upon two Zorbax-SIL adsorption chromatographic steps, the first using 10% isopropanol in hexane, the second using 3% methanol in methylene chloride.

Table 2. Chromatographic properties of important vitamin D metabolites

Metabolites	Retention time (in minutes) Zorbax-CN HIM 94-5-1 1.3 ml/min	Zorbax-SIL HIM 91-7-2 1.5 ml/min	Zorbax-CN M/MC 0.5-99.5 1.0 ml/min	Zorbax-SIL M/MC 2.5-97.5 1.5 ml/min
Vitamin D_3	3.87	3.60	6.70	4.51
25-OH-D_2	6.66	6.28	-	-
25-OH-D_3	7.11	6.78	7.92	6.70
24-oxo-25-OH-D_3	9.78	7.54	7.40	5.77
24(R),25-$(OH)_2D_2$	11.30	10.39	10.30	8.13
24(R),25-$(OH)_2D_3$	11.43	11.08	13.26	9.34
24-oxo-23,25-$(OH)_2D_3$	14.82	11.32	10.40	6.73
25(S),26-$(OH)_2D_3$	14.09	14.49	19.66	13.03
1α,25-$(OH)_2D_3$	18.66	21.90	19.24	16.71
25-OH-D_3-26,23-lactone	20.71	11.71	10.88	5.98

Adapted from Jones (1983) J. Chromatogr. 276:69-75
HIM = Hexane : Isopropanol : Methanol
M/MC = Methanol : Methylene Chloride

The ability to vary solvent composition in order to bring about changes in resolution can be rationalised by examination of the resolution equation:

$$R = \frac{1}{4}\left[\frac{\alpha-1}{\alpha}\right]\left[\sqrt{N}\right]\left[\frac{K'}{1+K'}\right]$$

where R = resolution
α = selectivity factor
N = plate count
k' = capacity factor.

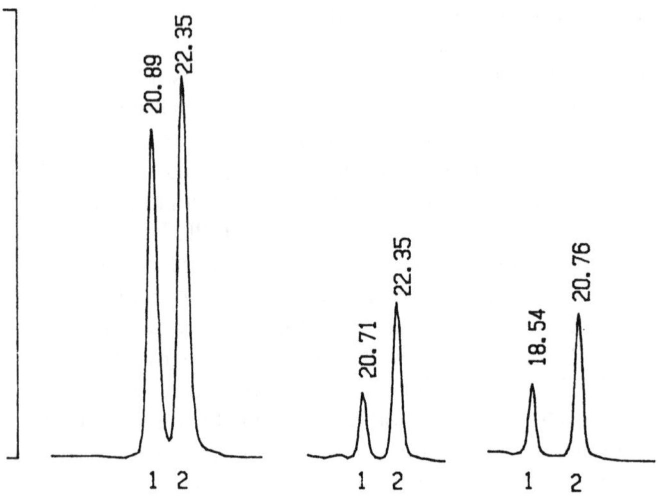

Fig. 13. HPLC separation of 1,25-$(OH)_2D_2$ and 1,25-$(OH)_2D_3$ using different solvent composition. (A) hexane-isopropanol, (85/15); 1.5 ml/min, (B) hexane/isopropanol/methanol, (87/10/3); 1.5 ml/min, (C) hexane-ethanol-chloroform, (80/10/10); 1.5 ml/min. All on Zorbax -SIL (6.2 mm × 25 cm); 254 nm. Only the relevant region of each chromatogram is reproduced in the figure. In all frames the first peak is 1,25-$(OH)_2D_2$ and the second, 1,25-$(OH)_2D_3$. From Jones (1980)

Changes in solvent composition modify the selectivity factor, α. Another example of the effect of the selectivity factor, α, is in the resolution of the hormone 1,25-$(OH)_2D_3$ from its D_2 counterpart, 1,25-$(OH)_2D_2$. Figure 13 illustrates the effect of varying the secondary solvents used with hexane for this separation. As can be observed, use of an halogenated hydrocarbon such as chloroform or methylene chloride increases this resolution. Presumably the halogenated hydrocarbon causes the side chains of 1,25-$(OH)_2D_2$ and 1,25-$(OH)_2D_3$ to be more exposed for interaction with silanol groups on Zorbax-SIL thereby accentuating the differences in the side chains.

Another way to influence the selectivity factor, α, is to change the column chemistry. By substitution of a Zorbax-CN column for a Zorbax-SIL column but using similar hexane: isopropanol-based solvent systems, Jones (1983) was able to effect a similar dramatic improvement in the resolution of 24,25-$(OH)_2D_3$ and 25-OH-D_3-26-23-lactone. In this case 25-OH-D_3-26-23-lactone is greatly retarded (Table 2) during Zorbax-CN chromatography presumably because of a C=O:C≡N interaction on the surface of the column packing material. This interpretation is further supported by the observation that 24-oxo-25-OH-D_3 and 24-oxo-23,25-$(OH)_2D_3$, both C=O containing compounds, are similarly retarded on Zorbax-CN relative to 24,25-$(OH)_2D_3$. Jongen et al. (1981) have used Nucleosil 10-NO_2 in HPLC and Kao and Heser (1984) have used NH_2-cartridges in pre-HPLC purification to modify separations achieved on conventional microparticulate silica packings.

Thus with a judicious use of different column packings and by manipulating solvent composition, the separation of most known dihydroxylated metabolites of vitamin D can be achieved.

Other metabolites of Vitamin D

These include several trihydroxy compounds: 1,24,25-$(OH)_3D_3$ (Holick et al., 1973), 23,24,25-$(OH)_3D_3$, 24,25,26-$(OH)_3D_3$ (Wichmann et al., 1981) and 1,25,26-$(OH)_3D_3$ (Reinhardt et al., 1981) as well as more polar compounds such as calcitroic acid (a 1-hydroxylated C_{23}-acidic vitamin D with loss of the end of the side chain C_{24}-C_{27} inclusive) (Esvelt et al., 1979) and the 25-OH-D-glucuronides (Levan et al., 1981). HPLC studies on these compounds have been scant and usually limited to HPLC steps involved in the isolation and identification of the compounds. In most cases the separations have been extensions of the strategy applied to other mono- and dihydroxylated vitamin D metabolites. In other cases some original approaches have been made towards exploiting the presence of additional functionalities in the vitamin D molecule.

In the case of the trihydroxylated metabolites the general approach to separations (Wichmann et al., 1981; Reinhardt et al., 1981) has been to use straight phase HPLC with increased strength solvent systems or to switch to reverse-phase HPLC and increase aqueous content to as much as 30% water in methanol.

In the case of the acidic vitamin D metabolites, such as calcitroic acid, an alternative approach to purification of the metabolites has been to methylate the carboxylic-acid group (Esvelt et al., 1979). A methylated version of this acidic metabolite is thereby rendered suitable for HPLC on the basis of its two remaining hydroxyl functions at positions 1α and 3β, and making it similar to 1α-OH-D_3. The chromatographic behaviour and solvent systems used for 1α-OH-D_3 (Jones and DeLuca, 1975) are shown in Figure 11.

In the case of 25-OH-D_2-glucuronide, Levan et al., 1981 used reverse-phase HPLC on Zorbax-ODS and the solvent system methanol-water (65:35) containing 10 mM ammonium bicarbonate to purify this charged metabolite from bile extracts. Levan et al. (1981) also devised a useful DEAE-Sephadex A-25 step to separate neutral from charged vitamin D species prior to HPLC.

Detection Techniques

Vitamin D and its metabolites all posses the same intense ultraviolet absorption spectrum with characteristic absorption maximum and minimum at 265 nm and 228 nm respectively. Molar extinction coefficient, ε_{265}, at the maximum is 18 200 for vitamin D_3. On-line detection techniques have thus been dominated by UV detection to the almost total exclusion of all else. Lack of fluorescence or of a ionic group in the vitamin D molecule have made other HPLC detectors virtually useless. Many biochemical studies utilise commercially available radioactive vitamin D preparations labelled in the ring or side chain with atoms of [^{3}H] or [^{14}C]. Such labelled metabolites have high specific radioactivity and allow for detection of picomolar quantities. For these metabolites off-line radioactivity scintillation counting was the method of choice following HPLC. But with the recent introduction of on-line radioactivity detectors with improved sensitivity there is now a second viable on-line alternative to UV detection.

In clinical studies or basic experiments where use of radioisotopically-labelled vitamin D metabolites is precluded or where concentrations of vitamin D metabolites fall below the limit of UV detectors (~ 1–5 ng), alternative means of detection must be sought. Though some have used HPLC-MS or HPLC with GC-MS with a degree of success most vitamin D biochemists have solved the detection problem by use of combinations of HPLC separation techniques with the classical tools of endocrinology, namely competitive-binding assays and radioimmunoassays. These tests afford a sensitivity and specificity rarely attained by even the most sensitive physico-chemical techniques. Some have even coupled HPLC to cytochemical and bioassay procedures to push sensitivity into the picomolar range.

Ultraviolet (UV) Detection

The most popular and sensitive type of UV detector is the mercury-lamp fixed-wavelength variety which allows detection at any mercury line including the intense band at wavelength 254nm. Absorbance of vitamin D metabolites at 254nm represents about 90% of that at the maximum at 265nm. Thus UV detection using standard fixed-wavelength detectors set at 254nm is virtually optimal. The gain in sensitivity offered by increasing the wavelength to the absorption maximum at 265nm is more than offset by the resulting loss in light intensity brought about by having to switch to a variable wavelength detector with a deuterium lamp.

However Optimising UV absorbance is not the only reason to switch detection from 254nm to 265nm. Many sample matrices, whether biological or artificial, contain UV interferences of an aromatic nature absorbing in the 254nm region e.g. ketones and benzene derivatives. Effects of these can be minimised by switching to 265nm or an even higher wavelength. Unfortunately, oestrogens several other compounds e.g. peptides, phenolics and often absorb in the 270–290nm range so that variable wavelength detectors offer no panacea.

UV detection has been hampered by its lack of specificity. In most cases the major criterion for peak identification has been on the basis of comparison of retention time with known standards. Recently, however, a new type of UV detector has been introduced based upon the diode-array scanning spectrophotometer (Miller et al., 1982). Such an instrument is able to make repeated 200–400nm scans of chromatographic effluent passing through the detector flow-cell in as little as 100 milliseconds or less. Using a microcomputer and associated data storage devices, vast amounts of chromatographic data can be acquired in a single 20 minute chromatographic run. Jones et al., (1983) and Jones et al., (1984) have applied this technique to the separation of vitamin D metabolites in extracts of kidney perfusates. As illustrated in Fig. 14A and B the diode-array scanning-spectrophotometric detector adds another dimension, wavelength, to the con-

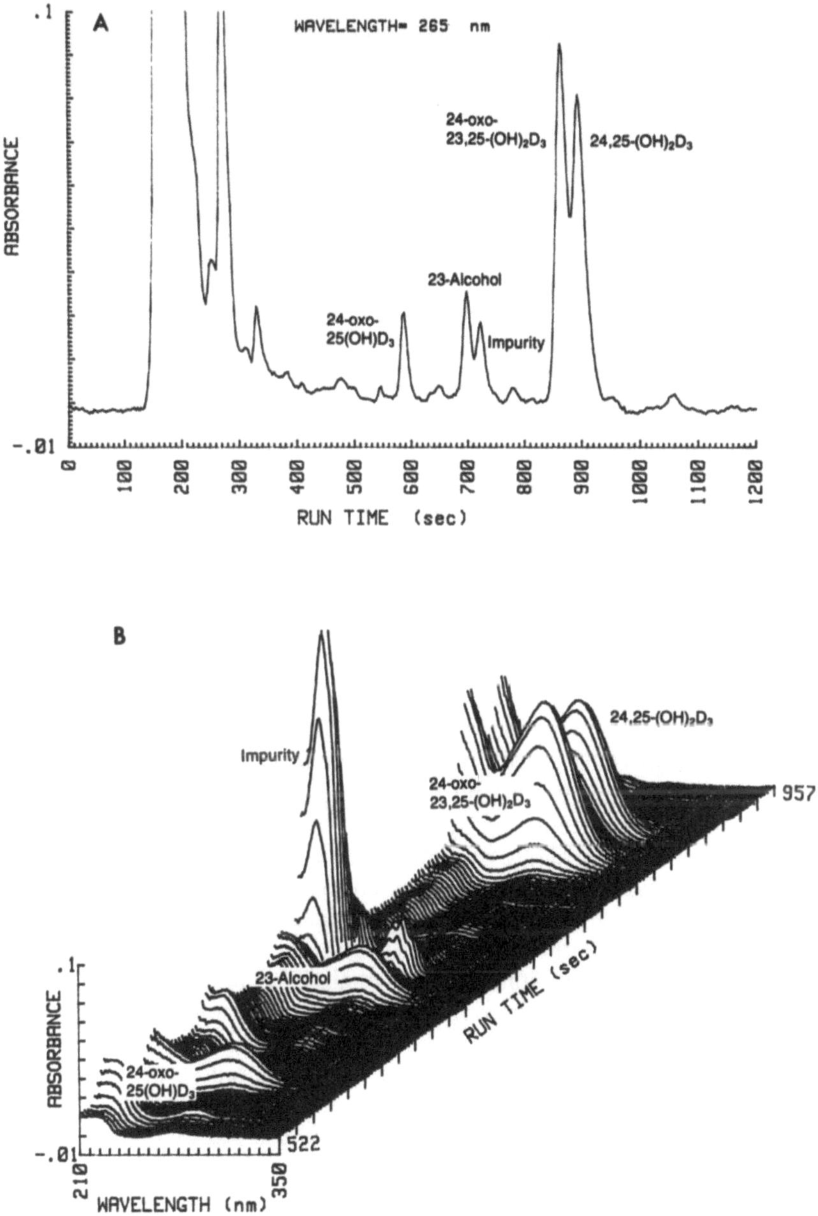

Fig. 14. HPLC separation of metabolites of 24,25-$(OH)_2D_3$ made in the perfused rat kidney. Extract of perfusate from kidneys perfused in the presence of 1250 nM 24,25-$(OH)_2D_3$:
(A) Absorbance at 265 vs run time.
(B) Absorbance vs wavelength vs run time.
(A) & (B) represent the same chromatographic run. Note in (A) that the peak marked as an impurity appears minor because the the wavelength used for monitoring is 265nm. The impurity peak absorbs strongly at 225nm and in the 275–280nm region. Note also that all metabolite peaks have the classical vitamin D absorption spectrum. Conditions: Zorbax-SIL, (6.2 mm × 25 cm); hexane-isopropanol-methanol (94/5/1); 1.5 ml/min. from Jones et al. (1983)

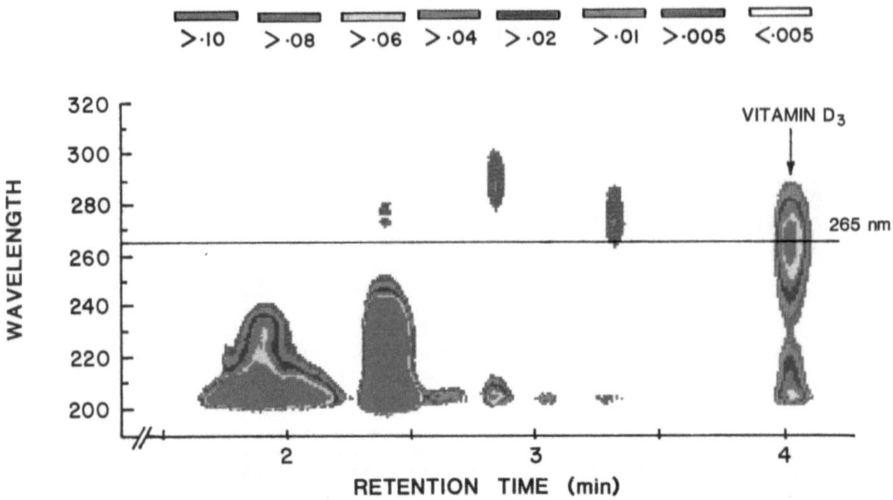

Fig. 15. HPLC of commercially available [^{3}H] vitamin D_3. Isoabsorbance plot showing full UV spectrum of each component in the mixture. [^{3}H] vitamin D_3 elutes at 4 min and shows a spectrum with maximum at 265nm and minimum around 228nm. The even gradient of color in the peak of vitamin D_3 suggests the absence of impurities.

ventional absorbance versus time plot. Vitamin D metabolite peaks can be recognized at 590, 700, 870 and 890 seconds from their λ_{max} = 265nm, λ_{min} = 228nm properties without the use of known standards. In fact, the technique allowed Jones et al. (1983) to recognize the existence of two new metabolites: 24-oxo-23,25-$(OH)_2D_3$ and 24,25,26,27-tetranor-23-$(OH)D_3$ formed in the perfused kidney *in vitro.* It should be noted however that mass spectrometry and comparison to chemically synthesised material provided the means for identification of the metabolites in these studies (Jones et al., 1983; Jones et al., 1984). In another example shown in Fig. 15, impurities in a batch of commercially-available radioisotopically labelled vitamin D_3 can be seen to differ in UV properties from the [^{3}H] vitamin D, when viewed in an isoabsorbance plot representation. Furthermore the [^{3}H] vitamin D peak shows the spectral properties and symmetry expected for a pure preparation. Irrespective of type of spectral representation we used, 3-D plot or isoabsorbance plot, the scanning spectrophotometry provides valuable information regarding the purity of vitamin D preparations. This is of tremendous value in predicting *when* during a metabolite purification scheme direct probe-MS will be feasible and will not be hampered by impurities (see Fig. 16).

Radioactivity Detectors

Several manufacturers now offer instruments that measure radioactivity in chromatographic effluent either by passing the effluent directly over glass beads

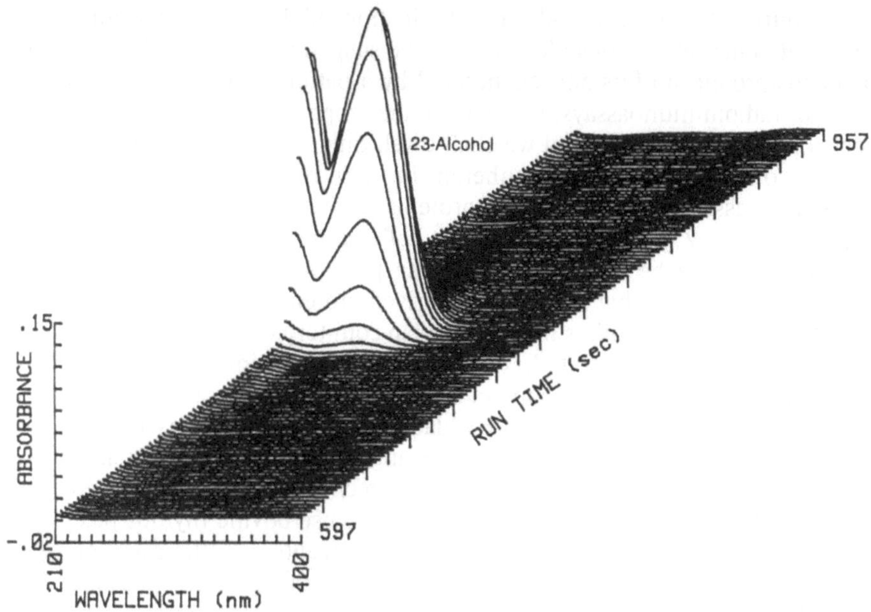

Fig. 16. Final HPLC of purified renal metabolite using diode-array spectrophotometry. Putative 24,25,26,27-tetranor-23-OH-D_3 on Zorbax-SIL (6.2 mm × 25 cm); hexane-isopropanol-methanol (94/5/1); 1.5 ml/min. Material from this column was used for mass spectrometry. from Jones et al., 1984

impregnated with scintillators and measuring resulting light emission with sensitive photomultiplier tubes or by mixing effluent with scintillation fluid using a post-column Y-arrangement, a mixing chamber and a flow scintillation counter. The former system preserves sharp HPLC peaks at the expense of sensitivity while the latter system causes peak broadening of HPLC peaks but sensitivities approach those of conventional stop-flow scintillation counters. Practically, the former system is more suitable for [^{14}C] labelled vitamin D metabolites, the latter system is more suitable for the weaker β-particles emitted from [^{3}H] vitamin D metabolites. Though Rosenthal et al., 1980 reported use of such a detector for screening effluents, this form of detection has found little use thus far. On the other hand, scintillation counting after fraction collection of effluent is one of the most popular forms of detection, particularly in basic biochemical studies.

Saturation Analysis or Radioimmunoassay as a means of detection

Workers in the field of vitamin D have been forced by limits of detection of conventional HPLC detectors to pioneer the blending of HPLC separation techniques with biological means of detection. The use of highly non-polar solvents during HPLC has made the elimination of these solvents quick and convenient. Vitamin D molecules can then be suspended in ethanolic solutions or aqueous-protein buffers and quantitated by a battery of competitive-protein binding or radioimmunoassays. It is beyond the scope of this chapter to cover all aspects of assay procedures and we shall dwell only on the details pertinent to discuss chromatographic strategy inherent in these methods. Competitive protein binding assays usually employ proteins derived from the blood-carrier protein, DBP for 25-OH-D_3 or the target cell receptor for the hormone 1,25-$(OH)_2D_3$. These can be mammalian or avian proteins. It has been recognized over the past few years that mammalian-derived proteins are less discriminating against vitamin D_2 metabolites than their avian counterparts (Belsey et al., 1974;a Reinhardt et al., 1984; Jones et al., 1980). This is of major importance in clinical assays since human blood can contain significant concentrations of vitamin D_2 metabolites. Belsey et al. (1971) and Haddad and Chyu (1971) devised assays for 25-OH-D based upon rat DBP while Brumbaugh et al. (1974) and Eisman et al. (1976) devised assays for 1,25-$(OH)_2D$ based upon chick intestinal cytosol. Recent advances have permitted Reinhardt et al. (1984) to use bovine thymus receptors and Manolagas et al. (1983) to use rat bone cell receptors in the assay of 1,25-$(OH)_2D$. Competitive-binding assays benefit from the greater specificity for ligand offered from use of the natural binding protein or receptor for 25-OH-D_3 or 1,25-$(OH)_2D_3$. Thus HPLC can be less rigorous and purification schemes are less susceptible to problems.

Radioimmunoassays were pioneered by O'Riordan's group working with antibodies directed to the immunogen: BSA-1,25-$(OH)_2D_3$-25-hemisuccinate (Clemens et al., 1978). Several others have reported successful generation of antibodies (Bouillon et al., 1980; Gray et al., 1981; Schaefer et al., 1989) but most of these antibodies are specific only for 1,25-$(OH)_2D_3$ and offer poor cross-reactivity against 1,25-$(OH)_2D_2$. Only one, generated in sheep, (S11, of Fraher et al., 1983) provides a binding with both 1,25-$(OH)_2D_3$ and 1,25-$(OH)_2D_2$ approaching that found in natural receptor proteins. Since cross-reactivity with other vitamin D metabolites is often significant, radioimmunoassays for 1,25-$(OH)_2D_3$ do not offer such good specificity as competitive protein binding techniques. Thus HPLC and separation techniques used in conjunction with radioimmunoassays must be well designed to rid 1,25-$(OH)_2D_3$ fractions of all other contaminating vitamin D metabolites; a difficult task! A recent study by Hummer et al. (1984) found that HPLC using methylene chloride-methanol mixtures gave a cleaner 1,25-$(OH)_2D_3$ fraction, suitable for radioimmunoassay, than do hexane: isopropanol mixtures.

Since HPLC techniques can be easily automated, one advantage of coupling HPLC to CPBA or RIA techniques is to reduce the high costs of technician time involved in these labour intensive assays. Though separation techniques other than HPLC can be used for purifying vitamin D metabolites these tend to be of the labour-intensive type.

Stern et al. (1978) devised a technique involving HPLC coupled with a sensitive bioassay for 1,25-$(OH)_2D_3$. The bioassay exploited the potent bone resorptive power of 1,25-$(OH)_2D_3$ releasing ^{45}Ca from deepseated ^{45}Ca in bone in organ culture. By measuring ^{45}Ca released in culture medium, the amount of 1,25-$(OH)_2D_3$ in HPLC effluents could be compared to a standard curve of crystalline

$1,25\text{-}(OH)_2D_3$ dissolved in ethanol. Manolagas et al., (1983) have used an ingenious technique to follow internalisation of $1,25\text{-}(OH)_2D_3$ into intact rat osteogenic sarcoma cells in culture and thereby measure $1,25\text{-}(OH)_2D_3$ concentrations in HPLC effluents. In this assay the bone cell envelope has the dual purpose of localising $1,25\text{-}(OH)_2D_3$ binding protein and obviating the separation of bound from free ligand. Such a method claims exquisite specificity that the requirement for HPLC in this assay has since been dropped (Manolagas et al., 1983). Prepurification steps involving cartridges are still retained however.

Application of HPLC to Vitamin D Research

Basic Research Studies

In Vivo Studies

Much of the definitive work on the *in vivo* metabolism of vitamin D_3 was carried out using open silicic acid-columns (Norman et al., 1964; Ponchon and DeLuca, 1969). Injection of pharmacological amounts of radioactive vitamin D preparations into rachitic animals was followed by distribution studies of vitamin D metabolites in blood and tissue extracts. Though the initial experiments were frequently well supported with rigorous physico-chemical techniques such as MS, NMR, GC, follow-up studies at physiological concentrations of vitamin D were forced to depend upon less rigorous methods. Most often identification of [^{3}H] metabolites was based upon comparison of their column elution volumes to known [^{3}H] standards run in a similar sized column side-by-side with the experimental sample.

When silicic acid chromatography was superseded by Sephadex LH-20 gel chromatography (Holick and DeLuca, 1971) in the early seventies elution positions became somewhat more reproducible but still metabolism studies (eg. Ribovich and DeLuca, 1978) depended heavily upon elution position for peak identification. Since peaks were often broad and assymmetrical it was difficult to state with any degree of certainty that metabolites were accurately identified. Later this problem was compounded when some workers reduced the column size recommended by Holick and DeLuca (1971) (1×60 cm) in order to speed up analyses. Performance was often compromised in order to achieve sample throughput.

With the advent of HPLC to the vitamin D field in the mid-seventies (Matthews et al., 1974; Jones and DeLuca, 1975; Koizumi and Ikekawa, 1976) researchers quickly applied HPLC as a tool to aid in the rechromatography and identification of Sephadex LH-20 peaks. One important blind spot of the Sephadex LH-20 column eluted with chloroform:hexane, 65:35 was its inability to separate $25,26\text{-}(OH)_2D_3$ and $1,25\text{-}(OH)_2D_3$. This separation was readily accomplished on Zorbax-SIL with 10% isopropanol in hexane (Jones and DeLuca, 1975). Later it became evident that HPLC could not only complement Sephadex LH-20 chromatography but in many cases HPLC could also replace it. Currently it is commonplace for blood extracts from animals injected with radioactive vitamin D

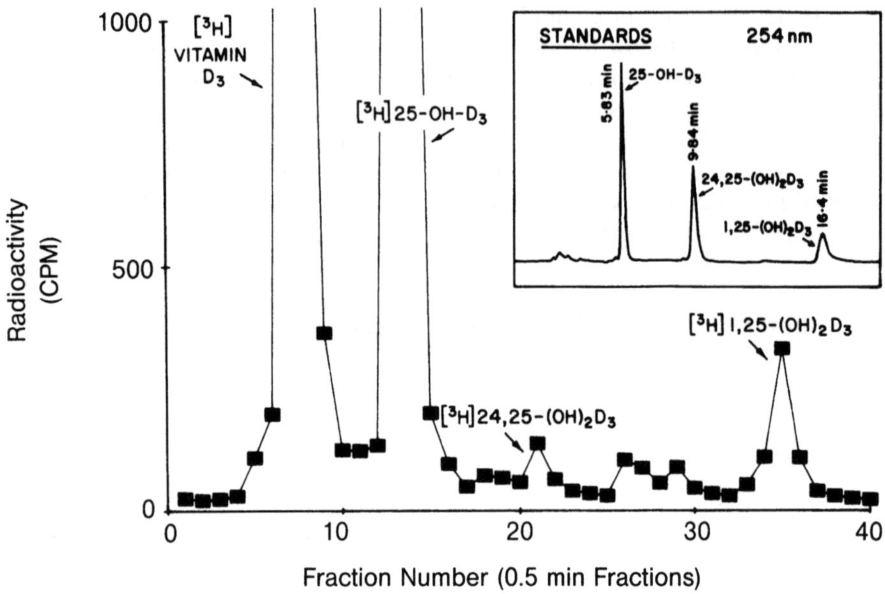

Fig. 17. HPLC of extract of plasma from vitamin D deficient rat which received [1,2-^{3}H] vitamin D_3. Rat received radioactive vitamin D_3 24 hr prior to sacrifice. Run on Zorbax-CN with hexane-isopropanol-methanol (94/5/1) at 1.5 ml/min

preparations to be run on HPLC without elaborate clean-up steps. The chromatographic profile shown in Figure 17 represents the direct HPLC analysis of a plasma total lipid extract from a vitamin D-deficient rat given 5 μg[1,2-^{3}H] vitamin D, subcutaneously 24 hr prior to being sampled. It shows the presence of peaks of vitamin D_3, 25-(OH)D_3, 24,25-$(OH)_2D_3$ and the hormone, 1,25-$(OH)_2D_3$. As with silicic acid and Sephadex LH-20 chromatography, identification of an unknown vitamin D metabolite is still based upon elution volume but now this parameter is more precise and can be checked in the very same column by injection of known standard. This can be achieved in the *same* run by cochromatography of the [^{3}H]-labelled unknown metabolite with a large amount of non-radioactive crystalline standard (internal standardisation) or in *back to back* runs (external standardisation). Dead space between UV detector and fraction collector (for scintillation counting of ^{3}H) can be precisely measured to offset UV and radioactivity traces. It is also feasible to overcome dead-space problems by co-chromatography of [^{3}H]-labelled unknown metabolites with [^{14}C] labelled-standards. Furthermore, HPLC allows for rapid rechromatography on alternative column packings or with other solvent systems. The putative [2-^{3}H] 1,25-$(OH)_2D_3$ peak from Figure 17 was rechromatographed using a different column packing, Zorbax-SIL instead of Zorbax-CN (Fig. 18).

Though, in earlier studies involving open column procedures rechromatography was occasionally used, the choice of packings and type of chromatography

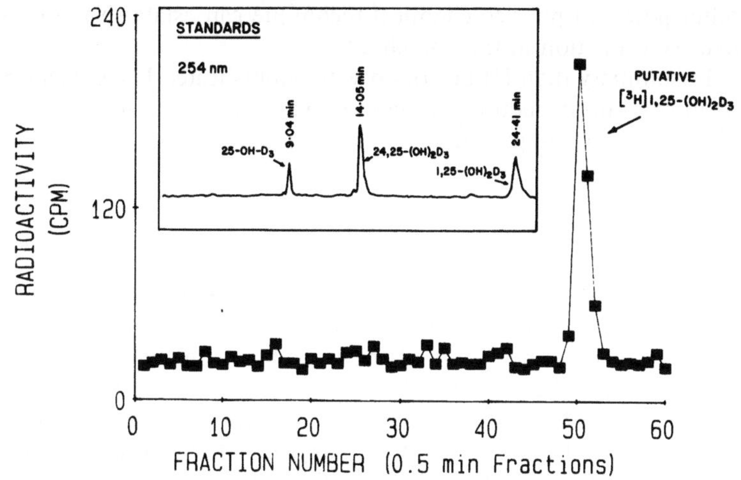

Fig. 18. Rechromatography of [2-^{3}H] 1,25-$(OH)_2D_3$ on Zorbax-SIL. Peak of putative [2-^{3}H] 1,25-$(OH)_2D_3$ from Fig. 12 rerun on second column system

were often limited and recovery of vitamin D metabolite poor. Thus, because of its improved resolving power, HPLC offers a greater likelihood that a metabolite can be correctly identified than with open-column procedures. However, it should always be bourne in mind that chromatographic behaviour alone does not establish unequivocal identification of an unknown peak.

Use of diode-array scanning spectrophotometry augments identification on the basis of retention time. Presence of the classical absorption spectrum with λ_{max} 265, λ_{min} 228, $\lambda_{max}/\lambda_{min} = 1.8$-2 along with a retention time identical to known standard is fairly powerful evidence that 25-OH-D_3 in hypervitaminotic D-rabbit plasma is correctly identified (Fig. 19). However, only mass spectroscopy or some

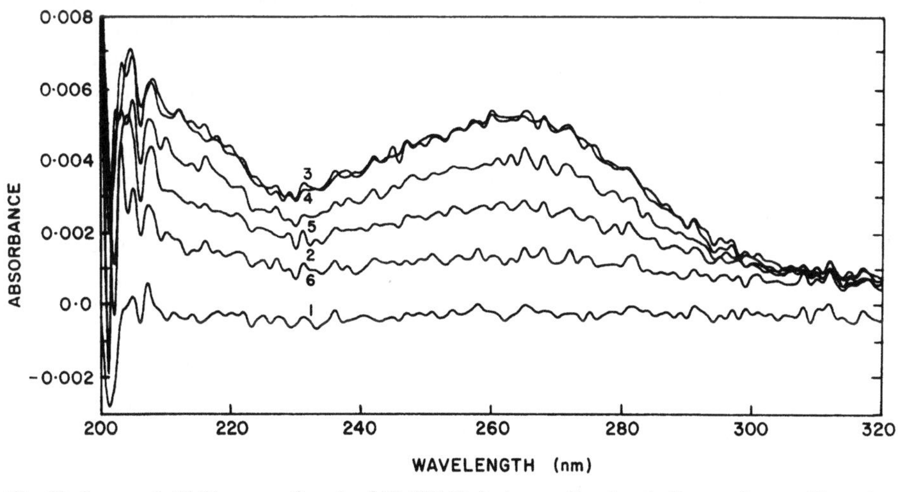

Fig. 19. Sequential UV scans of peak of 25-OH-D_3 in hypervitaminotic D_3 rat plasma. Note the typical vitamin D spectrum in these scans from a peak of only 50 nanograms of metabolite

other powerful physico-chemical technique can establish the presence of a 25-hydroxyl function in this metabolite.

In summary then HPLC, often with sophisticated UV detection, has replaced open column procedures as the method of choice for tentatively ascribing peak identity in routine physiological *in vivo* studies.

In Vitro Studies

There has been tremendous interest in the elucidation of the physiological factors which control the hydroxylations of the vitamin D molecule prior to its action as a hormone. Study of the basic biochemistry of the complex enzymes which execute these hydroxylations has also been intense. With the advent of liver and renal *in vitro* enzyme models, the need for routine separation tools and tentative identification of products has been strong. HPLC has virtually replaced Sephadex LH-20 chromatography in this application too. Here higher concentrations of vitamin D metabolites can be used than *in vivo* situations and UV absorption assay becomes a viable alternative to replace or augment use of [^{3}H] labelled substrates. Several problems encountered in analysis of *in vitro* enzyme extracts are similar to those with *in vivo* experiments and need not be discussed again.

However, a few new problems arise. The range of vitamin D metabolites encountered *in vitro* is often greater than *in vivo.* For example, a vitamin D-replete rat kidney will convert 25-(OH)D_3 into several metabolites besides 24,25-$(OH)_2D_3$ which never reach significantly high concentrations in the blood *in vivo.* These include 24-oxo-25-(OH)D_3, 24-oxo-23,25-$(OH)_2D_3$ and 24,25,26,27-tetranor-23-(OH)D_3. In addition, the D-replete kidney is capable of the synthesis of several metabolites such as 25,26-$(OH)_2D_3$, 23(S),25-$(OH)_2D_3$ and 25-OH-D_3-26,23-lactone which never reach a high concentration in the blood (above 1 ng/ml) unless the animal is given pharmacological amounts of vitamin D_3. Furthermore, many renal *in vitro* systems (Vieth and Fraser, 1979; Tanaka and DeLuca, 1981) require large amounts of substrate in order to reach conditions of Zero-order kinetics, conditions which favour production of some of these minor metabolites. These metabolites can be partially resolved on one HPLC system eg. Zorbax-SIL with isopropanol-hexane mixtures (as in Fig. 11) and then be rerun on a second HPLC system such Zorbax-CN with isopropanol-hexane mixtures or Zorbax-SIL with methylene chloride-methanol mixtures. Thus 24,25-$(OH)_2D_3$ produced *in vitro* can be resolved from several other metabolites which have very similar chromatographic mobility and which have been included with 24,25-$(OH)_2D_3$ in earlier studies using Sephadex LH-20 chromatography or in early HPLC separations.

The second problem unique to *in vitro* studies is the purity of radioactive 25-OH-D_3 commonly used with kidney extracts. Commercially available radioactive preparations, though often 95-98% pure as claimed, contain radioactive impurities which co-migrate with enzymic products. These must be removed by HPLC prior to study. Though one might expect *in vivo* studies to be plagued by similar problems, in practice rarely does this happen presumably because these non-D radioactive impurities are rapidly excreted and do not remain in blood and tissues.

Metabolite Identification on a Semi-Preparative Scale

Many workers in the vitamin D field have used HPLC as a tool to augment physiological studies with the purpose of providing tentative identification of the vitamin D metabolite on the basis of comparison to known standards. In such cases, the choice of *in vivo* or *in vitro* model usually provides an indicator as to the types of metabolites to be expected. Rarely has HPLC been used in conjunction with physico-chemical techniques as a step in the initial identification of metabolites. But times are changing. The major metabolites of vitamin D: 25-(OH)D_3, 25,26-$(OH)_2D_3$, 1,25-$(OH)_2D_3$ and 24,25-$(OH)_2D_3$ were identified (Blunt et al., 1968, Suda et al., 1970; Holick et al., 1971; Holick et al., 1972b) prior to the introduction of HPLC. The identification schemes for the minor metabolites: 25-OH-D_3-26,23-lactone, 23(S),25-$(OH)_2D_3$, 1,25-$(OH)_2D_2$, 24(R), 25-$(OH)_2D_2$, calcitroic acid (Wichmann et al., 1979; Tanaka et al., 1981; Jones et al., 1975; Jones et al., 1980; Esvelt et al., 1979) all used a stage of HPLC.

Initial volumes of plasma or weights of tissue are large in these identification studies and open-column techniques are frequently employed to reduce the weight of lipid to a manageable size. In the extreme case of Holick et al., (1971), it is difficult to see how HPLC could have been used until well into the purification scheme of 1,25 $(OH)_2D_3$. They extracted 1450 chick intestines weighing 12.5 kg to yield 210 ml of lipid extract which was later reduced to 22 g by a solvent partition step. A combination of open column techniques gave about 8 μg of pure 1,25-$(OH)_2D_3$ suitable for UV absorption spectroscopy and mass spectrometry. A large preparative HPLC of the type used by Levan et al. (1981) (9.4 mm × 25 cm PARTISIL column) may have aided here.

Even in the case of 24,25-$(OH)_2D_3$, identified by Holick et al. (1972b) from chick kidney homogenates, 5 column steps were required to recover approximately 10 μg of metabolite from 1 g of lipid. Holick et al. used a combination of gel-chromatography on Sephadex LH-20 with two solvent mixtures and molecular sieving using Biobeads SX-8 and Sephadex LH-20 in methanol. Compare this to the recent scheme of Jones et al., (1983) for purifying 24,25-$(OH)_2D_3$ and a number of other metabolites from renal perfusates where no open column steps were used (Fig. 20). Preparative HPLC on 9.4 mm × 25 cm Zorbax-SIL reduced lipid content in an analogous way to gel-chromatography on Sephadex LH-20 chromatography but in a fraction of the time. Subsequent steps in Jones et al. (1983) procedure took advantage of the solvent selectivity and column selectivity strategies presented earlier whereas Holick et al. (1972) used relatively inefficient molecular sieving effects. The resultant reduction in steps from 5 to 3 between Holick et al. (1972) and Jones et al. (1983) although partly the result of knowledge gained from newer metabolites in the 10 years between the studies, is also an indicator of the tremendous improvement in resolution offered by HPLC. There are corresponding reductions in time.

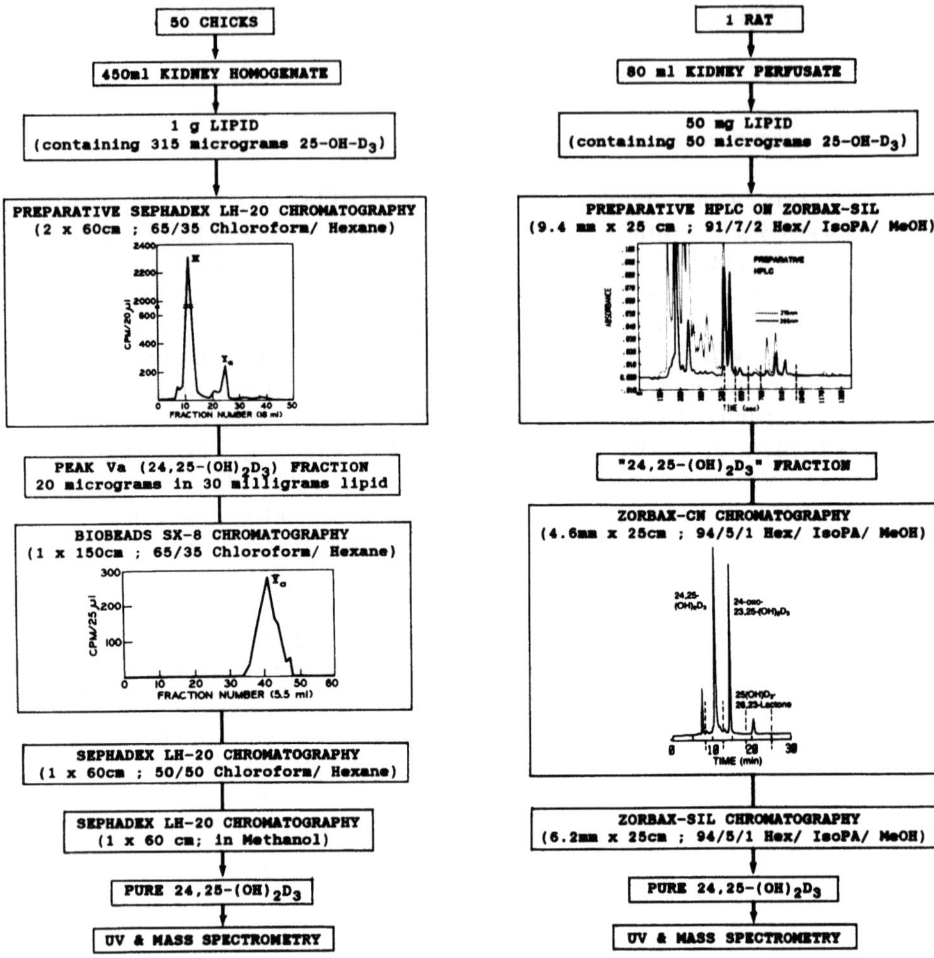

Fig. 20. Comparison of purifications schemes of Holick et al. (1972b) and Jones et al. (1983). Comparison of open column purification to one involving HPLC

The impressive purification and identification schemes reported by DeLuca and coworkers for a variety of metabolites reported in the early 1970s made it clear that ultraviolet spectroscopy provided an important criterion for assessing purity of an isolated metabolite. Usually on the basis of this technique the decision was made to sacrifice valuable nanograms of metabolite to the important identification technique, mass spectrometry. With the advent of scanning spectrophotometry this assessment has been made simpler and more sensitive (Jones et al., 1983; 1984). It also appears that the tenet "when UV pure proceed with MS" still applies since Jones et al. (1983) have utilized the same approach used by Holick et al., (1972b) (Fig. 20). In a more elaborate way than stop-flow UV spec-

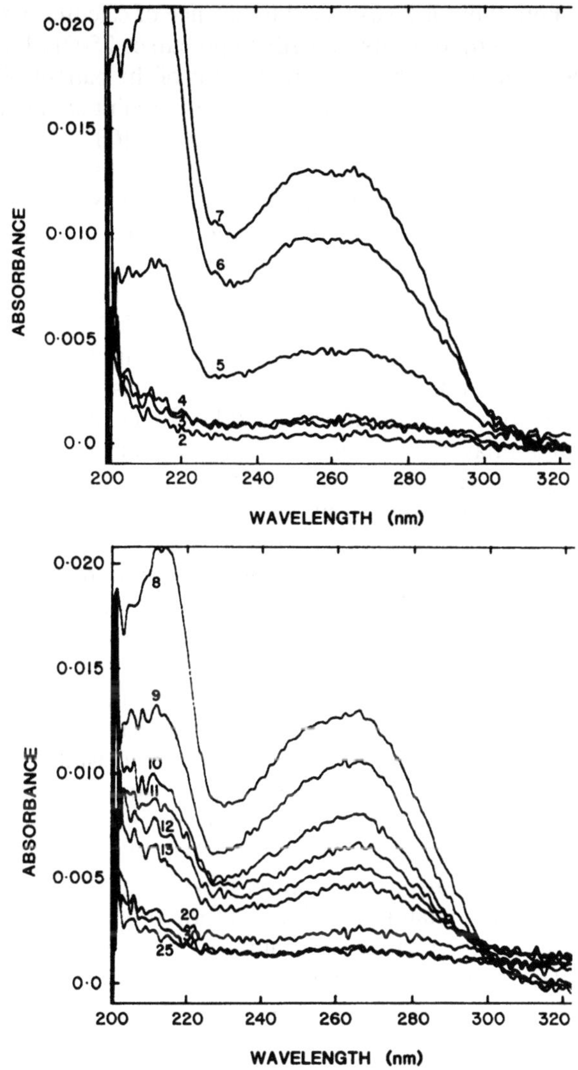

Fig. 21. Sequential UV scans of 24,25-$(OH)_2D_3$ peak from perfused kidney extracts. From work of Jones et al. (1983). Figure not published. Note this is the product of purification scheme shown in Fig. 20. Note the distorted spectra present in scans 6,7,8,9 with intense absorbance around 210 nm and a shoulder around 254 nm. The clean spectra 10 onwards show that the tail of the peak is devoid of UV contaminants and suitable for mass spectrometry

trometry, scanning spectrophotometry can be used to assess the leading and tailing sides chromatographic peaks thereby indicating which portion of the peak to apply to mass spectrometry (Fig. 21).

Diagnosis, Pathogenesis and Treatment of Endocrine Disorders Involving Vitamin D

Assay of Blood Levels of Vitamin D Metabolites

Since 1,25-$(OH)_2D_3$ is the hormonal form of vitamin D it is the *blood level of this metabolite* which is central to any endocrinological investigation of a patient where there is a suspected imbalance of calcium metabolism. It appears that the measurement of its immediate precursor 25-OH-D_3, is also valuable because its level indicates the vitamin D status of the patient. Without adequate supplies of vitamin D, an individual is unable to maintain adequate levels of 25-OH-D and this in turn leads to a deficiency of 1,25-$(OH)_2D$. It is possible to measure other metabolites such as vitamin D itself, 24,25-$(OH)_2D_3$, 25,26-$(OH)_2D_3$ or 25-OH-D_3-26,23-lactone but at this point in time there seems to be little clinical reason to do so since no basic or clinical studies have yet defined a specific function for a metabolite other than 1,25-$(OH)_2D_3$.

Both vitamin D_2 and D_3, have been measured by competitive binding assay with (Horst et al., 1981b) or without HPLC (Belsey et al., 1971). These methods work better for the low levels: 0.5–5 ng/ml found in normal individuals receiving adequate intakes of vitamin D. Other assays based upon HPLC and which are able to resolve vitamins D_2 and D_3 (Jones, 1978 and Horst et al., 1981a) are more appropriate for the higher plasma vitamin D levels found in patients receiving high dosage vitamin D therapy. Though some (Kooh et al., 1979) have argued that measurement of plasma vitamin D levels may aid in the elucidation of the cause of bone disease in some hepatobiliary diseases of childhood, the absence of impaired 25-hydroxylation in these conditions makes it unlikely that high plasma vitamin D levels will be observed. Thus aside from being a gross indicator of vitamin D dosage in treated patients the measurement of plasma vitamin D is of little clinical value.

The use of DBP in the assay of the principal circulating metabolite, 25-OH-D has already been discussed. Of equal value and gaining in popularity is the direct measurement of 25-OH-D after HPLC (Eisman et al., 1977, Jones, 1978). This type of assay not only provides a measure of the total 25-OH-D substrate available to the patient but it can also make use of the type of chromatography described in (V) to separate 25-OH-D_2 and 25-OH-D_3 thereby providing an assessment of relative contribution of vitamins D_2 and D_3. Where the major dietary sources of vitamin D are in the form of vitamin D_2 (e.g. some areas of U. S. and Canada) the relative contributions of skin and diet can be assessed (Arnaud et al., 1977). In a similar way Jones et al., 1979 used 25-OH-D_2/25-OH-D_3 ratios to assess the amount of skin synthesis of vitamin D_3 in TPN patients receiving 200 IU of vitamin D_2/day parenterally as their sole source and in whom gut absorption of vitamin D was insignificant. Present day 25-OH-D assays utilise some combination of reverse phase and straight phase chromatography, a concept used by Jones (1978) but speeded up by others by adapting one of the chromatographic steps to a cartridge step (Dabek et al., 1981; Turnbull et al., 1982; Kao and Heser, 1984). The advent of HPLC to the vitamin D field in the mid 1970s led to a decline in the use of non-ochromatographic assays for 25-OH-D (Belsey et al., 1974b; Garcia-Pascual et al., 1976; Hollis and Conrad, 1976; Morris and Peacock, 1976). It is clear that these "direct" assays gave erroneously high values for 25-OH-D, a point which could be proven with the introduction of 25-OH-D assay with UV detection. It is also interesting to note that there has not been a re-emergence of nonchromatographic assays to measure 25-OH-D.

Assays for the hormone 1,25-$(OH)_2D$ continue to proliferate. Over the years these have simplified considerably from the method of Brumbaugh et al. (1974)

or Hughes et al. (1976). The assay of Eisman et al. (1976) provided the basis which led many laboratories to develop a permutation of extraction, purification and radioreceptor assay steps.

Eisman et al. (1976) used only 4 steps in their 1,25-$(OH)_2D$ assay. With the interest in other metabolites having waned somewhat the complex assay schemes developed by many groups (Shepard et al., 1979; Aksnes, 1980a and b; Horst et al., 1981b; Parvainen et al., 1981; and Imawari et al., 1982) have also lost favour. Instead, newer clinical assays measure only one or two metabolite(s), are simple, have only 3 steps instead of 4 and some have even forsaken HPLC altogether.

Present versions (Reinhardt et al., 1984; Manolagas et al., 1983; Kao and Heser, 1984) though not involving HPLC steps, have gained from having been tested by comparison to HPLC assays. It will be interesting to see if this newer generation of assays will in turn be replaced by even simpler versions without purification steps. Though the streamlined, widely available, 1,25-$(OH)_2D$ assay of today is aimed at the routine endocrinological laboratory, newer research assays for 1,25-$(OH)_2D$ still utilise HPLC in order to resolve 1,25-$(OH)_2D_2$ and 1,25-$(OH)_2D_3$ (Fraher et al., 1983; Horst et al., 1982). This resolution achieved by the use of solvent selectivity is based upon the concepts referred to (Jones, 1980) in a previous section of this chapter.

Plasma vitamin D metabolite levels in health and disease

Plasma vitamin D metabolite measurements have one serious drawback that befalls all plasma hormone determinations. Plasma concentrations reflect both the synthesis and tissue uptake (whether target cells or degradative pathways) of each metabolite in a pathway. Plasma levels measured at one moment in time do not reflect the dynamics of the biological system. Thus when we define normal ranges for vitamin D metabolites we must consider this serious drawback. Furthermore the level of a highly regulated hormone such as 1,25$(OH)_2D_3$ is determined by the physiologic state of the organism. A normal value for 1,25$(OH)_2D_3$ is not meaningful unless PTH, calcium status, phosphorous level and mineralization state is specified. Nevertheless, plasma vitamin D metabolite measurements, particularly of 25-OH-D and 1,25-$(OH)_2D$, have been especially useful in helping to define the pathogenesis and etiology of some calcium related syndromes.

Normal ranges for the four commonly assayed metabolites are presented in Table 3. It can be readily appreciated that 25-OH-D is present in the blood at a concentration 500–1000 times that of its product, the hormone 1,25-$(OH)_2D$. While the normal human can tolerate a wide range of 25-OH-D levels from ~ 5 ng/ml to ~ 100 ng/ml without showing any ill effects, 1,25-$(OH)_2D$ levels are much more closely controlled and rarely fluctuate outside of the range 20–60 pg/ml under normal circumstances.

The various calcium-related disease states can be generally categorised under those causing:-

- an excess of 1,25-$(OH)_2D_3$
- a deficiency of 1,25-$(OH)_2D_3$.

Table 3. Reference Ranges for Plasma Concentrations of Important vitamin D metabolites

		Mean (± SD)	Range
Vitamin D		5.7 ± 2.9 nmol/L[a]	2.1 - 12.2 nmol/L[a]
25-OH-D	Summer	55.3 ± 17.8 nmol/L[b]	20.0 - 82.5 nmol/L[b]
	Winter	40.0 ± 9.8 nmol/L[a]	22.8 - 59.8 nmol/L[a]
1,25-$(OH)_2D$		88.9 ± 30.5 pmol/L[c]	40.9 - 141.8 pmol/L[c]
24.25$(OH)_2D_3$		5.8 ± 2.6 nmol/L[d]	1.9 - 10.3 nmol/L[d]

[a] From Jones, G. (1978)
[b] From Duthie, D., Fraser, D., unpublished observations
[c] From Jones, G., unpublished observations
[d] From Dreyer, B., Goodman, D. B. P. (1981)

These are summarised in Table 4. High plasma 1,25-$(OH)_2D_3$ can result from excessive stimulation of the kidney by normal physiological factors such as low plasma Ca (Hughes et al., 1975) or low plasma phosphate (Tanaka and DeLuca, 1973) or by excessive amounts of PTH (hyperparathyroidism) (Garabedian et al., 1972). During the early stages of the onset of rickets, falls in blood calcium and phosphate levels, are accompanied by a rise in plasma 1,25-$(OH)_2D_3$. Similarly the

Table 4a. Physiological, Disease and Therapeutic States Associated with High Vitamin D Metabolite Levels

High Plasma Vitamin D (more than 10 ng/ml)	High Plasma 25-OH-D (more than 50 ng/ml)	High Plasma 1,25-$(OH)_2D$ (more than 60 pg/ml)	Cause
Post-prandial transient increase Excessive dietary intake of D Excessive exposure to UV light High dosage therapy with D	Excessive dietary intake D Excessive exposure to UV light High dosage therapy with D or 25-$(OH)D_3$	High dosage therapy with 1,25-$(OH)_2D_3$	High intakes
		Low Ca or PO_4 diet Pregnancy	Physiological
	Vitamin D dependency rickets Type I?		Metabolic block
		Hyperparathyroidism Sarcoidosis Neprolithiasis	Regulatory Defect
		Vitamin D dependency rickets Type II	Target cell Receptor Defect

Table 4b. Diseases States Associated with Low Vitamin D Metabolite Levels

Low Plasma Vitamin D (less than 1 ng/ml)	Low Plasma 25-(OH)D (less than 4 ng/ml)	Low Plasma $1,25\text{-}(OH)_2D_3$ (less than 15 pg/ml)	Cause
Nutritional Deficiency Rickets Malabsorption of dietary D (e. g. Primary Biliary Atresia)	Nutritional Deficiency Rickets Malabsorption of dietary D	Nutritional Deficiency Rickets Malabsorption of dietary D	Lack of D intake
	Primary Biliary Cirrhosis?	Primary Biliary Cirrhosis Vitamin D dependency rickets Type I Renal Osteodystrophy	Metabolic Blocks
		Hypoparathyroidism Pseudohypopara-thyroidism Hypophosphatemic Rickets Tumour-induced Hypophosphatemia	Regulatory Defects
	Anticonvulsant-induced rickets Neprotic syndrome	Total Parenteral Nutrition associated Osteomalacia Osteoporosis?	Other

administration of vitamin D_3 to a vitamin D deficient patient results in a high blood level of $1,25\text{-}(OH)_2D_3$ during the healing period (Papapoulos et al., 1980). High plasma $1,25\text{-}(OH)_2D$ levels, about twice the normal level, are found in pregnancy (Kumar et al., 1979; Reddy et al., 1983).

In sarcoidosis, plasma $1,25\text{-}(OH)_2D_3$ levels rise above normal (Bell et al., 1979; Papapoulos et al., 1979; Chesney et al., 1981) due to the unique synthesis of $1,25\text{-}(OH)_2D_3$ by sarcoid tissue (Adams et al., 1983).

Low plasma $1,25\text{-}(OH)_2D$ result from a variety of acquired and genetically inherited diseases some of which involve specific defects in calcium homeostasis while others are secondary complications of more general problems. Low plasma $1,25\text{-}(OH)_2D$ can result from inadequate supply of vitamin D (eg. malnutrition or vitamin D deficiency rickets) or defective metabolism of vitamin D_3 to the hormone, $1,25\text{-}(OH)_2D$ (eg. hepatobiliary disease or vitamin D dependency rickets, Type I (VDDR, Type I). In all cases, the low plasma level of $1,25\text{-}(OH)_2D$ results from an inadequate synthesis of $1,25\text{-}(OH)_2D$ in the kidney and this produces a similar pattern of symptoms. These include low blood Ca, low blood PO_4, elevated PTH and unmineralised bone. In children, $1,25\text{-}(OH)_2D$ deficiency manifests as a thickening of the unmineralised cartilage matrix at the growth plate and a weakening of the shaft of the bone; defects that are collectively termed rickets. In adults the unmineralised bone is confined to the shaft of the bone, results in wide osteoid seams and the condition is called osteomalacia.

All types of rickets or osteomalacia which result from an inadequate synthesis of 1,25-$(OH)_2D$ can be cured simply by exogenous supplies of physiological amounts of the hormone. In most types of rickets, vitamin D will also cure the condition but pharmacological amounts must often be supplied to overcome poor absorption or metabolic blocks.

Chronic renal failure often results in a type of bone disease known as renal osteodystrophy (Coburn et al., 1982). Here, the reduction in renal tissue causes a loss in amounts of renal 1-hydroxylase enzyme and this in turn causes a low plasma 1,25-$(OH)_2D_3$ level. The extensive data of Gray et al., (1979) suggests that a fall in glomerular filtration rate below 25 is the critical point below which the kidney-1-hydroxylase cannot maintain a normal circulating concentration of 1,25-$(OH)_2D_3$. But in renal osteodystrophy low plasma 1,25-$(OH)_2D$ level is accompanied by uremia and secondary hyperparathyroidism; the latter resulting from an inability to excrete fragments of PTH and exacerbated by continued hypocalcemic stimulation of PTH release. Aluminum has also been implicated as a complicating factor in chronic renal failure patients receiving haemodialysis (Parkinson et al., 1979). It is because of these complications of renal osteodystrophy that it is not always successfully treated with 1,25-$(OH)_2D_3$. Slatopolsky et al. (1985) have presented evidence recently that intravenous 1,25-$(OH)_2D_3$ administration markedly suppresses secondary hyperparathyroidism in patients with chronic renal failure without causing the concomitant hypercalcemia associated with oral 1,25-$(OH)_2D_3$ therapy.

Perhaps the best example of the usefulness of plasma 1,25-$(OH)_2D$ assay comes in the differential diagnosis of vitamin D dependency rickets. Here a lack of the metabolic effects of 1,25-$(OH)_2D$ can result from an inadequate synthesis of the hormone (VDDR, Type I) (Fraser et al., 1973) or a defective hormone response in the target cell (VDDR, Type II) (Rosen et al., 1979). As might be expected, 1,25-$(OH)_2D$ levels in VDDR, Type I are below normal and in VDDR, Type II above normal. With the development of HPLC methods for the purification of the 1,25-$(OH)_2D_3$ receptor (Pike et al., 1982), it is likely that methods will soon be developed to pinpoint any possible subtypes of VDDR, Type II in human patients. Indeed using other procedures Liberman and Marx (1982) have suggested that the known VDDR Type II patients can be subdivided into those with receptor and postreceptor defects.

Another group of diseases which result in low 1,25-$(OH)_2D$ levels are those in which there is a lack of one of the normal physiological factors which modulates 1-hydroxylase activity (eg. PTH in hypoparathyroidisms, Lund et al., 1980; or pseudohypoparathyroidism, Drezner et al., 1976) or inappropriate response to these modulating factors (eg. low PO_4 in hypophosphatemic rickets).

Though Scriver et al. (1978), Chesney et al. (1980), Lyles et al. (1982) reported low or low normal levels of plasma 1,25-$(OH)_2D$ in hypophosphatemic states the deficiencies were not dramatic. In fact, Delvin et al. (1981) were unable to demonstrate a low 1,25-$(OH)_2D$ in this condition. However, when one bears in mind that a low plasma phosphate produces a dramatic increase in the plasma 1,25-$(OH)_2D$ level in normal animals (Hughes et al., 1975), it is evident that the level of 1,25-$(OH)_2D$ in hypophosphatemic rickets is inappropriately low (Chesney et al., 1980). Another rare condition which is accompanied by a very low 1,25-$(OH)_2D_3$ is tumour-induced hypophosphatemia. Here, it is possible that a PTH-like peptide produced by the tumour causes a number of renal problems including phosphate loss and suppression of 1-hydroxylase (Lyles et al., 1980; Sweet et al., 1980).

Thus far disease states have been considered where there is a rise or fall in plasma 1,25-$(OH)_2D$ often without a discernable change in other plasma metabolites. Only in diseases where there is a problem of vitamin D delivery is a low 1,25-$(OH)_2D$ accompanied by a fall in plasma 25-OH-D (eg. vitamin D deficiency rickets, malabsorption syndromes). However, there are disease states in which normal plasma 1,25-$(OH)_2D$ levels are maintained along with very low plasma 25-OH-D levels. These include rickets due to anticonvulsant drug therapy (Hahn et al., 1972) and nephrotic syndrome (Lambert et al. 1981).

Monitoring Vitamin D Therapy

Despite the fact that 1,25-$(OH)_2D$ assays have provided much useful information to help in the diagnosis and pathogenesis of a variety of calcium related diseases, these assays are rarely used in monitoring therapy. Part of the reason for this can be understood by considering the half life of exogeneously administered 1,25-$(OH)_2D_3$ which is of the order of 4–8 hrs. Thus the daily measurement of plasma 1,25-$(OH)_2D$ is of little value in the assessment of the effects of exogenous 1,25-$(OH)_2D_3$ treatment. Administered 1,25-$(OH)_2D_3$ has essentially disappeared from the blood stream by 4–6 hrs, even though its biological effects continue for up to 24 hrs. This of course makes 1,25-$(OH)_2D_3$, a superior therapeutic agent because its hypercalcemic actions disappear within 24 hr after drug withdrawal while hypercalcemia persists for days or even weeks when cased by large doses of vitamin D. Aside from the problem of when to measure plasma 1,25-$(OH)_2D_3$ after the exogenous dose we have the problems of assay turn around time and lack of reference data as to what levels of plasma 1,25-$(OH)_2D$ should reach in a patient being healed with 1,25-$(OH)_2D_3$. Bell and Stern (1978) report transient 1,25-$(OH)_2D$ levels as high as 60–70 pg/ml whereas others (Papapoulos et al., 1980) have reported sustained levels of several hundred pg/ml in the healing of osteomalacia with vitamin D. Thus the indicators of the biological action of 1,25-$(OH)_2D$, namely plasma calcium and phosphate, are more convenient and reliable indicators of the amounts of 1,25-$(OH)_2D_3$ used in therapy.

Monitoring of exogenous vitamin D_2 or D_3 can be very useful in the management of the patient. Though vitamin D itself, and the other metabolites of vitamin D: 25-OH-D_3, 24,25-$(OH)_2D_3$ are being rapidly replaced by 1,25-$(OH)_2D_3$ as the drug of choice in some calcium-related diseases (eg. hypophosphatemic rickets), treatment with pharmacological amounts of vitamin D_2 or D_3 remains a popular form of therapy. It is cheap, but the dangers of overdosage are great and correction of overdosage is difficult. In contrast, the hypercalcemia resulting from overdosage of 1,25-$(OH)_2D_3$ is of short duration due to the short half life of 1,25-$(OH)_2D_3$ in the body.

It is believed that vitamin D has its biological effects by generating a pharmacological level of 25-OH-D_3 which can mimic the actions of physiological amounts of 1,25-$(OH)_2D_3$ (Stern, 1981). Thus it is the level of 25-OH-D_3, necessary to maintain the appropriate plasma Ca and PO_4, that should be monitored in these patients.

Several groups (Hughes et al., 1975 and Shepard et al., 1979) have reported that at times of hypervitaminosis D_3 (toxicity of vitamin D_3), 1,25-$(OH)_2D_3$ levels are normal or even suppressed. Thus, some other metabolite such as 25-OH-D_3 or its products must play a role in the toxicological/healing effects of vitamin D_3. Vitamin D_2 is commonly used as the form of vitamin D in drug formulations and though it is generally believed D_2 and D_3 are equipotent in man there is some work (Christensen et al., 1974 and Horst et al., 1982) to suggest otherwise. The main advantage of the use of the vitamin D_2 analog in treatment is that it represents an exogenous molecule different from the natural compound which can be monitored during the healing period, but this has not been exploited.

References

Abe E, Miyaura C, Sakagami H, Takeda M, Konno K, Yamazaki T, Yoshiki S, and Suda T (1981) Proc Natl Acad Sci USA 78:4990–4994

Abe E, Miyaura C, Tanaka H, Shiina Y, Kuribayashi T, Suda S, Nishii Y, DeLuca HF, and Suda T (1983) 1,25-Dihydroxyvitamin D_3 promotes fusion of mouse alveolar macrophages both by a direct. Proc Natl Acad Sci USA 80:5583–5587

Adams JA, Clemens TL, Holick MF (1981) Silica Sep-Pak preparative chromatography for vitamin D and its metabolites. J Chromatogr 226:198–201

Adams JS, Sharma OP, Singer FR (1983) Metabolism of 25-hydroxy vitamin D_3 by pulmonary alveolar macrophages in sarcoidosis. Am Soc Bone Mineral Res A53 (abstract)

Aksnes L (1980a) Quantitation of the main metabolites of vitamin D in a single serum sample. I. Extraction, separation and purification of metabolites. Clin Chim Acta 104:133–146

Aksnes L (1980b) Quantitation of the main metabolites of vitamin D in a single serum sample. II. Determination by UV-absorption and competitive protein binding assays. Clin Chim Acta 104:147–159

Arnaud SB, Matthusen M, Gilkinson JB, Goldsmith RS (1977) Components of 25-OH-D in serum of young children in upper midwestern United States. Am J Clin Nutr 30:1082–1086

Bell NH, Stern PH (1978) Hypercalcemia and increases in Serum hormone value during prolonged administration of 1α,25-dihydroxyvitamin D. New Engl J Med 298:241–1243

Bell NH, Stern PH, Pantzer E, Sinha TK, DeLuca HF (1979) Evidence that increased circulating 1,25-$(OH)_2D$ is the probable cause for abnormal calcium metabolism in sarcoidosis. J Clin Invest 64:218–225

Belsey R, DeLuca HF, Potts JT (1971) Competitive binding assay for vitamin D and 25-OH vitamin D. J Clin Endocrinol Metab 33:554–557

Belsey R, DeLuca HF, Potts JT (1974a) Selective binding properties of the vitamin D transport protein in chick plasma in vitro. Nature (Lond) 247:208–209

Belsey RE, DeLuca HF, Potts JT (1974b) A rapid assay for 25-OH-vitamin D_3 without preparative chromatography. J Clin Endocrinol Metab 38:1046–1051

Bills CE (1954) Chemistry. In: "The Vitamins" (Sebrell WH Jr. & Harris RS, eds), Chapter 6, pp 132-203. Academic Press, NY

Bishop JE, Norman AW, Coburn JW et al (1980) Studies on the metabolism of calciferol. XVI. Determination of the concentration of 25-OH-D, 24,25-$(OH)_2D$, 1,25-$(OH)_2D$ in a single 2 ml plasma sample. Miner Electrolyte Metab 3:181–189

Björkhem I, Holmberg I, Oftebro H, Pedersen JI (1980) Properties of a reconstituted vitamin D_3 25-hydroxylase from rat liver mitochondria. J Biol Chem 255:5244–5249

Bligh EG, Dyer WJ (1959) A rapid method of total lipid extraction and purification. Can J Biochem Physiol 37:911–917
Bloch K (1965) The biological synthesis of cholesterol. Science 150:19–22
Blunt JW, DeLuca HF, Schnoes HK (1968) 25-Hydroxycholecalciferol. A biologically active metabolite of vitamin D_3. Biochemistry 7:3317–3322
Bouillon R, DeMoor P, Baggiolini EG, Uskokovic MR (1980) A radioimmunoassay for 1,25-$(OH)_2D_3$. Clin Chem 26:562–567
Boyle IT, Gray RW, DeLuca HF (1971) Regulation by calcium of *in vivo* synthesis of 1,25-dihydroxycholecalciferol and 21,25-dihydroxycholecalciferol. Proc Natl Acad Sci USA 68:2131–2134
Boyle IT, Miravet L, Gray RW, Holick MF, DeLuca HF (1972) The response of intestinal calcium transport to 25-hydroxy and 1,25-dihydroxy vitamin D in nephrectomized rats. Endocrinology 90:605–608
Brommage R, Jarnagin K, DeLuca HF, Yamada S, Takayama H (1983) 1-But not 24-hydroxylation of vitamin D is required for skeletal mineralization in rats. Am J Physiol 244:E298–E304
Brumbaugh PF, Haussler DH, Bursac KM, Haussler MR (1974) Filter assay for 1,25-dihydroxyvitamin D_3. Utilisation of the hormone's target receptor. Biochemistry 13:4091–4097
Brumbaugh PF, Haussler MR (1975) Specific binding of 1α,25-dihydroxycholecalciferol to nuclear components of chick intestine. J Biol Chem 250:1588–1594
Brunette MG, Chan M, Ferriere C, Roberts KD (1978) Site of 1,25-dihydroxyvitamin D_3 synthesis in kidney. Nature 276:287–289
Burns DT, MacKay C, Tillman J (1980) Rationalisation of the chromatographic behaviour of vitamin D_2/D_3 and related compounds in adsorption HPLC. J Chromatogr 190:141–144
Chesney RW, Hamstra AJ, DeLuca HF et al (1981) Elevated serum 1,25-dihydroxyvitamin D concentrations in the hypercalcemia of sarcoidosis: correction by glucocorticoid treatment. J Pediatr 98:919–922
Chesney RW, Mazess RB, Rose P, Hamstra AJ, DeLuca HF (1980) Supranormal 25-OH-D and subnormal 1,25-$(OH)_2D$. Their role in X-linked hypophosphatemic rickets. Am J Dis Child 134:140–143
Christiansen C, Rodbro P (1974) Initial and maintenance doses of vitamin D_2 in the treatment of anticonvulsant osteomalacia. Acta Neurol Scand 50:631–641
Clemens TL, Hendy GN, Graham RF, Baggiolini EG, Uskokovic MR, O'Riordan JLH (1978) A radioimmunoassay of 1,25-$(OH)_2D_3$. Clin Sci 54:329–332
Coburn JW, Sherrard DJ, Ott SM, Hodsman AB, DiDomenico NC, Alfrey AC (1982) Bone disease in uremia: a reappraisal. In: Norman AW, Schaefer K, Herrath D, Grigoleit HG (eds), Vitamin D, Chemical, Biochemical and Clinical Endocrinologoy of Calcium Metabolism, de Gruyter, Berlin pp 827–831
Dabek JT, Harkonen M, Wahlroos O, Aldercreutz H (1981) Assay for plasma 25-OH-D_2 and 25-OH-D_3 by HPLC. Clin Chem 27:1346–1351
DeLuca HF (1972) The functional metabolism of vitamin D_3. In: Calcium, Parathyroid Hormone and the Calcitonins (Talmage RV & Munson PL, eds), pp 221–235. Excerpta Medica, Amsterdam
DeLuca HF (1974) Vitamin D: The vitamin and the hormone. Fed Proc 33:2211–2219
DeLuca HF (1984) The metabolism and function of vitamin D. In: "Biochemistry of Steroid Hormones" (Makin HLJ, ed), Second edition, pp 71–116. Blackwell Scientific Publications, Boston, MA
DeLuca HF (1985) The vitamin D-calcium axis--1983. In: Calcium in Biological Systems" (Rubin RP, Weiss GB & Putney JW, Jr, eds), pp 491–511. Plenum Publishing Corp, New York
DeLuca HF (1988) The vitamin D story: a collaborative effort of basic science and clinical medicine. FASEB J 2:224–236
DeLuca HF, Franceschi RT, Halloran BP, Massaro ER (1982) Molecular events involved in the 1,25-dihydroxyvitamin D_3 stimulation of intestinal calcium transport. Fed Proc 41:66–71
DeLuca HF, Schnoes HK (1983) Vitamin D: Recent advances. Ann Rev Biochem 52:411–439
DeLuca HF, Schnoes HK (1984) Vitamin D: Metabolism and mechanism of action. In: Annual Reports in Medicinal Chemistry (Bailey DM, ed), Chapter 18, vol 19, pp 179–190. Academic Press, New York
Delvin EE, Glorieux FH (1981) Serum 1,25-dihydroxyvitamin D concentration in hypophosphatemic vitamin D-resistant rickets. Calcif Tissue Int 33:173–175

Dreyer BE, Goodman DBP (1981) A simple direct spectrophotometric assay for 24,25-$(OH)_2D_3$. Analyt Biochem 114:37–41
Drezner MK, Neelon FA, Haussler MR, McPherson HT, Lebovitz HE (1976) 1,25-Dihydroxycholecalciferol deficiency the probable cause of hypocalcemia and metabolic bone disease in pseudohypoparathyroidism. J Clin Endocrinol Metab 42:621–628
Eisman JA, Hamstra AJ, Kream BE, DeLuca HF (1976) A sensitive precise and convenient method for determination of 1,25-dihydroxyvitamin D in human plasma. Arch Biochem Biophys 176:235–243
Eisman JA, Shepard RM, DeLuca HF (1977) Determination of 25-OH-D_2 and 25-OH-D_3 in human plasma using HPLC. Analyt Biochem 80:298–305
Esvelt RP, Schnoes HK, DeLuca HF (1979) Isolation and characterisation of 1-alpha-hydroxy-23-carboxytetranorvitamin D: a major metabolite of 1,25-dihydroxyvitamin D_3. Biochemistry 18:3977–3983
Fraher LJ, Adami S, Clemens TL, Jones G, O'Riordan JLH (1983) Radioimmunoassay of 1,25-dihydroxyvitamin D_2: studies on the metabolism of vitamin D_2 in man. Clin Endocrinol 18:151–165
Fraser D, Kooh SW, Kind P, Holick MF, Tanaka Y, DeLuca HF (1973) Pathogenesis of hereditary vitamin D dependancy rickets. N Engl J Med 289:817–822
Fraser DR, Kodicek E (1970) Unique biosynthesis by kidney of a biologically active vitamin D metabolite. Nature 228:764–766
Garabedian M, Holick MF, DeLuca HF, Boyle IT (1972) Control of 25-hydroxycholecalciferol metabolism by the parathyroid glands. Proc Natl Acad Sci USA 69:1673–1676
Garabedian M, Tanaka Y, Holick MF, DeLuca HF (1974) Response of intestinal calcium transport and bone calcium mobilization to 1,25-dihydroxyvitamin D_3 in thyroparathyroidectomized rats. Endocrinology 94:1022–1027
Garcia-Pascual B, Peytremann A, Courvoisier B, Lawson DEM (1976) A simplified competitive protein-binding assay for 25-hydroxycalciferol. Clin Chim Acta 68:99–105
Ghazarian JG, Jefcoate CR, Knutson JC, Orme-Johnson WH, DeLuca HF (1974) Mitochondrial cytochrome P_{450}: A component of chick kidney 25-hydroxycholecalciferol-1α-hydroxylase. J Biol Chem 249:3026–3033
Gilbertson TJ, Stryd RP (1977) HPLC assay for 25-OH-D_3 in serum. Clin Chem 23:1700–1704
Gray RW, Lemann J, Adams ND (1979) The regulation of plasma 1,25-$(OH)_2D$ concentrations in healthy adults. ibid pp 545–551
Gray RW, Omdahl JL, Ghazarian JG, DeLuca HF (1972) 25-Hydroxycholecalciferol-1-hydroylase: Subcellular location and properties. J Biol Chem 247:7528–7532
Gray TK, McAdoo T, Pool D, Lester GE, Williams ME, Jones G (1981) A modified radioimmunoassay for 1,25-$(OH)_2D_3$. Clin Chem 27:458–462
Haddad JG, Chyu KJ (1971) Competitive protein-binding assay for 25-hydroxycholecalciferol. J Clin Endocrinol Metab 33:992–995
Hahn TJ, Hendin BA, Scharp CR, Haddad JG (1972) Effect of chronic anticonvulsant therapy on serum 25-hydroxycalciferol levels in adults. New Engl J Med 287:900–904
Halloran BP, Barthell EN, DeLuca HF (1979) Vitamin D metabolism during pregnancy and lactation in the rat. Proc Natl Acad Sci USA 76:5549–5553
Havinga E (1973) Vitamin D, example and challenge. Experientia 29:1181–1316
Henry HL, Norman AW (1978) Vitamin D: Two dihydroxylated metabolites are required for normal chicken egg hatchability. Science 201:835–837
Henry HL, Taylor AN, Norman AW (1977) Response of chick parathyroid glands to the vitamin D metabolites 1,25-dihydroxyvitamin D_3 and 24,25-dihydroxyvitamin D_3. J Nutr 107:1918–1926
Holick MF (1981) The cutaneous photosynthesis of previtamin D_3: A unique photoendocrine system. J Invest Dermatol 76:51–58
Holick MF, Clark MB (1978) The photobiogenesis and metabolism of vitamin D. Fed Proc 37:2567–2574
Holick MF, Garabedian M, DeLuca HF (1972a) 1,25-Dihydroxycholecalciferol: Metabolite of vitamin D_3 active on bone in anephric rats. Science 176:1146–1147
Holick MF, DeLuca HF (1971) A new chromatographic system for vitamin D_3 and its metabolites: Resolution of a new vitamin D_3 metabolite. J Lipid Res 12:460–465

Holick MF, Kleiner-Bossaller A, Schnoes HK, Kasten PM, Boyle IT, DeLuca HF (1973) 1,24,25-Trihydroxyvitamin D_3. A metabolite of vitamin D_3 effective on intestine. J Biol Chem 248:6691–6696

Holick MF, Schnoes HK, DeLuca HF, Suda T, Cousins RJ (1971) Isolation and identification of 1,25-dihydroxycholecalciferol. A metabolite of vitamin D active in intestine. Biochemistry 10:2799–2804

Holick MF, Schnoes HK, DeLuca HF, Gray RW, Boyle IT, Suda T (1972b) Isolation and identification of 24,25-dihydroxycholecalciferol: A metabolite of vitamin D_3 made in the kidney. Biochemistry 11:4251–4255

Hollis BW, Conrad HR (1976) Measurement of 25-hydroxycholecalciferol in the bovine. J Dairy Sci 59:1522–1524

Horst R (1979) 25-OH-D_3-26,23-Lactone: a metabolite of vitamin D_3 that is 5 times more potent than 25-OH-D_3 in the rat plasma competitive protein binding radioassay. Biochem Biophys Res Comm 89:286–293

Horst RL, Littledike ET, Riley JL, Napoli JL (1981a) Quantitation of vitamin D and its metabolites and their plasma concentrations in five species of animals. Analyt Biochem 116:189–203

Horst RL, Napoli JL, Littledike ET (1982) Discrimination in the metabolism of orally dosed ergocalciferol and cholecalciferol by the pig, rat and chick. Biochem J 204:185–189

Horst RL, Reinhardt TA, Beitz DC, Littledike ET (1981b) A sensitive competitive protein binding assay for vitamin D in plasma. Steroids 37:581–591

Horst RL, Reinhardt TA, Pramanik BC, Napoli JL (1983) 23-Keto-25-hydroxyvitamin D_3: A vitamin D_3 metabolite with high affinity for the 1,25-dihydroxyvitamin D specific cytosol receptor. Biochemistry 22:245–250

Horst RL, Wovkulich PM, Baggiolini EG, Uskokovic MR, Engstrom GW, Napoli JL (1984) (23S)-1,23,26-Trihydroxyvitamin D_3: Its biologic activity and role in 1α,25-dihydroxyvitamin D_3 26,23-lactone biosynthesis. Biochemistry 23:3973–3976

Hughes MR, Baylink DJ, Jones PG, Haussler MR (1976) Radioligand receptor assay for 25-hydroxyvitamin D_2/D_3 and 1,25-dihydroxyvitamin D_2/D_3. Application to hypervitaminosis D. J Clin Invest 58:61–70

Hughes MR, Brumbaugh PF, Haussler MR, Wergedal JE, Baylink DJ (1975) Regulation of serum 1-alpha-25-dihydroxyvitamin D_3 by calcium and phosphate in the rat. Science 190:578–580

Hummer L, Tjellesen L, Christiansen C (1984) Discrepancy between serum 1,25-$(OH)_2D_3$ measured by antisera and by cytosol receptor protein in epileptic patients treated with vitamin D_3. Am Soc Bone Mineral Ress A67 (abstract)

Ikekawa N, Koizumi N (1976) Separation of vitamin D metabolites and their analogues by HPLC. J Chromatogr 119:227–232

Imawari M, Kozawa K, Yoshida T, Osuga T (1982) A simple and sensitive assay for 25-hydroxyvitamin D, 24,25-dihydroxyvitamin D and 1,25-dihydroxyvitamin D in human serum. Clin Chim Acta 124:63–73

Jarnagin K, Brommage R, DeLuca HF, Yamada S, Takayama H (1983) 1- But not 24-hydroxylation of vitamin D is required for growth and reproduction in rats. Am J Physiol 244:E290–E297

Jones G (1977) In: Norman AW, Schaefer K, Coburn JW, DeLuca HF, Fraser D, Grigoleit HG, Herrath D. "Vitamin D Biochemical, Chemical and Clinical Aspects Related to Calcium Metabolism. De Gruyter, Berlin pp 495–500. Application of HPLC for Assay of Vitamin D Metabolites

Jones G (1978) Assay of vitamins D_2 and D_3,25-OH-D_2 and 25-OH-D_3 in human plasma by HPLC. Clin Chem 24:287–298

Jones G (1980) Ternary solvent mixtures for improved resolution of hydroxylated metabolites of vitamin D_2 and vitamin D_3 during HPLC. J Chromatogr Biomedical Applications 221:27–37

Jones G (1983) Chromatographic separation of 24(R),25-dihydroxyvitamin D_3 and 25-hydroxyvitamin D_3-26,23-lactone using a cyano-bonded phase packing. J Chromatogr 276:69–75

Jones G, Byrnes B, Jeejeebhoy KN (1979): Contribution of skin vitamin D_3 synthesis in patients receiving total parenteral nutrition. In: Norman AW, Schaefer K, Herrath D, Grigoleit HG, Coburn JW, DeLuca HF, Mawer B, Suda T (eds), Vitamin D: Basic Research and its Clinical Applications, de Gruyter, Berlin pp 169–172

Jones G, Byrnes B, Palma F, Segev D, Mazur Y (1980) Displacement potency of vitamin D_2 analogs in competitive protein-binding assays for 25-OH-D_3, 24,25-$(OH)_2D_3$, and 1,25-$(OH)_2D_3$. J Clin Endocrinol Metab 50:773–775

Jones G, DeLuca HF (1975) HPLC: Separation of the metabolites of vitamin D_2 and D_3 on small particle silica columns. J Lipid Res 16:448–453

Jones G, Kano K, Yamada S, Furusawa T, Takayama H, Suda T (1984) Identification of 24,25,26,27-tetranor 23-$(OH)D_3$ as a product of renal metabolism of 24,25-$(OH)_2D_3$. Biochemistry 23:3749–3754

Jones G, Kung M, Kano K (1983) The isolation and identification of two new metabolites of 25-hydroxyvitamin D_3 produced in the kidney. J Biol Chem 258:12920–12928

Jones G, Schnoes HK, DeLuca HF (1975) Isolation and identification of 1,25-dihydroxyvitamin D_2. Biochemistry 14:1250–1256

Jones G, Schnoes HK, DeLuca HF (1976) An *in vitro* study of vitamin D_2 hydroxylases in the chick. J Biol Chem 251:24–28

Jones G, Schnoes HK, Levan L, DeLuca HF (1980) Isolation and identification of 24-hydroxyvitamin D_2 and 24,25-dihydroxyvitamin D_2. Arch Biochem Biophys 202:450–457

Jongen MJM, van der Vijgh WJF, Willems HJJ et al (1981) Simultaneous determination of 25-OH-D and 1,25-$(OH)_2D$ in plasma or serum. Clin Chem 27:1757–1760

Kandutsch AA (1964) XXV. Sterol metabolism in skin and epidermis. In: "The Epidermis" (Montagna W & Lobitz WC, eds), Chapter 25, pp 493–510. Academic Press, New York

Kao PC, Heser DW (1984) Simultaneous Determination of 25-hydroxy- and 1,25-dihydroxyvitamin D from a single sample by dual-cartridge extraction. Clin Chem 30:56–61

Kobayashi Y, Taguchi T, Kanuma N, Ikekawa N, Oshide J-I (1980) Synthesis of 26,26,26,27,27,27-hexafluoro-25-hydroxyvitamin D_3. J Chem Soc Chem Commun 10:459–460

Kobayashi Y, Taguchi T, Terada T, Oshida J, Morisaki M, Ikekawa N (1982) Studies on organic fluorine compounds. Part 37. Studies on steroids. Part 78. Synthesis of 24,24-difluoro- and 24-fluoro-25-hydroxyvitamin D_3. JCS Perkin I:85–91

Kooh SW, Jones G, Reilly BJ, Fraser D (1979) Pathogenesis of rickets in chronic hepatobiliary disease in children. J Pediatr 94, 870–874

Koshy KT, Van Der Slik AL (1976) HPLC method for the determination of 25-OH-D_3 in cow plasma. Analyt Biochem 74:282–291

Kream BE, Reynolds RD, Knutson JC, Eisman JA, DeLuca HF (1976) Intestinal cytosol binders of 1,25-dihydroxyvitamin D_3 and 25-hydroxyvitamin D_3. Arch Biochem Biophys 176:779–787

Kumar R, Cohen WR, Silva P, Epstein FH (1979) Elevated 1,25-dihydroxyvitamin D levels in normal pregnancy and lactation. J Clin Invest 63:342–344

Lambert P, DeOreo P, Siegal B, Fu I, Hollies B, Roos B (1981) Vitamin D metabolism in the nephrotic syndrome. Calcif Tissue Int 33:338 (abstract)

Lester GE, Gray TK, Williams ME (1981) *In vitro* 1α-hydroxylation of ^{3}H-25-hydroxyvitamin D_3 by isolated cells from rat kidneys and placentae. In: "Hormonal Control of Calcium Metabolism" (Cohn DV, Talmage RV, Matthews JL, eds), p 376. Excerpta Medica, Amsterdam

Levan LW, Schnoes HK, DeLuca HF (1981) Isolation and identification of 25-hydroxyvitamin D_2 25-glucuronide: a biliary metabolite of vitamin D_2 in the chick. Biochemistry 20:222–226

Liberman UA, Eil C, Marx SJ (1982) Defects in target cell response to 1,25-dihydroxyvitamin D. Evaluation of cellular basis with cultured skin fibroblasts. In: Norman AW, Schaefer K, Herrath D, Grigoleit HG (eds) Vitamin D, Chemical, Biochemical and Clinical Endocrinology of Calcium Metabolism, de Gruyter, Berlin pp 73–78

Lohnes D, Jones G (1987) Side chain metabolism of vitamin D_3 in osteosarcoma cell line UMR-106. J Biol Chem 262:14394–14401

Lohnes D, Makin G, Ray R, Jones G (1988) Calcitroic acid production from 1,25-dihydroxyvitamin D_3 in perfused rat kidney and cultured bone cells. J Bone Min Res inpress

Lund BJ, Sorensen OH, Lund BI, Bishop JE, Norman AW (1980) Vitamin D metabolism in hypoparathyroidism. J Clin Endocrinol Metab 51:606–610

Lyles KW, Berry WR, Haussler MR, Harrelson JM, Drezner MK (1980) Hypophosphatemic osteomalacia: association with prostatic carcinoma. Annals Int Med 93:275–278

Lyles KW, Clark AG, Drezner MK (1982) Serum 1,25-$(OH)_2D$ levels in subjects with X-linked hypophosphatemia and osteomalacia. Calcif Tissue Int 34:125–130

Madhok TC, DeLuca HF (1979) Characteristics of the rat liver microsomal enzyme system converting cholecalciferol into 25-hydroxycholecalciferol. Evidence for the participation of cytochrome P-450. Biochem J 184:491–499
Manolagas SC, Culler FL, Howard JE, Brinkman AS, Deftos LJ (1983) The cytoreceptor assay for 1,25-$(OH)_2D$ and its application to clinical studies. J Clin Endocrinol Metab 56:751–760
Matthews EW, Byfield PGH, Colston KW, Evans IMA, Galante LS, MacIntyre I (1974) Separation of hydroxylated derivatives of vitamin D_3 by high speed liquid chromatography. FEBS Lett 48:122–125
Mayer E, Bishop JE, Chantraratna RAS, Okamura WH, Kruse JR, Popjak G, Ohnuma N, Norman AW (1983) Isolation and identification of 1,25-dihydroxy-24-oxo-vitamin D_3 and 1,23,25-trihydroxy-24-oxo-vitamin D_3. New metabolites of vitamin D_3 produced by a C-24 oxidation pathway of metabolism for 1,25-dihydroxyvitamin D_3 present in intestine and kidney. J Biol Chem 258:13458–13465
McCollum E v, Simmonds N, Becker JE, Shipley PG (1922) Studies on experimental rickets. XXI. An experimental demonstration of the existence of a vitamin which promotes calcium deposition. J Biol Chem 53:293–312
Mellanby E (1919) An experimental investigation on rickets. Lancet 1:407–412
Miller JC, George SA, Willis BG (1982) Multichannel detection in HPLC. Science 218:241–246
Miyaura C, Abe E, Kuribayashi T, Tanaka H, Konno K, Nishii Y, Suda T (1981). 1,25-Dihydroxyvitamin D_3 induces differentiation of human myeloid leukemia cells. Biochem Biophys Res Commun 102:937–943
Morris JF, Peacock M (1976) Assay of plasma 25-hydroxyvitamin D. Clin Chim Acta 72:383–391
Nakada M, Tanaka Y, Schnoes HK, DeLuca HF, Kobayashi Y, Ikekawa N (1985) Biological activities and binding properties of 23,23-difluoro-25-hydroxyvitamin D_3 and its 1α-hydroxy derivative. Arch Biochem Biophys, in press
Napoli JL, Fivizzani MA, Schnoes HK, DeLuca HF (1979) 1-Fluorovitamin D_3: A vitamin D_3 analog more active on bone-calcium mobilization than intestinal calcium transport. Biochemistry 18:1641–1646
Napoli JL, Horst RL (1983a) (23)-1,23,25-Trihydroxycholecalciferol, an intestinal metabolite of 1,25-dihydroxycholecalciferol. Biochem J 214:261–264
Napoli JL, Horst RL (1983b) C(24)- and C(23)-Oxidation, converging pathways of intestinal 1,25-dihydroxyvitamin D_3 metabolism: Identification of 24-keto-1,23,25-trihydroxyvitamin D_3. Biochemistry 22:5848–5853
Norman AW, Henry HL, Malluche HH (1980) 24R,25-Dihydroxyvitamin D_3 and 1α,25-dihydroxyvitamin D_3 are both indispensable for calcium and phosphorus homeostasis. Life Sci 27:229–237
Norman AW, Lund J, DeLuca HF (1964) Biologically active forms of vitamin D_3 in kidney and intestine. Arch Biochem Biophys 108:12–21
Ohnuma N, Norman AW (1982) Identification of a new C-23 oxidation pathway of metabolism for 1,25-dihydroxyvitamin D_3 present in intestine and kidney. J Biol Chem 257:8261–8271
Okamoto S, Smith C, DeLuca HF, Yamada S, Takayama H (1983) Biological activity of 24,24-difluoro-25-hydroxycholecalciferol in chicks. J Nutr 113:1607–1612
Okamoto S, Tanaka Y, DeLuca HF, Yamada S, Takayama H (1981) 24,24-Difluoro-25-hydroxyvitamin D_3-enhanced bone mineralization in rats. Comparison with 25-hydroxyvitamin D_3 and vitamin D_3. Arch Biochem Biophys 206:8–14
Olson EB Jr, Knutson JC, Bhattacharyya MH, DeLuca HF (1976) The effect of hepatectomy on the synthesis of 25-hydroxyvitamin D_3. J Clin Invest 57:1213–1220
Onisko BL, Schnoes HK, DeLuca HF, Glover RS (1979) Metabolism and biological activity of 25-fluorocholecalciferol, 24-dehydrocholecalciferol and 25-dehydrocholecalciferol in the rat. Biochem J 182:1–9
Ornoy A, Goodwin D, Noff D, Edelstein S (1978) 24,25-Dihydroxyvitamin D is a metabolite of vitamin D essential for bone formation. Nature 276:517–519
Papapoulos SE, Clemens TL, Fraher LJ, Gleed J, O'Riordan JLH (1980) Metabolites of vitamin D in human vitamin D deficiency. Effect of vitamin D_3 or 1,25-$(OH)_2D_3$. Lancet ii:612–615
Papapoulos SE, Fraher LJ, Sandler LM, Clemens TL, Lewin IG, O'Riordan JLH (1979) 1,25-Dihydroxycholecalciferol in the pathogenesis of the hypercalcemia of sarcoidosis. Lancet i:627–630

Parfitt AM, Mathews CHE, Brommage R, Jarnagin K, DeLuca HF (1984) Calcitriol but no other metabolite of vitamin D is essential for normal bone growth and development in the rat. J Clin Invest 73:576–586
Parkinson IS, Feest TG, Ward MR et al (1979) Fracturing dialysis osteodystrophy and dialysis encephalopathy: An epidemiological survey. Lancet i:406–409
Parvainen MT, Savolainen KE, Korhonen PH, Alhava EM, Visakorpi JK (1981) An improved method for routine determination of vitamin D and its hydroxylated metabolites in serum from children and adults. Clin Chim Acta 114:233–247
Pedersen JI, Ghazarian JG, Orme-Johnson NR, DeLuca HF (1976) Isolation of chick renal mitochondrial ferredoxin active in the 25-hydroxyvitamin D_3-1α-hydroxylase system. J Biol Chem 251:3933–3941
Pike JW, Kelly MA, Haussler MR (1982) HPLC of the Receptor for 1,25-dihydroxyvitamin D_3: its application toward resolution of receptor-monoclonal antibody complexes. Biochem Biophys Res Commun 109:902–907
Ponchon G, DeLuca HF (1969) The role of the liver in the metabolism of vitamin D. J Clin Invest 48:1273–1279
Ponchon G, Kennan AL, DeLuca HF (1969) "Activation" of vitamin D by the liver. J Clin Invest 48:2032–2037
Rasmussen H, Bordier P (1978) Vitamin D and bone. Metab Bone Dis Rel Res 1:7–13
Reddy GS, Norman AW, Willis DM, et al (1983) Regulation of vitamin D metabolism in normal human pregnancy. J Clin Endocrinol Metab 56:363–370
Reeve L, Tanaka Y, DeLuca HF (1983) Studies on the site of 1,25-dihydroxyvitamin D_3 synthesis in vivo. J Biol Chem 258:3615–3617
Reinhardt TA, Horst RL, Orf JW, Hollis BW (1984) A microassay for 1,25-$(OH)_2D$ not requiring HPLC: application to clinical studies. J Clin Endocrinol Metab 58:91–98
Reinhardt TA, Napoli JL, Praminik B, Littledike BT, Beitz DC, Partridge JJ, Uskokovic MR, Horst RL (1981) 1α,25,26-Trihydroxyvitamin D_3: An in vivo and in vitro metabolite of vitamin D_3. Biochemistry 20:6230–6235
Ribovich ML, DeLuca HF (1978) Effect of dietary calcium and phosphorus on intestinal calcium absorption and vitamin D metabolism. Arch Biochem Biophys 188:145–156
Rosen JF, Fleischman AR, Finberg L, Hamstra A, DeLuca HF (1979) Rickets with alopecia: an inborn error of metabolism. J Pediatr 94:729–735
Rosenthal A, Jones G, Kooh SW, Fraser D (1980) 25-Hydroxyvitamin D_3 metabolism by isolated perfused kidney. Am J Physiol 239 (Endorcrinol Metab):E12–E20
Schaefer PC, Goldsmith RS (1979) Radioimmunoassay of vitamin D metabolites. In: Norman AW, Schaefer K, Herrath D, Grigoleit HG, Coburn JW, DeLuca HF, Mawer B, Suda T (eds), Vitamin D: Basic Research and its Clinical Applications, de Gruyter, Berlin pp 205–212
Scriver OR, Reade TM, DeLuca HF, et al (1978) Serum 1,25-dihydroxyvitamin D levels in normal subjects and in patients with hereditary rickets or bone disease. N Engl J Med 299:976–979
Shepard RM, Horst RL, Hamstra AJ, DeLuca HF (1979) Determination of vitamin D and its metabolites in plasma from normal and anephric man. Biochem J 182:55–69
Shultz TD, Fox J, Heath H, III., Kumar R (1983) Do tissues other than the kidney produce 1,25-dihydroxivitamin D_3 *in vivo?* A reexamination. Proc Natl Acad Sci USA 80:1746–1750
Simpson RU, Wichmann JK, Paaren HE, Schnoes HK, DeLuca HF (1984) Metabolism of 25-hydroxyvitamin D_3 by rat kidney cells in culture: isolation and identification of cis- and trans-19-nor-10-oxo-25-hydroxyvitamin D_3. Arch Biochem Biophys 230:21–29
Slatopolsky E, Hilker S, Chan Y, Morrissey J, Martin K (1985) Parathyroid hormone secretion in renal failure. Proceedings of the 6th Workshop on Vitamin D, Merano, Italy, p 374
Steenbock H (1924) The induction of growth promoting and calcifying properties in a ration by exposure to light. Science 60:224–225
Stern PH (1981) A monolog on analogs. In vitro effects of vitamin D metabolites and consideration of the mineralisation question. Calcif Tissue Int 33:1–4
Stern PH, Hamstra AJ, DeLuca HF, Bell NH (1978) A bioassay capable of measuring 1 picogram of 1,25-$(OH)_2D_3$. J Clin Endocrinol Metab 46:891–896
Stumpf WE, Sar M, DeLuca HF (1981) Sites of action of 1,25$(OH)_2$ vitamin D_3 identified by thaw-mount autoradiography. In: "Hormonal Control of Calcium Metabolism" (Cohn DV, Talmage RV, Matthews JL, eds), pp 222–229. Excerpta Medica, Amsterdam

Stumpf WE, Sar M, Narbaitz R, Reid FA, DeLuca HF, Tanaka Y (1980) Cellular and subcellular localization of 1,25-$(OH)_2$-vitamin D_3 in rat kidney: Comparison with localization of parathyroid hormone and estradiol. Proc Natl Acad Sci USA 77:1149–1153
Stumpf WE, Sar M, Reid FA, Tanaka Y, DeLuca HF (1979) Target cells for 1,25-dihydroxyvitamin D_3 in intestinal tract, stomach, Kidney, skin. pituitary and parathyroid. Science 206:1188–1190
Suda T, DeLuca HF, Schnoes HK, Tanaka Y, Holick MF (1970) 25,26-Dihydroxyvitamin D_3, a metabolite of vitamin D_3 with intestinal transport activity. Biochemistry 9:4776–4780
Sutton RAL, Dirks JH (1978) Renal handling of calcium. Fed Proc 37:2112–2119
Sweet JA, Males JL, Hamstra AJ, DeLuca HF (1980) Vitamin D metabolite levels in oncogenic osteomalacia. Annals Int Med 93:279–280
Taguchi T, Mitsuhashi S, Yamanouchi Y, Kobayashi Y, Sai H, Ikekawa N (1984) Synthesis of 23,23-difluoro-25-hydroxyvitamin D_3. Tetrahedron Lett 25:4933–4937
Takasaki Y, Suda T, Yamada S, Takayama H, Nishii Y (1981) Isolation, identification, and biological activity of 25-hydroxy-24-oxovitamin D_3: A new metabolite of vitamin D_3 generated by in vitro incubations with kidney homogenates. Biochemistry 20:1681–1686
Tanaka Y, DeLuca HF (1973) The control of 25-hydroxyvitamin D metabolism by inorganic phosphorus. Arch Biochem Biophys 154:566–574
Tanaka Y, DeLuca HF (1981) Measurement of mammalian 25-hydroxyvitamin D_3 24R- and 1-alpha-hydroxylase. Proc Natl Acad Sci USA 78:196–199
Tanaka Y, DeLuca HF (1983) Stimulation of 1,25-dihydroxyvitamin D_3 production by 1,25-dihydroxyvitamin D_3 in the hypocalcaemic rat. Biochem J 214:893–897
Tanaka Y, DeLuca HF, Ikekawa N, Morisaki M, Koizumi N (1975) Determination of stereochemical configuration of the 24-hydroxyl group of 24,25-dihydroxyvitamin D_3 and its biological importance
Tanaka Y, DeLuca HF, Kobayashi Y, Taguchi T, Ikekawa N, Morisaki M (1979a) Biological acitivity of 24,24-difluoro-25-hydroxyvitamin D_3. Effect of blocking of 24-hydroxylation on the functions of vitamin D. J Biol Chem 254:7163–7167
Tanaka Y, DeLuca HF, Schnoes HK, Ikekawa N, Eguchi T (1981a) 23,25-Dihydroxyvitamin D_3: A natural precursor in the biosynthesis of 25-hydroxyvitamin D_3-26,23-lactone. Proc Natl Acad Sci USA 78:4805–4808
Tanaka Y, Halloran B, Schnoes HK, DeLuca HF (1979b) In vitro production of 1,25 dihydroxyvitamin D_3 by rat placental tissue. Proc Natl Acad Sci USA 76:5033–5035
Tanaka Y, Pahuja DN, Wichmann JW, DeLuca HF, Obayashi Y, Taguchi T, Ikekawa N (1982) 25-Hydroxy-26,26,26,27,27,27-Hexafluorovitamin D_3: Biological activity in the rat. Arch Biochem Biophys 218:134–141
Tanaka Y, Wichmann JK, Schnoes HK, DeLuca HF (1981b) Isolation and identification of 23,25-dihydroxyvitamin D_3, an in vivo metabolite of vitamin D_3. Biochemistry 20:3875–3879
Tucker G, III, Gagnon RE, Haussler MR (1973) Vitamin D_3-25-hydroxylase: Tissue occurrence and apparent lack of regulation. Arch Biochem Biophys 155:47–57
Turnbull H, Trafford DJH, Makin HLJ (1982) A rapid and simple method for the measurement of plasma 25-OH-D_2 and 25-OH-D_3 using Sep-Pak C_{18} cartridges and a single HPLC step. Clin Chim Acta 120:65–76
Wichmann JK, DeLuca HF, Schnoes HK, Horst RL, Shepard RM, Jorgenson NA (1979). 25-Hydroxyvitamin D_3-26,23-lactone: a new in vivo metabolite of vitamin D. Biochemistry 18:4713–4780.
Wichmann JK, Schnoes HK, DeLuca HF (1981) 23,24,25-Trihydroxyvitamin D_3, 24,25,26-trihydroxyvitamin D_3, 24-keto-25-hydroxyvitamin D_3 and 23-dehydro-25-hydroxyvitamin D_3: new in vivo metabolites of vitamin D_3. Biochemistry 20:7385–7391
Velluz L, Amiard G (1949) Chimie organique-equilibre de reaction entre precalciferol et calciferol. Compt Red 228:853–855
Velluz L, Amiard G, Petit A (1949) Le precalciferol: Ses relations d'equilibre avec le calciferol. Bull Soc Chim France 16:501–507
Vieth R, Fraser D (1979) Kinetic behavior of 25-hydroxyvitamin D_3-1-hydroxylase and 24-hydroxylase in rat kidney mitochondria. J Biol Chem 254:12455–12460
Yamada S, Ohmori M, Takayama H (1979) Synthesis of 24,24-difluoro-25-hydroxyvitamin D_3. Tetrahedron Lett 21:1859–1862
Yamada S, Ohmori M, Takayama H, Takasaki Y, and Suda T (1983) Isolation and identification of 1α- and 23-hydroxylated metabolites of 25-hydroxy-24-oxovitamin D_3 from *in vitro* incubates of chick kidney homogenates. J Biol Chem 258:457–463

High Performance Liquid Chromatography in the Analysis of Prostaglandins, Thromboxanes and Leukotrienes

I. A. Tavares

Introduction

The history of prostaglandins goes back to the 1930s when it was observed that lipid fractions isolated from semen induced contractions or relaxations of smooth muscle (Kurzrok and Lieb, 1930; von Euler 1934, 1935; Goldblatt, 1935). The lack of suitable chromatographic and analytical techniques hampered further investigation until the 1950s when Bergstrom et al. using conventional chromatographic methods of open columns, thin layer chromatography and gas liquid chromatography, isolated and then structurally identified and synthesized different pure prostaglandins (Bergstrom and Sjovall, 1957, 1960a, 1960b; Bergstrom et al., 1959, 1962, 1963). Since then, our knowledge of prostaglandin chemistry and chromatography have progressed rapidly. High performance liquid chromatography (HPLC) is now an important tool, and on it has depended the characterization of newly identified leukotriene metabolites of arachidonate.

Biochemistry and Chemistry

The naturally occurring, commonly encountered prostaglandins (PG), thromboxanes (TX), leukotrienes (LT) and other substances biosynthesized from arachidonate are shown in Fig. 1 (for reviews see Samuelsson et al., 1978 and Hammarstrom 1983). These products are called eicosanoids because their precursors are eicosa-enoic acids. Prostaglandins (PG) are a group of closely related 20-carbon unsaturated fatty acids whose structures comprise a cyclopentane ring with two aliphatic chains, a double bond at C13–C14, and a C15-hydroxyl group which is important for their biological activity (Fig. 2). Ten prostaglandins have been isolated to date. These have been classified according to the functional groups attached to the cyclopentane ring and are referred to by letters A to J. In addition they have numerical subscripts to indicate the number of double bonds in the side chains (determined by the precursor fatty acids). However, subscript 2 compounds arising from arachidonate tend to predominate. The subscript α given to the F class of prostaglandins refers to the configuration of the hydroxyl group attached to C-9.

Thromboxanes (TX) have an oxacyclohexane ring in place of the cyclopentane ring. The thromoxanes formed from arachidonate are TXA_2 and its more stable metabolite TXB_2, (Fig. 2). Prostaglandins and thromboxanes are formed by the

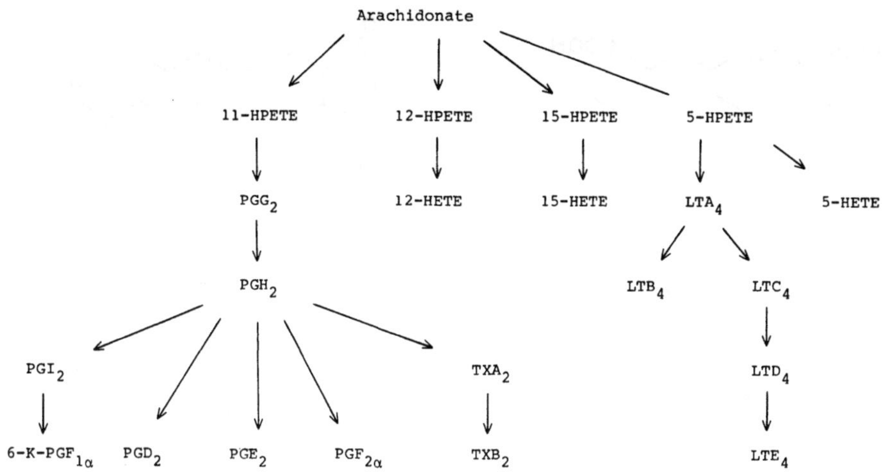

Fig. 1. Arachidonic acid cassade

action of a cyclo-oxygenase enzyme via the cyclic endoperoxides (PGG_2 and PGH_2 Fig. 1) formed from aracidonate.

Lipoxygenases transform arachidonate into straight chain metabolites, the hydroxy-eicosatetraenoic acids (HETE) and leukotrienes (LT). Several hydroxy fatty acids have now been isolated and found to contain 20 carbon atoms and four double bonds, two of which are conjugated (Fig. 3). A common feature of the leukotrienes is that three of their four double bonds are conjugated. The unstable epoxide leukotriene (LTA_4 is formed from 5-hydroperoxy-eicosatetraenoate (5-HPETE) and gives rise to LTB_4 via one pathway, and to LTC_4, LTD_4 and LTE_4 (peptide derivatives) by a second pathway (Fig. 3).

The analysis and quantitation of arachidonate metabolites is difficult not only because they comprise a very small fraction of total lipid but also because they constitute a large family of structurally related compounds with similar physical and chemical properties. Chromatographic separation of prostaglandins cannot depend on chain length for separation but has to rely on the differing polarity of the functional groups and the degree of unsaturation. Moreover the prostaglandins do not have a strong chromophore and hence, detection following chromatography has been difficult.

UV spectral scans of the prostaglandins demonstrates that all prostaglandins have a wavelength maximum at 192.5 nm (Terrango et al., 1981). Only fairly recently have UV detectors capable of working at low UV wavelengths been available, making it possible to detect nanogram quantities of underivatized prostaglandins. The leukotrienes have characteristic absorbance spectra with an absorbance maxima between 270 and 280 nm (Murphy et al., 1979) so making the UV detection and quantitation of leukotrienes relatively easy. Hydroxy fatty acid metabolites of arachidonate have conjugated double bonds which absorb maximally at 235 nm.

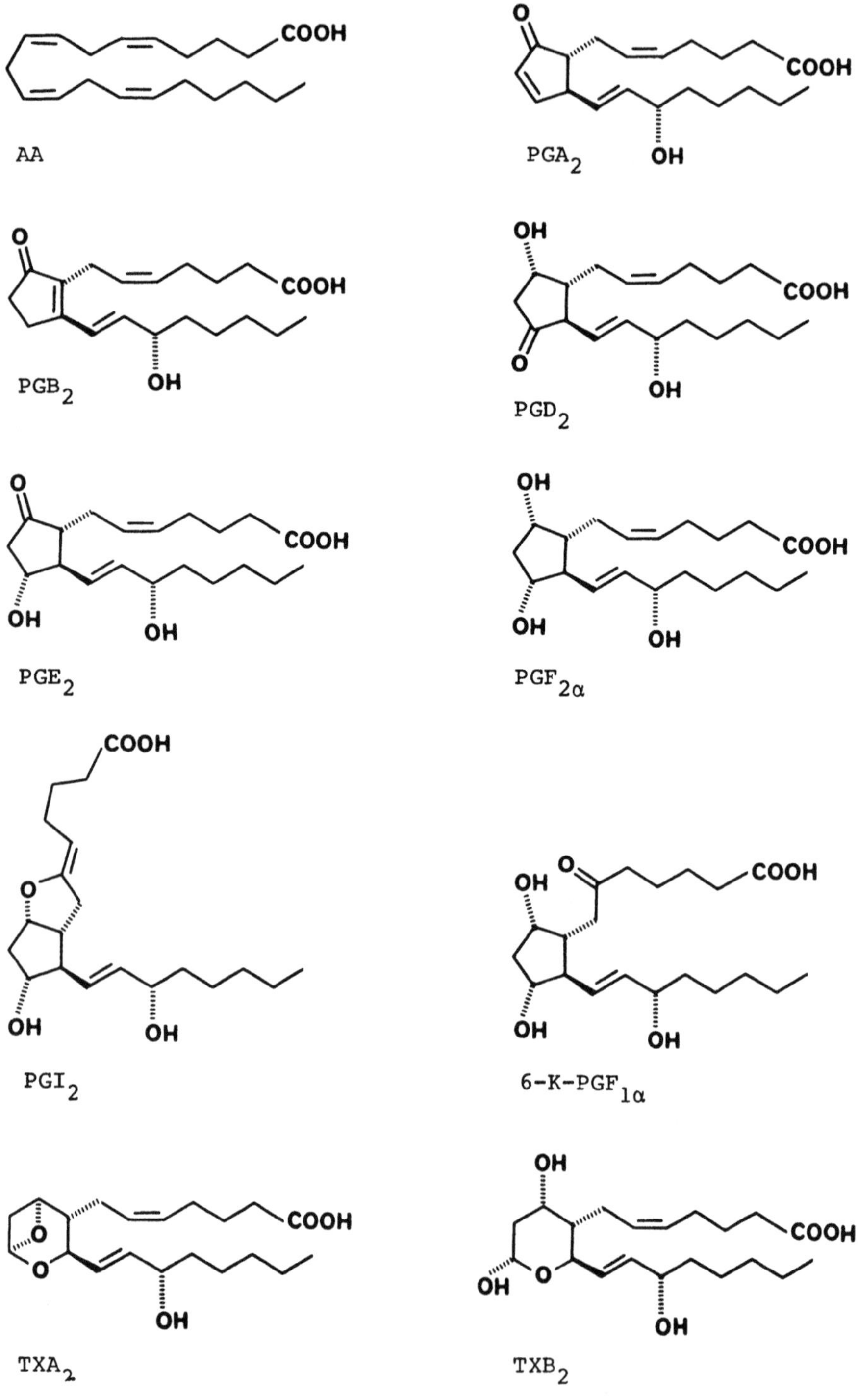

Fig. 2. Arachidonate metabolites : thranboxanes and prostaglandins

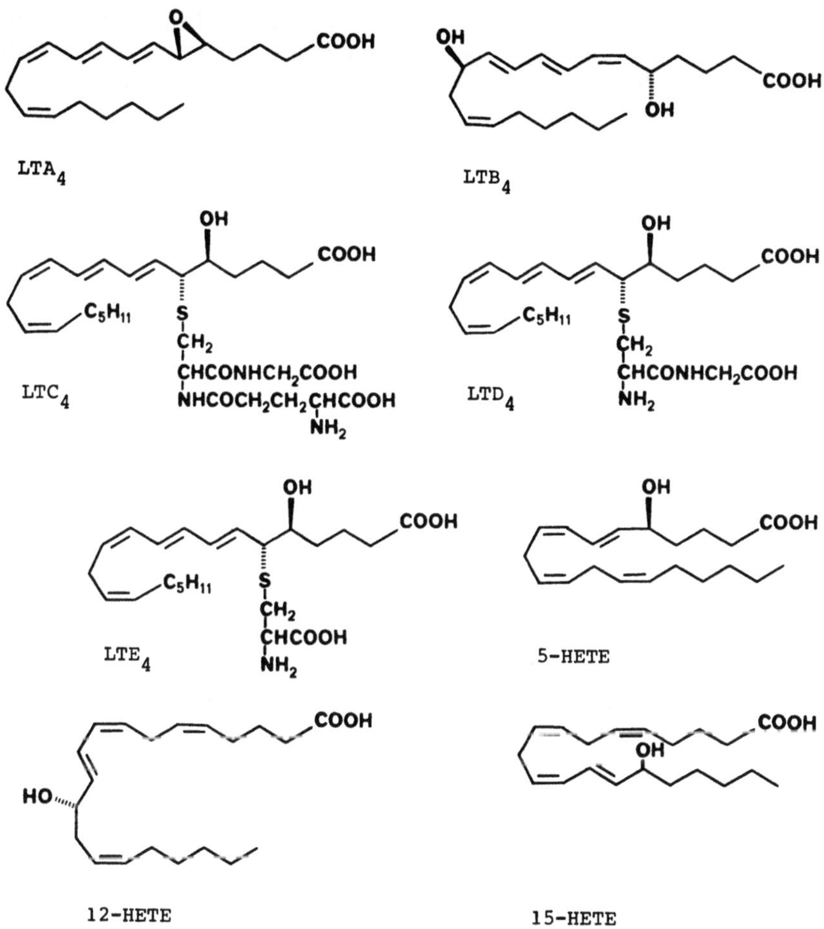

Fig. 3. Arachidonate metabolites Leukotrienes

Several different approaches have been employed in the detection and separation of prostaglandins using HPLC. Initially prostaglandin mixtures were separated by HPLC followed by time consuming fraction collection and scintillation counting of radiolabel. Nowadays, on-line radioactivity detectors eliminate the need for fraction collection. An alternative approach is to derivatize prostaglandins prior to HPLC for ultraviolet of fluorescence detection. More recently reproducible, rapid and sensitive HPLC methods have been developed which separate and detect underivatized eiosanoids. Apart from quantitation, HPLC has been used preparatively for further analysis by other methods.

Extraction and Sample Preparation

Eicosanoids are relatively soluble in water because of their polar groups. Their extraction into organic solvents are greatly improved by prior acidification of the aqueous phase to about pH3. The acids used range from hydrochloric acid (Bygdeman and Samuelsson, 1966) to formic acid (Unger et al., 1971). Chloroform, diethyl ether and ethyl acetate are the most frequently used organic solvents for eicosanoid extraction. With plasma, precipitation of proteins greatly reduces the prostaglandin yield. Non-precipitating amounts of ethanol (40–50%) are therefore recommended (Unger et al., 1971). Prostaglandins can also be extracted from biological fluids by percolation through Amberlite XAD-2, a neutral cross-linked polystyrene polymer, followed by elution with ethanol (Samuelsson et al., 1971).

Eicosanoids are unstable and are susceptible to either acid or alkaline attack (Shaw and Ramwell, 1969). Moreover most biological samples contain enzymes which can modify the eicosanoid content. If extractions cannot be performed immediately it is essential that the samples be frozen at minus 20° C or lower. Extracted samples are best stored in absolute ethanol at minus 20° C (Srivastava and Clausen, 1973).

Extraction for eicosanoids also recovers other lipids which need to be removed prior to HPLC. This can be achieved by chromatography on silicic acid or as shown more recently, on reverse phase octadecylsilyl (ODS) silica. Silicic acid (1 g, activated at 120° C) is used for column chromatography. The dry residue from extractions dissolved in diethyl ether/hexane (8/92, v/v) is applied to the columns. The column is eluted with 30 ml of each of the following solvents: diethyl ether/hexane (8/92, v/v) eluting unreacted arachidonate and other unpolar lipids, diethyl ether/hexane (25/75, 45/55, v/v) the latter eluting 5-HETE, and ethyl acetate eluting more polar metabolites of arachidonate (Borgeat and Samuelsson, 1979a). If the whole spectrum of arachidonate metabolites needs to be analysed by HPLC in the same run, the silicic acid column can be eluted by diethyl ether/hexane (8/92, v/v) followed by ethyl acetate.

Rapid extraction of arachidonate metabolites from biological samples using octadecylsilyl (ODS) silica has been described by Powel (1982). Cartridges, containing ODS silica are pre-eluted with ethanol (20 ml) and water (20 ml). Biological samples (homogenates, incubation mixtures, body fluids) acidified to pH 3 are passed through the cartridge which is then eluted with 20 ml of the following solvents: ethanol/water (15/85, v/v) (eluting polar lipids), water, petroleum ether, petroleum ether/chloroform (65/35, v/v) (eluting fatty acids and monohydroxy fatty acids), methyl formate (eluting prostaglandins and thromboxanes) and ethanol/water (80/20, v/v) (eluting moderately polar hydrophilic substances). Leaving out the petroleum ether/chloroform (65/35, v/v) would cause all arachidonate metabolites to be eluted by methyl formate. This method is also employed as a cleanup procedure following conventional extraction.

HPLC Columns and Detectors

HPLC of eicosanoids has been achieved using adsorption, argentation-cation and partition modes of chromatography. Adsorption by liquid-solid chromatography has been achieved using silica. Retention and selectivity are caused primarily by the interaction of polar functional groups of the eicosanoid and the silica. The small particles in the columns give a high effiency and permit the separation of the closely related eicosanoids. In argentation - HPLC, where a cation exchange column is treated with $AgNO_3$, this interaction is extended to double bonds. These form a loose complex with the silver, cis double bonds being more susceptible than trans. Partition chromatography, particularly reverse-phase chromatography has been widely used for HPLC of eicosanoids. The most common bonded phase is octadecylsilane which has C-18 carbon chains attached to the silica. The separation is based on the relative association of the hydrophobic part of the eicosanoid with the nonpolar C-18 groups versus their solubility in a polar mobile phase.

The earliest detectors employed measured the difference in refractive index between the solution and the pure solvent. However these refractive index detectors are not very sensitive and they are unsuitable for gradient elution. Ultraviolet detectors, which are among the most sensitive available for HPLC, measure UV absorption of the solute. Most prostaglandins have a UV absorption at 192 nm and only recently have UV detectors sensitive enough in this range been available. This has led to the use of derivatives to enhance detection by UV absorption or by fluorescence assay. Fluorescence detectors are said to be even more sensitive than UV detectors. They depend on excitation of an eicosanoid derivative at one wavelength followed by detection at a higher wavelength. The advent of online radioactivity detectors has eliminated the need for time-consuming fraction collection. Two types of radioactive detectors have been developed: homogenous and heteregenous. Homogenous radioactivity detection requires the addition of scintillation fluid to the eluant from the column via a secondary pump prior to monitoring. A disadvantage of this method is peak-broadening. Heterogenous detection uses solid phase scintillation which permits the direct monitoring of column eluant in smaller flow cells, and gives better resolution.

Another development is the availability of sophisticated, integrated microprocessor-controlled high performance liquid chromatographs which permit a large number of samples to be analysed for eicosanoids with little technical assistance.

HPLC of Prostaglandins and Thromboxanes

Early investigations of the applications of HPLC to prostaglandin analysis had the major problem of detector sensitivity, especially with assays of biological materials. These investigations employed straight phase silicic acid columns in conjunction with a refractive index detector and hence could only operate in the microgram region. Weinshenker et al. (1972) separated 15-epi-$PGF_{2\alpha}$ from $PGF_{2\alpha}$ by eluting a silica gel column with ethyl acetate and acetic acid, and monitoring with a refractive index detector. This approach was extended by Andersen and Leovey (1974) to study a wide range of PGE and PGF compounds.

Dunham and Anders (1973) used the observation that 9-keto prostaglandins form a strongly absorbing chromophore, PGB (λ max = 278 nm), following treatment with base, to analyse prostaglandins by HPLC and UV detection. However this method did not allow differentiation between the prostaglandins.

Carr et al. (1976) successfully separated several radioactive prostaglandins using a silicic acid column eluted with a gradient of chloroform, methanol and acetic acid, and monitored radioactivity in collected fractions. They were unable to separate PGE_2, PGE_1, 6-K-$PGF_{1\alpha}$ and TXB_2 using the silicic acid column, but were successful using a C18 reverse phase column eluted isocratically with water / acetonitrile / benzene / acetic acid (76.6/23/0.2/0.2), v/v/v/v) and monitoring the radioactivity in collected fractions. Russell and Deykin (1979) used a reverse phase column and radioactive monitoring to separate arachidonate metabolites by elution with mixtures of tetrahydrofuran, acetonitrile, water and acetic acid.

Since fraction collection is time-consuming, direct detection methods for prostaglandin analysis were sought. Several studies have used prostaglandin derivatization to enhance the UV absorption of the compounds and increase the sensitivity of the analytical methods. Morozowich and Douglas (1975) rapidly and quantitatively converted prostaglandins to p-nitrophenacyl esters without any degradation or epimerisation. They used a straight phase silica gel column eluted isocratically with methylene chloride/hexane/methanol (55/45/5 v/v/v) for the PGF series, and a less polar solvent system of methylene chloride/acetonitrile/dimethylformamide (160/40/1 v/v/v) for the PGE series, both monitored at 254 nm. Ten PGF series analogues were separated in 20 minutes while eight PGE series analogues were separated in 35 minutes, the lower limit of detection being about 1 nanogram. Roseman et al. (1976) also prepared the nitrophenacyl esters of dinoprost and its 15-epimer prior to chromatography on silica using a mobile phase of acetonitrile/dichloromethane/water (50/50/1 v/v/v). HPLC determination of PGs $F_{2\alpha}$, E_2 and D_2 as p-bromophenacyl esters was done by Fitzpatrick (1976). Detection and simultaneous quantitation of less than 3 micrograms was achieved using a reverse phase C18 column eluted with acetonitrile/water (50/50, v/v) monitored at 254 nm. Watkins and Peterson (1982) studied fluorescent and ultraviolet-absorbing ester derivatives by HPLC and achieved greater sensitivity. The esters of p-(9-anthroyloxy) phenacyl bromide were synthesized and following removal of excess phenacyl bromide the eicosanoid derivatives were chromatographed on reverse phase columns (C18) eluted isocratically with acetonitrile/water/acetic acid (68/22/0.02, v/v/v). Using ultraviolet absorption (254 nm) the lower limit of eicosanoid detection was 280 picogram while with fluorescence detection (249 nm emission, 413 nm cut off) the sensitivity was 50 picogram. The adequate separation of 7 common eicosanoids was obtained with the exception of TXB_2 which coeluted with $PGF_{2\alpha}$. Previously, fluorescent esters of the prostaglandins D_2, E_2, $F_{2\alpha}$, and 6-keto-$F_{1\alpha}$ and TXB_2 had been prepared using the reagent 4-bromomethyl-7-methoxycoumarin (Turk et al., 1978). HPLC was performed, eluting with a gradient of chloroform/iso-octane/methanol (35/65/1, v/v/v) and chloroform/methanol (100/1.35 v/v) using a CN-10 Micropak column, and monitored in a flow-through fluorimeter (excitation at 325 nm and cut off at 385 nm). All the eicosanoid derivatives were separated in a single run with the limit of detection being 20 nanogram. Identification of TXB_2 produced by human platelets and of 6-keto-$PGF_{1\alpha}$ produced by bovine aortic microsomes has been achieved with this method. Tsuchiya et al. (1982) using a similar reagent, 4-bromomethyl-7-acetoxycoumarin were able to detect prostaglandins in human seminal fluid in the 1 nanomole to 5 picomole range.

Another method for the separation of prostaglandins is argentation HPLC. Merritt and Bronson (1977) prepared a silver ion-loaded microparticulate cation-exchange resin column for HPLC separation of the p-nitrophenacyl esters of several series of closely related prostaglandins, eluted with dioxane/acetonitrile (99.94/0.06 v/v). Powell (1981) extended the study to radioactive, underivatized prostaglandins on a cation exchange column treated with silver nitrate solution. The mobile phase contained various proportions of methanol and acetonitrile in chloroform/acetic acid and the PGs were monitored using a radiocolumn detector. The system gave a good separation of 6-keto-$PGF_{1\alpha}$, PGE_2, PGD_2 and $PGF_{2\alpha}$ as well as several monohydroxy fatty acids. The retention times of the PGs were very susceptible to the acetonitrile content of the mobile phase which interacted with the silver ions of the stationary phase.

Recently, Van Rollins et al. (1980) and Terrango et al. (1981) have described a fast and reliable method for the separation of underivatized cyclo-oxygenase and lipoxygenase metabolites of arachidonic acid using a C18 reverse phase column. Using gradient elution with acetonitrile/water/phosphoric acid mix (30.5/69.5 pH2 to 95/5 pH2) over 150 minutes they separated all the known major cyclo-oxygenase and lipoxygenase products from platelets (van Rollins et al., 1980). The compounds were monitored at 192 nm and characterised by comparison with retention times of standards. They also collected fractions for radioactivity determinations. Terrango et al. (1981) separated the cyclooxygenase products (6-keto-$PGF_{1\alpha}$, 6-keto-PGE_1, TXB_2, $PGF_{2\alpha}$, $PGF_{1\alpha}$, PGE_2, PGE_1, PGD_2, and 15-keto-PGE_2) isocratically with 0.017 M H_3PO_4/acetonitrile (67.2/32.8), and monitored at 192.5 nm. The separation took 20 minutes and the minimum detection limit was 30 nanograms (Fig. 4). This HPLC system was used to study the stimulatory effect of comfrey on prostaglandin synthesis by rat isolated stomach (Stamford and Tavares, 1983) and also in a study on the effect of isoxicam on arachidonate metabolism (Bennett et al., 1984). More recently this HPLC system has been further refined by the use of narrow bore (2 mm) C18 columns, permitting the picogram detection of eicosanoids by ultraviolet absorbance (Rydzik et al., 1984). Lisboa and Ohm (1983) used similar columns and eluting systems but had an on-line radioactive flow monitor to measure the prostaglandins. With these systems, the TXB_2 peak broadened consistently, possibly because of its oxacyclohexane structure. Using non-silica reverse phase support (PRP-1) and eluting with a gradient of acetonitrile in triethylamine in water adjusted to pH 7.5 with acetic acid, Moonen et al. (1983) eluted TXB_2 as well as 6-K-$PGF_{1\alpha}$, $PGF_{2\alpha}$ and PGE_2 with little peak-broadening. Monitoring was done using an online radioactive detector.

HPLC of Hydroxyeicosatetraenoic Acids and Leukotrienes

The development of HPLC analysis for arachidonate lipoxygenase products has been somewhat easier because they can easily be monitored by UV absorption at 235 and 280 nm. Both straight phase and reverse phase HPLC systems have been

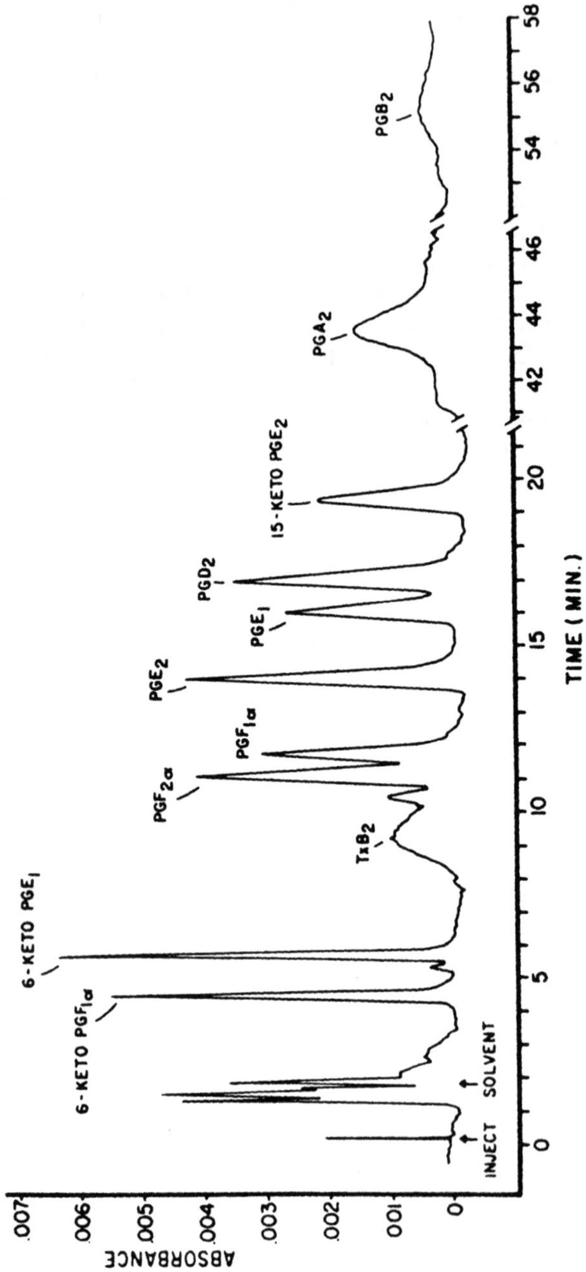

Fig. 4. HPLC of Cyclooxygenase products. Reverse-phase C_{18} column. 0.017 M H_3PO_4/Acetonitrile (67.2/32.8). UV absorbance at 192.5 nm. (From Terrango et al., 1981, with permission of authors and publisher)

employed in the analysis of hydroxy fatty acids. Using a silica gel column eluted with hexane/ethanol/acetic acid (993/6,1, v/v/v), monitored at 235 nm, Porter et al. (1979) separated chemically synthesized hydroxy and hydroperoxy derivatives of arachidonic acid. Boeynams et al. (1980) used a normal phase column with a linear gradient from hexane/acetic acid (99.2/0.8, v/v) to chloroform/acetic acid (99.2/0.8, v/v) to separate several HETEs. Straight phase HPLC for HETEs was extended to biological samples by Goetzl and Sun (1979) in studies on homogenates of human neutrophils and by Rabinovitch et al. (1979) who studied HETEs in mouse peritoneal macrophages.

Reverse phase is also more widely used than straight phase in the HPLC analysis of hydroxy acids. Most of these systems depend on a C18 column eluted with methanol/water/acetic acid (75/25/0.01 v/v/v) and monitored by a UV detector as described by Borgeat and Samuelsson (1979a). Studying the metabolism of arachidonic acid in human polymorphonuclear leucocytes, they obtained base line separations LTB_4 (dihydroxy acid, monitored at 280 nm) as well as 15-HETE, 12-HETE and 5-HETE (monohydroxy acids, monitored at 235 nm) within 35 minutes, with a limit of detection of about 20 nanogram. Using the above system, steroisomers of LTB_4 co-eluted and required further straight phase HPLC to separate them (Lindren et al., 1981).

HPLC of leukotrienes has been developed primarily using reverse phase C18 columns eluted isocratically with methanol/water/acetic acid (65/35/0.01 v/v/v) with the pH adjusted to 5.4 using NH_4OH (Hammarstrom et al., 1979). Lewis et al. (1980) using the above system eluted LTC_4 followed by LTD_4 within 36 minutes. The 11-trans isomers eluted directly after their respective 11-cis isomers. LTA_4, the unstable intermediate, has been chromatographed as the methoxy derivative using a C18 column and methanol/water/acetic acid (70/30/0.01, v/v/v) (Hammarstrom and Samuelsson, 1980). HPLC estimation of intact LTA_4 was achieved using a reverse phase column and a basic eluting system of acetonitrile/0.01 M borate buffer (40/60, v/v, pH 10) (Wynalda et al., 1982). The methyl ester of LTA_4 was purified on a straight phase HPLC column of aluminium oxide eluted with hexane/isopropyl alcohol/triethylamine (100/0.15/0.01, v/v/v) (Radmark et al., 1980). Osborne et al. (1983) prepared several HETEs and leukotrienes and separated them using a C18 reverse phase column eluted isocratically with methanol/water/acetic acid (65/35/0.06, v/v/v) ph 5.3 adjusted with ammonia solution (Fig. 5). They compared the separation achieved on several reverse phase packing materials and solvents systems. The method was used to identify lipoxygenase products released from mouse macrophage cell.

Using a variable wavelength UV detector both leukotrienes and monohydroxy acids can be studied in a single chromatography step (Borgeat and Samuelsson, 1979a). Prostanoids and hydroxy acids have also been analysed in a single run by reverse phase HPLC (Van Rollins et al., 1980). Peters et al. (1983) achieved separation of major prostaglandins, leukotrienes and mono-HETEs by HPLC using a gradient of acetonitrile/water/acetic acid over 100 minutes, and applied the method to the analysis of arachidonate metabolites produced by human lung macrophages. Birkle and Bazan (1984) employed similar systems to study the metabolism of radiolabelled arachidonic acid by intact bovine retina. The active synthesis of prostaglandins and HETEs as well as the incorporation of arachidon-

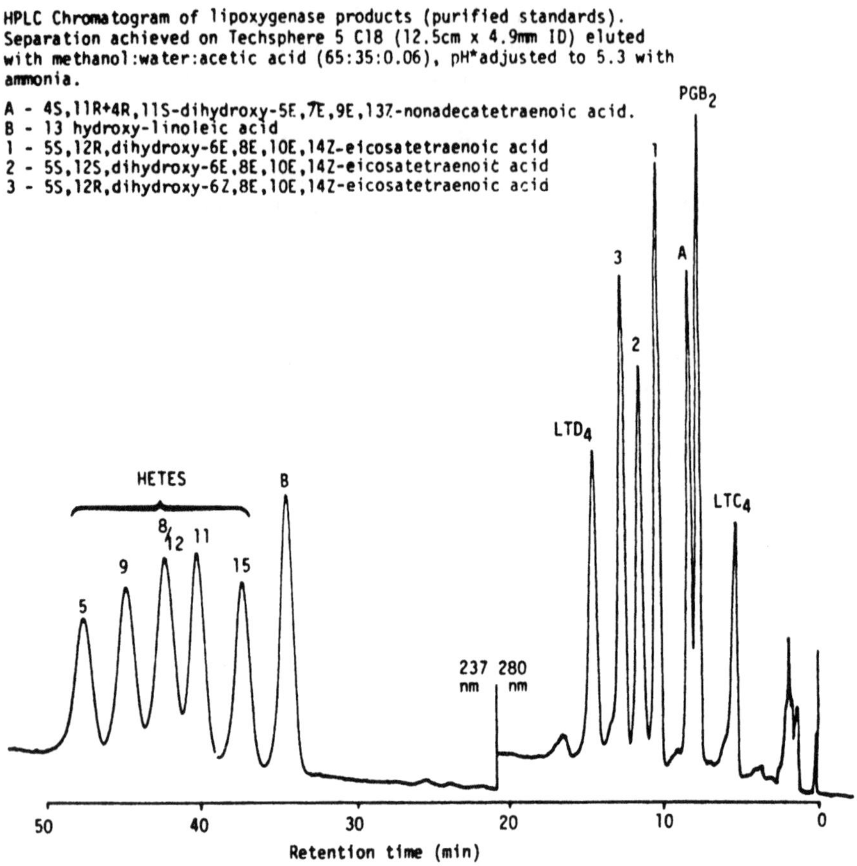

Fig. 5. HPLC of lipoxygenase products (from Osborne et al., 1983 with permission of authors and publisher).

ic acid into glycerolipids was measured by reverse-phase HPLC (Radial-pak C18 column) with both, UV and flow scintillation detection. They confirmed their results using straight phase HPLC (Porter et al., 1980).

Radioimmunossay and Sample Preparation by HPLC

Radioimmunossay (RIA) of prostaglandins was first carried out by Levine and van Vunakis (1970). Since then, RIAs for the quantitation of most eicosanoids including leukotrienes are now commercially available and have widespread use. The preparation of specific antibody with a low cross-reactivity usually involves the injection into a rabbit of an eicosanoid-protein conjugate for production of the antibody. RIAs are based upon the competition of standard amounts of tritiated antigen with unknown amounts of unlabelled antigen for specific antibody sites. By measuring either the radioactivity bound to the antibody of the radioactivity in the supernatant, the unlabelled antigen can be quantitated by comparison with a standard curve. In order to increase the specificity and sensitivity of RIA, some sample preparation is usually necessary, involving extraction and sometimes chromatographic purification by HPLC prior to RIA.

Early studies of RIA in conjunction with HPLC were performed to analyse the specificity of the system. Using HPLC and RIA, Alam et al. (1979) observed that their PGF and PGE antibodies did not distinguish between the 1 and 2 series, and that their 6-keto-$PGF_{1\alpha}$ antibody significantly cross-reacted with $PGF_{1\alpha}$. Levine and Alam (1979) measured prostanoids produced by several cell types in culture by RIA before and after HPLC. They observed that HPLC purification consistently reduced both PGE_2 and $PGF_{2\alpha}$ levels, presumably by removing cross reactivity from the 1-series PGs. Alam et al. (1980) extended the HPLC - RIA approach to analysis of cyclo-oxygenase products in plasma of rabbits bearing the VX_2 carcinoma. On comparing these data with studies using direct RIA of extracted plasma, they concluded that direct assay with their antisera was a satisfactory procedure. Sraer et al. (1981, 1982) in a series of studies on prostaglandin synthesis by kidney after incubation with [^{14}C] arachidonate also used radiometric HPLC followed by RIA for 6-keto-$PGF_{1\alpha}$, TXB_2, $PGF_{1\alpha}$ and PGE_2. Siess and Dray (1982) measured 6-keto-$PGF_{1\alpha}$ and 6,15-diK-$PGF_{1\alpha}$ in human plasma and urine by RIA after separation on reverse phase HPLC. They found that previous purification of the samples on HPLC markedly enchanced the specificity of the RIA determinations. HPLC has also been used in the purification of prostaglandins metabolites prior to antisera production. These techniques have been well documented by Granstrom et al. (1982a, b).

HPLC in conjunction with RIA has also been employed to lipoxygenase products. Morgan and Levine (1982) used both straight phase and reverse phase HPLC prior to RIA for 12-HETE to remove other cross-reacting hydroxy fatty acids. Salmon et al. (1982) used reverse phase HPLC to purify [^3H] LTB_4 as well as to validate their direct LTB_4 assay in studies of human neutrophil incubations. Subsequently, HPLC-RIA was used to study the release of LTB_4 during experimental inflammation in rats (Simmons et al., 1983). In RIA of LTC_4 and LTD_4 the antibodies currently available do not differentiate among the various 6-sulphidopeptide containing leukotrienes. Levine et al. (1982) used reversed phase HPLC to resolve these compounds prior to RIA.

The advantages that RIA offers over other methods of measurements of eicoasnoids are the high sensitivity and the potential for handling large numbers of samples rapidly. HPLC purification prior to RIA greatly increases handling time and should probably be used only to validate the technique. A better solution might be the production of better more specific antibodies.

Mass Spectrometry and Sample Preparation by HPLC

Gas chromatography - mass spectrometry (GC-MS) is also widely used in the analysis of eicosanoids. Deuterium-labelled standards of the compound to be measured are added to act as carriers and internal standards during the mass spectrometry. Selectivity and sensitivity are achieved by selected ion monitoring (SIM). Quantitation is based on the proportion of the deuterated form and the natural form of one or more selected ions (Hubbard et al., 1978). Eicosanoids have been purified by HPLC prior to GC-MS usually as the free acid or the methyl ester, sometimes with a radioactive tracer to facilitate monitoring.

Hubbard and Watson (1976) were among the first to apply HPLC purification to GC-MS of eicosanoids. They used straight phase silicic acid columns to purify methyl derivatives of 15-keto-13,14-dihydro-metabolites of PGE_2 and $PGF_{2\alpha}$ in plasma. Prior to analysis by GC-MS, the methyl esters were further derivatized to the bis-O-methyloxime-trimethylsilyl ether derivatives. Other studies of prostaglandins metabolites, particularly in urine, have employed HPLC purification prior to GC-MS. Normal phase HPLC has been used prior to GC-MS to quantitate PGD_2 metabolites (Ellis et al., 1979; Roberts, 1982). Several PGD ring metabolites were identified. Quantitation of the major urinary metabolite of TXB_2, the 2,3-dinor-TXB_2 has been reported (Roberts et al., 1977, 1978; Maas et al., 1982). HPLC, both normal and reverse phase, were used in the latter stages of purification. Rosenkranz et al. (1980, 1981) studied the metabolism of prostacyclin in human urine and plasma. Purification was by chromatography using several systems including normal phase HPLC. Four dinor metabolites were identified in urine, while in plasma 6 keto-$PGF_{1\alpha}$ was the main degradation product. The study of prostacyclin in humans was extended by Brash et al. (1983) who identified 10 metabolites by HPLC and GC-MS, all of which had retained the basic 6-keto-PGF structure. Fischer et al. (1983) also studied urinary profiles of prostacyclin metabolites.

HPLC purification prior to GC-MS has also been applied to the primary prostaglandins. These include PGE_1 (Whorton et al., 1979), PGE_2 (Seyberth et al., 1977) and PGD_2 (Oelz et al., 1977). Several prostaglandins were studied by Danon et al. (1978). The measurement of 5-HETE in biological fluids by GC-MS has recently been detailed by Ogletree et al. (1982), and employs reverse phase HPLC purification.

Another application of HPLC is in the purification of standards. Powell (1983) used argentation HPLC to separate unlabelled arachidonic acid metabolites from their deuterium and tritium labelled analogues and analysed the results by mass spectrometry. This method can be used to obtain labelled compounds of high isotopic purity which are suitable for use as standards in GC-MS. HPLC has played an important part in purifying and elucidating the structure of slow-reacting substances (SRS). SRS generated by murine mastocytoma cells treated with the calcium ionophore A23187 and L-cysteine, was semi-purified chromatographically followed by two steps of reverse phase HPLC prior to partial structural analysis by GC-MS (Murphy et al., 1979). Further analysis identified this SRS as LTC_4. The SRS from guinea-pig lung was extracted and following partial purification the SRS was also further purified by two reverse phase HPLC systems. The SRS was derivatized and identified by mass spectrometry (electron impact mode) as LTD_4 (Morris et al., 1978, 1980a, b), and LTB_4 was also identified. LTB_4 formed by rabbit peritoneal polymorphonuclear leucocytes was purified by HPLC prior to characterization by GC-MS analysis. Several other hydroxy acid have been structurally identified by MS following HPLC (Borgeat and Samuelsson, 1979b). Fast atom bombardment mass spectrometry (FAB-MS) of underivatized LTC_4 following HPLC has also been reported by Murphy et al. (1982).

Mass spectrometry, although unsurpassed in precision and accuracy as an analytical tool, requires fairly rigorous sample preparation. HPLC, because of the resolution that it offers, is ideally suited to fill this role.

Bioassay and Sample Preparation by HPLC

Bioassays have been used extensively for the detection and quantitation of eicosanoids. The types of prostaglandins can also be determined since there are both quantitative and qualitative differences in their effects on smooth muscle from different animal organs (Bergstrom et al., 1968; Bennett, 1972; Main, 1973). Tissue superfusion and cascade techniques which involves several different muscle preparations superfused in turn by the same sample are particularly useful for qualitative analysis (Vane, 1969). The choice of isolated tissues depends on the substances to be assayed, and specificity can be increased using selective antagonists (Moncada et al., 1978).

Recently, HPLC purification has been performed prior to bioassay of leukotrienes on guinea-pig ileum, the classical tissue for bioassay of SRS-A (Orange and Austen, 1969). The method has been applied to LTC_4 (Murphy et al., 1979) LTD_4 (Morris et al., 1979; Orning et al., 1980) LTE_4 and LTF_4 (Bernstrom and Hammarstrom, 1981, 1982).

Conclusions

HPLC offers a rapid, reliable and reproducible method for the separation of arachidonic acid metabolites. Using reverse-phase columns and UV spectrophotometers, sensitivity in the nanogram range has been achieved for underivatized metabolites. This range is suitable for studies in several biological systems. However, because arachidonate metabolites are not stored, but produced as required, usually in picogram amounts, quantitation can sometimes only be achieved by derivatization for UV or fluorescent detection. HPLC has proved a useful tool in sample preparation prior to further analysis by mass spectrometry and has also been employed prior to RIA determinations to increase the specificity and sensitivity of the RIA.

References

Alam I, Ohuchi K, Levine L (1979) Determination of cyclooxygenase products and prostaglandin metabolites using high-pressure liquid chromatography and radioimmunoassay. Anal Biochem 93:339–345

Alam I, Voelkel EF, Tashjian AH-JR, Levine L (1980) Analysis by high performance liquid chromatography of cyclooxygenase products of arachidonic acid metabolism in plasma of rabbits bearing the VX2 carcinoma. Prostaglandins Med 4:227–237

Anderson NH, Leovey EM (1974) Identification and quantitative determination of prostaglandins by high pressure liquid chromatography. Prostaglandins 6:361–374
Bennett A (1972) Effects of prostaglandins on the gastrointestinal tract. In: The Prostaglandins: Progress in Research. (SMM Karim, editor) 205–221 (Oxford and Lancaster: Medical and Technical Publishing Co)
Bennett A, Carrol MA, Collins PO, Gaffen JD, Melhuish PB, Stamford IF, Tavares IA (1985). The effect of isoxicam on arachidonate metabolism. Roy Soc Med, International Congress and Symposium Series 75:3–11
Bergstrom S, Carlson LA, Weeks JR (1968) The Prostaglandins: A family of biologically active lipids. Pharmacological Reviews 20:1–48
Bergstrom S, Eliasson R, von Euler US, Sjovall J (1959) Some biological effects of two crystalline prostaglandin factors. Acta Physiol Scand 45:133–144
Bergstrom S, Ryhage R, Samuelsson B, Sjovall J (1962) The structure of prostaglandin E, F_1 and F_2. Acta Chem Scand 16:501–502
Bergstrom S, Ryhage R, Samuelsson B, Sjovall J (1963) Prostaglandins and related factors 15. The structures of prostaglandin E_1 $F_{1\alpha}$ and $F_{1\beta}$. J Biol Chem 238:3555–3564
Bergstrom S, Sjovall J (1957) The isolation of prostaglandin. Acta Physiologica Scandanavia 11:1086
Bergstrom S, Sjovall J (1960a) The isolation of prostaglandin F from sheep prostate glands. Acta Chem Scand 14:1701–1705
Bernstrom K, Hammarstom S (1981) Metabolism of leukotriene D by porcine kidney. J Biol Chem 256:9579–9582
Bernstrom K, Hammarstrom S (1982) A novel leukotriene formed by transpeptidation of leukotriene E. Biochem Biophys Res commun 109:800–804
Bernstrom S, Sjovall J (1960b) The isolation of prostaglandin E from sheep prostate glands. Acta Chem Scand 14:1701–1705
Birkle DL, Bazan NG (1984) Lipoxygenase and cyclo-oxygenase reaction products and incorporation into glycerolipids of radiolabelled arachidonic acid in the bovine retina. Prostaglandins 27:203–216
Boeynaems JM, Brash AR, Oates JA, Hubbard WC (1980) Preparation and assay of monohydroxy-eicosatetraenoic acid. Analyt Biochem 104:259–267
Borgeat P, Samuelsson B (1979a) Arachidonic acid metabolism in polymorphonuclear leukocytes: Effects of ionophore A23187. Proc Natl Acad Sci USA 76:2148–2152
Borgeat P, Samuelsson B (1979b) Metabolism of arachidonic acid in polymorphunuclear leukocytes. Structural analysis of novel hydroxylated compounds. J Biol Chem 254:7865–7869
Brash AR, Jackson EK, Saggese CA, Lawson JA, Oates JA, Fitzgerald GA (1983) Metabolic disposition of prostacyclin in humans. Pharmacol Exp Ther 226:78–87
Bygdeman M, Samuelsson B (1966) Analyses of prostaglandins in human semen. Prostaglandins and related factors 44. Clin Chim 13:465–474
Carr K, Sweetman BJ, Frolich JC (1976) High performance liquid chromotography of prostaglandins: biological applications. Prostaglandins 11:3–14
Danon A, Knapp HR, Oelz O, Oates JA (1978) Stimulation of prostaglandin biosynthesis in the renal papilla by hypertonic mediums. Am J Physiol 234:F64–F67
Dobson P, Aharony D, Krell RD (1983) Development of a sensitive method for measurement of leukotriene production by isolated cells. Res Commun Chem Pathol Pharmacol 42:3–23
Dunham EW, Anders MW (1973) High speed liquid chromatographic analysis of prostaglandins in rat kidney. Prostaglandins 4:85–92
Ellis CK, Smigel MD, Oates JA, Oelz O, Sweetman BJ (1979) Metabolism of prostaglandin D_2 in the monkey. J Biol Chem 254:4152–4163
Fischer S, Scherer B, Weber PC (1983) Prostacyclin metabolism in adults and neonates. Urinary profiles of 6-ketoprostaglandin F_1 and 2,3-dinor-6-ketoprostaglandin F_1 alpha studied by gas chromatography-mass spectrometry. Biochem Biophys Acta 750:127–133
Fitzpatrick FA (1976) High performance liquid chromatographic determination of Prostaglandins $F_{2\alpha}$, E_2 and D_2 from in vitro enzyme incubations. Analyt Chem 48:499–502
Gotzl EJ, Sun FF (1979) Generation of unique mono-hydroxyeicosatetraenoic acids from arachidonic acid by human neutrophils. J Exp Med 150:406–411

Goldblatt MW (1935) Properties of human seminal plasma. J Physiol (London) 84:208–218
Granstrom E, Kindahl H (1982a) Radioimmunoassay for the Major Plasma Metabolite of $PGF_{2\alpha}$ 15-Keto-13,14-dihydro-PGF_2. Methods Enzymol 86:320–338
Granstrom E, Kindahl H (1982b) Radioimmunoassay of $5_\alpha,7_\alpha$-dihydroxy-11-keto-tetranorprostane-1,16-dioic acid a major prostaglandin F metabolite in blood and urine. Methods Enzymol 86:339–353
Hammarstrom S, Murphy RC, Samuelsson B (1979) Structure of leukotriene C. Identification of the aminoacid part. Biochem Biophys Res Commun 91:1266–1272
Hammarstrom S (1983) Leukotrienes. Ann Rev Biochem 52:355–377
Hammarstrom S, Samuelsson B (1980) Detection of leukotriene A_4 as an intermediate in the biosynthesis of leukotriene C_4 and D_4. FEBS Lett 122:83–86
Hubbard WC, Watson JT (1976) Determination of 15-keto 13,14-dihydro-metabolites of PGE_2 and $PGF_{2\alpha}$ in plasma using high performance liquid chromotography and gas chromotography mass spectrometry. Prostaglandins 12:21–35
Hubbard WC, Watson JC, Sweetman BJ (1979) Prostaglandin analysis: The role of high performance liquid chromatography in sample processing. In: Biological Applications of Liquid Chromotography (Hawk G ed) Marcel Dekker New York 10:31–55
Kurzrok R, Lieb CC (1983) Biochemical studies of human semen. II. The action of semen on the human uterus. Proceedings of the Society for Experimental Biology and Medicine 28:268–272
Levine L, Alam I (1979) Archidonic acid metabolism by cells in culture: analyses of culture fluids for cyclooxygenase products by radioimmunoassay before and after separation by high pressure liquid chromatography. Prostaglandins Med 3:295–304
Levine L, Lewis RA, Austen KF, Corey EJ (1982) Radioimmunoassay of the 6-sulfido-peptide-leukotrienes and serologic specificity of the anti-leukotriene D4 plasma. Methods Enzymol 86:252–258
Levine L, van Vunakis H (1970) Antigenic activity of prostaglandins. Biochem Biophys Res Commun 41:1171–1177
Lewis RA, Drazen JM, Austen KF, Clark DA, Corey EJ (1980) Identification of the C(6)-S-conjugate of leukotriene A with cysteine as a naturally occurring slow reacting substance of anaphylaxis (SRS-A). Importance of the 11-cis-geometry for biological activity. Biochem Biophys Res Commum 96:271–277
Lindgren JA, Hansson G, Samuelsson B (1981) Formation of novel hydroxylated eicosatetracnoic acids in preparations of human polymorphonuclear leukocytes. FEBS Lett 128:329–335
Lisboa BP, Ohm K (1983) Separation of underivatized prostaglandins by reverse-phase high pressure liquid chromatography (HPLC). Acta Endocrinol Legica 102:113–114
Maas RL, Taber DF, Roberts LJ (1982) Quantitative assay of urinary 2,3-dinor thromboxane B_2 by GC-MS. Methods Enzymol 86:592–603
Main IHM (1973) Prostaglandins and the gastrointestinal tract. In: The Prostaglandins: Pharmacological and Therapeutic Advances (Cutherbert MF (ed) London: Heinemann. 287–323
Mathews WR, Rokach J, Murphy RC (1981) Analysis of leukotrienes by high pressure liquid chromatography. Anal Biochem 118:96–101
Merritt MV, Bronson GE (1977) High performance liquid chromotography of p-nitrophenacyl esters of selected prostaglandins on silver ion-loaded microparticular cation-exchange resin. Anal Biochem 80:392–400
Moncada S, Korbut R, Bunting S, Vane J (1978) Prostacylcin is a circulating hormone. Nature 273–767
Moonen P, Klok G, Keirse MJ (1983) An improved method for separation of thromboxane B2 by reversed phae liquid chromatography. Prostaglandins 26:797–803
Morgan RA, Levine L (1982) Radioimmunoassay and Immunochrommatography of 12-L-Hydroxyeicosatetraenoic acid. Methods Enzymol 86:246–252
Morozowich W, Douglas SL (1975) Resolution of prostaglandin p-nitrophenacyl esters by liquid chromotography and conditions for rapid quantitative p-nitrophenacylation. Prostaglandins 10:19–40
Morris HR, Taylor GW, Piper PJ, Sirois P,Tipins JR (1978) Slow reacting substance of anaphylaxis. Purification and characterization. FEBS Lett 87:203–206

Morris HR, Taylor GW, Piper PJ, Samhoun NM, Tippins JR (1980a) Slow reacting substances (SRSs): The structure identification of SRSs from rat basophilic leukemia (RBL-1) cells. Prostaglandins 19:185–201

Morris HR, Taylor GW, Piper PJ, Tippins JR (1980b) Structure of slow reacting substance of anaphylaxis from guinea pig lung. Nature 285:104–106

Murphy RC, Mathews WR, Rokach J, Fenselau C (1982) Comparison of biological-derived and synthetic leukotriene C_4 by fast atom bombardment mass spectrometry. Prostaglandins 23:210–206

Murphy RC, Hammarstrom S, Samuelsson B (1979) Leukotriene C: A slow reacting substance from murine mastocytoma cells. Proc Natl Acad Sci USA 76:4275–4279

Oelz O, Oelz R, Knapp HR, Sweetman BJ, Oates JA (1977) Biosynthesis of prostaglandin D_2. 1. Formation of prostaglandin D_2 by human platelets. Prostaglandins 13:225–235

Ogletree ML, Schlesinger K, Nettleman M, Hubbard WC (1982) Measurements of 5-Hydroxyeicosatetraenoic Acid (5-HETE) in Biological Fluids by GC-MS. Methods Enzymol 86:607–612

Orange RP, Austen KF (1969) Slow reacting substance of anaphylaxis. Adv Immunol 10:104–144

Orning L, Hammarstrom S, Samuelsson B (1980) Leukotriene D: A slow reacting substance from rat basophilic leukemia cells. Proc Natl Acad Sci USA 77:2014–2017

Osborne DJ, Peters BJ, Meade CJ (1983) The separation of leukotrienes and hydroxyeicosatetraenoic acid metabolites of arachidonic acid by high performance liquid chromatography (HPLC). Prostaglandins 26:817–832

Peters SP, Schulman ES, Liu MC, Hayes EC, Lichtenstein LM (1983) Separation of major prostaglandins, leukotrienes, and monoHETEs by high performance liquid chromatography. J Immunol Methods 64:335–343

Porter NA, Wolf RA, Pagels WR, Macnett LJ (1980) A text for the intermediacy of 11-Hydroperoxyeicosa- 5,8,12 14-tetaenoic Acid (11-HPETE) In: Prostaglandin Biosynthesis. Biochem Biophys Res Commun 92:349–355

Porter NA, Wolf RA, Yarbro EM, Weenen H (1979) The autoxidation of arachidonic acid: formation of the proposed SRS-A intermediate. Biochem Biophys Res Commun 89:1058–1064

Powell WS (1981) Separation of icosenoic acids, monohydroxyicosenoic acids and prostaglandins by high pressure liquid chromotography on a silver ion loaded cation exchange column. Anal Biochem 115:267–277

Powell WS (1982) Rapid extraction of arachidonic acid metabolites from biological samples using Octadecylsilyl silica Methods Enzymol 86:467–477

Powell WS (1983) Separation of unlabeled metabolites of arachidonic acid from their deuterium- and tritium-labeled analogs by argentation high-pressure liquid chromatography. Anal Biochem 128:93–103

Rabinovitch H, Durand J, Rigaud M, Mendy F, Breton JC (1981) Transformation of arachidonic acid into monohydroxyeicosatetraenoic acid by mouse peritoneal macrophages. Lipids 16:518–524

Radmark O, Malmsten C, Samuelsson B, Goto G, Marfat A, Corey EJ (1980) Leukotriene. A. Isolation from human polymorphonuclear leukocytes. J Biol Chem 255:11828–11831

Roberts LJ (1982) Quantification of the PGD_2 urinary metabolite 9 alpha-hydroxy-11,15-dioxo-2,3,18,19-tetranorprost 5-ene-1,10 dioic acid by stable isotope dilution mass spectrometric assay. Methods Enzymol 86:559–570

Roberts J, Sweetman B, Oates J (1978) Metabolism of thromboxane B_2 in the monkey. J Biol Chem 253:5305–5318

Roberts J, Sweetman B, Payne N, Oates J (1977) Metabolism of thromboxane B_2 in man. Identification of the major urinary metabolite. J Biol Chem 252:7415–7417

Roseman TJ, Butler SS, Douglas SL (1976) High-pressure liquid chromatographic determination of the 15-epimer of dinoprost in bulk drug. J Pharm Sci 65:673–676

Rosenkranz B, Fischer C, Frolich JC (1981) Prostacyclin metabolites in human plasma. Clin Pharmacol Ther 29:420–424

Rosenkranz B, Fischer C, Weimer KE, Frolich JC (1980) Metabolism of prostacyclin and 6-keto-prostaglandin F1 alpha in man. J Biol Chem 255:10194–10198

Russell FA, Deykin D (1979) The use of high pressure liquid chromatography (HPLC) for the separation of radiolabeled arachidonic acid and its metabolities produced by thrombin-tested human platelets. I. The validation of the technique. Prostaglandins 18:11–18

Rydzik R, Terragno A, Tackett R (1984) Picogram detection of eicosanoids by ultraviolet absorbance after narrow-bore high-performance liquid chromatography. Comparison with conventional bore column. J Chromatogr 308:31–41

Salmon JA, Simmon PM, Palmer RMJ (1982) A radioimmunoassay for leukotriene B_4. Prostaglandins 24:225–235

Samuelsson B, Goldyne M, Granstrom E, Amberg M, Hammarstrom S, Malmsten C (1978) Prostaglandins and thromboxanes. Ann Rev Biochem 47:997–1029

Samuelsson B, Granstrom E, Green K, Hamberg M (1971) Metabolism of prostaglandins. In: Prostaglandins (Ramwell PW, Shaw JE eds) Ann NY Acad Sci 180:138–163

Serhan CN, Lundberg U, Lindgren JA, Weissman G, Samuelsson B (1984) Formation of leukotriene C4 by human leukocytes exposed to monosodium urate crystals. FEBS Lett 167:109–112

Seyberth HW, Hubbard WC, Oelz O, Sweetman BJ, Watson JT, Oates JA (1977) Prostaglandin-mediated hypercalcemia in the VX_2 carcinoma-bearing rabbit. Prostaglandins 14:319–331

Shaw JE, Ramwell PW (1969) Separation, identification and estimation of prostaglandins. Methods Biochem Anal 17:325–371

Siess W, Dray F (1982) Very low levels of 6-keto-prostaglandin $F_{1\alpha}$ in human plasma. J Lab Clin Med 99:388–398

Siess W, Dray F, Seillan C, Ody C, Russo-Marie F (1981) Prostanoid synthesis by vascular slices and cultured vascular cells of piglet aorta. Biochem Biophys Res Commun 99:608–616

Simmons PM, Salmon JA, Moncada S (1983) The release of leukotriene B4 during experimental inflammation. Biochem Pharmacol 32:1353–1359

Sraer JD, Moulonguet-Doleris L, Delarue F, Sraer J, Ardaillou R (1981) Prostaglandin synthesis by glomeruli isolated from rats with glycerol-induced acute renal failure. Circ Res 49:775–783

Sraer J, Ardaillou N, Sraer JD, Ardaillou R (1982) In vitro prostaglandin synthesis by human glomeruli and papillae. Prostaglandins 23:855–864

Srivastava KC, Clausen J (1973) Stability of prostaglandin E compounds in solution. Lipids 8:592–594

Stamford IF, Tavares IA (1983) The effect of an aqueous extract of comfrey on prostaglandin synthesis by rat isolated stomach. J Pharm Pharmacol 35:816–817

Terrango A, Rydzik R, Terragno NA (1981) High performance liquid chromatography and UV detection for the separation and quantitation of prostaglandins. Prostaglandins 21:101–112

Tsuchiya H, Hayashi T, Naruse H, Takagi N (1982) Sensitive high performance liquid chromatographic method for prostaglandins using a fluorescence reagent, 4-bromomethyl-7 acetoxy-coumarin. O J Chromatogr 231:247–254

Turk J, Weiss SJ, David JE, Needleman P (1978) Fluorescent derivatives of prostaglandins and thromboxane for liquid chromatography. Prostaglandins 16:291–309

Unger WG, Stamford I, Bennett A (1971) Extraction of prostaglandins from human blood. Nature (London) 233:336–337

Van Rollins M, Ho SHK, Greenwald JE, Alexander M, Dorman NJ, Wong LK, Horrocks LA (1980) Complete separation by high performance liquid chromotography of metabolites of arachidonic acid from incubations with human and rabbit platelets. Prostaglandins 20:571–577

Vane JR (1969) The release and fate of vaso-active hormones in the circulation. Br J Pharmacol 35:209–242

von Euler US (1934) An adrenaline-like action in extracts from prostatic and related glands. J Physiol (London) 81:102–112

von Euler US (1935) A depressor substance in the vesicular gland. J Physiol (London) 84:21P

Watkins WD, Peterson MB (1982) Fluorescent/ultraviolet absorbing ester derivative formation and analysis of eicosanoids by high-pressure liquid chromatography. Anal Biochem 125:30–40

Weinshenker NM, Longwell A (1972) Quantitative determination of 15-epi-$PGF_{2\alpha}$ in PGF_2. Prostaglandins 2:207–211

Whorton AR, Sweetman BJ, Oates JA (1979) Application of high performance liquid chromatography and gas chromatography-mass spectrometry to analysis of prostaglandin E_1 in biological media. Anal Biochem 98:455–463

Wynalda MA, Morton DR, Kelly RC, Fitzpatrick FA (1982) Liquid chromatographic determination of intact leukotriene A_4. Anal Chem 54:1079–1082

High-Performance Liquid Chromatography of Gastrointestinal Hormones

CLARE SELDEN, P. N. MATON, and V. S. CHADWICK

General Principles

Role of HPLC

The rapid expansion of knowledge in gut endocrinology can be attributed, in part, to the development of laboratory techniques for isolation, purification, characterization and synthesis of gastrointestinal peptides and for their sensitive and specific assay in body fluids permitting physiological studies.

High-performance liquid chromatography (HPLC) has proved to be one of the more important of the techniques instrumental in this progress. In particular it has superceded many of the conventional chromatographic techniques, for purification of known and unknown peptides in addition permitting their characterization by peptide mapping, amino-acid composition and sequence analysis.

Radioimmunoassay (RIA) is the conventional technique for measurement of gastrointestinal hormones reaching sensitivities in the fmol range (Bloom & Long, 1982). There are, however, problems with RIA in some instances because of the existance of several molecular forms of a particular hormone, e.g. cholecystokinin (CCK); substances in tissues and plasma interfering with normal assay, and structural homology among hormones such that antibody binding is non-specific. For example, gastrin and CCK share a common C-terminal pentapeptide, so that antibodies raised to CCK cross-react with gastrin resulting in spurious levels unless multiple region specific antibodies are used. RIA also relies on competition between radiolabelled tracer and sample and frequently, commerical preparations of synthetic peptides used for tracers are impure.

HPLC has provided an adjunct to radioimmunoassay by obviating most of the above problems. The technique can reliably and reproducibly separate multimolecular forms from each other and from other similar hormones with a recovery not significantly different from 100%, while removing interfering substances in biological tissues. Pre-purification of peptides prior to radiolabelling, separation of peptide-bound from free iodine after radio-iodination, together with separation of labelled and unlabelled peptide have served to markedly improve tracer preparation and hence sensitivity and specificity of radioimmunoassay and radioreceptor assays (Paul et al. 1984, Pingoud & Trautschold, 1984).

HPLC Modes

The technique of high performance liquid chromatography encompasses each of the conventional liquid chromatographic procedures or modes. Gel permeation chromatography, separating on the basis of size uses gel columns of aqueous-compatible or non-aqueous compatible support media. These enable separation of peptides and proteins in the range of a few thousand daltons to several hundred thousand daltons. Ion-exchange chromatography is achieved using cross-linked polystyrene divynyl benzene supports linked to cation or anion exchangers. Medium pressure (ca 1000 psi) ion-exchange chromatography can achieve separations of peptides with excellent recoveries and more recently has been used for polypeptides and proteins (Mikes, 1979). Normal and reverse-phase chromatography occurs on a silica based support. The former relies on direct interaction with the silanol (Si-OH) groups, and is rarely used in gut hormone analyses. Reverse-phase supports, however, have carbon chains of differing length attached to the silanol groups and peptides are separated on the basis of their hydrophobicity. This last technique is the most commonly used HPLC mode for gastrointestinal hormone analysis and is used in combination with gel permeation chromatography and conventional chromatography, during purification procedures.

Reverse-phase columns are commonly constructed of C3, C8 or C18 bonded silica. In each of these only a certain proportion, usually 5–15% of the silanol groups are covered by these carbon chains. The remaining silanyl groups may be uncapped, so that during chromatography retention of solutes is due to both hydrophobic and normal phase reaction with the column; alternatively each of the silanol groups can be end-capped with a methyl group. In this case the solute retention is based entirely on apparent hydrophobicity. For small peptides ($<$ 15 residues) the predicted retention times can be calculated from the individual hydrophobicities of each amino acid; as the peptide chain length increases this prediction is no longer reliable because of the steric hindrance induced by the carbon chain length of the support and the "tertiary" structure of the peptide. Cyanopropylsilyl (CN) and n-propyldimethylsilyl(C3) supports have been used for the separation of larger peptides and proteins. The advantage of these supports is the larger size of peptides separated ranging from 5 amino acid peptides to proteins of 68 kilodaltons in size (Tempst et al., 1984).

Mobile Phases

HPLC may employ only one set of solvent conditions, i.e. isocratic conditions: in which case samples to be separated each interact differently with the column under those conditions and are thus resolved. Alternatively, gradient elution can be used. In this case all the samples are loaded in an aqueous or polar media, each is retained on the column until a second solvent, an organic modifier, alters the polarity of the solid phase-mobile phase interaction. The samples to be separated are eluted sequentially.

When considering buffer systems, those useful in gastrointestinal hormone separations should exhibit four properties. Ideally they should be volatile, e.g. trifluoroacetic acid (TFA) and acetonitrile (MeCN), if further characterization studies are required so that primary structure and sequencing are not interfered with by high salt concentrations: if the detection system to be used is radioimmunoassay, the immuno-reactivity of the peptides should be retained, and if required for physiological studies the bioactivity should also be preserved. For example, physiological saline at an acid pH together with acetonitrile as organic modifier results in retention of both bio- and immunoactivity of cholecystokinin peptides. Prior to any bioassay such samples should be freeze dried in silanised glass containers to remove all acetonitrile. Lastly, if samples are to be analysed by UV detection at low wavelength (peptide bond 206–225 μm) solvent systems must be UV transparent, hence only high quality (Aristar) reagents and HPLC grade solvents should be used.

Ion-pairing reagents such as heptane sulphonic acid can markedly alter the characteristics of peptide interaction with the columns, providing more versatile systems for the separation of hormones (Terabe et al., 1979). The ion-pair is formed between a non-charged solute and its ionic counterpart in the mobile phase providing the basis for hydrophobic-ionic interaction.

Detection Systems

Detection systems for HPLC of hormones can be non-destructive of samples, e.g. UV and fluorescence detection, allowing for fraction collection and further characterisation. If, however, physiological concentrations are to be measured neither of these detection modes offer the required sensitivity so radioimmunoassay is employed. For peptide mapping UV detection at 206–214 nm is frequently sufficient, although peptides may be dansylated to increase sensitivity and may then be viewed by UV or fluorescent light. Radioactivity detectors may be used for hormone/receptor studies and in vivo biosynthesis of peptide hormones. Amino acid analysis and peptide sequence analyses rely on derivatization procedures to identify the resultant amino acids. For complete acid hydrolysis either dansylation for nmol quantities of amino acids, o-phthalaldehyde (OPA) or dimethylaminoazobenzenesulphonyl chloride (DABS-Cl) pre-column derivatization for pmol quantities of amino acids may be used with reverse-phase supports. Ion-exchange of the free amino acids followed by post-column OPA derivatization is acceptable when quantities greater than 1 nmol are present.

Advantage of Reverse Phase HPLC

One of the major advantages of reverse phase HPLC is the versatility of separation achieved by altering mobile phase conditions (Knip, 1984). This allows for separations of peptide hormones from crude tissue homogenates to homogeneity in a matter of hours whereas conventional chromatography would take days.

Another most important aspect is the recovery inherent in reverse phase chromatography, so that for peptides, and even large proteins, using larger pore reverse phase supports (300 Å), recovery is usually greater than 90%. In purification procedures this is a great advance over conventional chromatography allowing the use of grams rather than kilograms of starting material. This is particularly relevant in gastrointestinal hormone purification from human tissues.

HPLC of Individual Gastrointestinal Hormones Gastrin

In 1905 Edkins demonstrated stimulation of gastric acid secretion by extracts of the antral region of the stomach after intravenous injection in cats. Gastrin has been demonstrated in several species, and has been shown to have several molecular forms. Gastrin 17, is either nonsulphated (gastrin I) or sulphated on tyrosine at position 12 (gastrin II); both amino and carboxyl groups are blocked by pyroglutamyl and phenylalanine amide groups respectively. Yalow & Berson (1970) identified the major circulating forms of gastrin as 34-amino acid peptides (Big Gastrin) both sulphated and non-sulphated; the extension of G17 is from the N-terminus. This portion may be cleaved by proteolytic enzymes to convert G34 to G17 gastrin. Small amounts of other gastrin-like molecules have been demonstrated by immunoreactivity with gastrin antibodies. Gastrin has been localised in the G-cell of the antrum of the stomach and in the duodenum; there is some evidence that gastrin-like peptides are found in the central nervous system (Vanderhaeghen et al., 1975) where they may act as neurotransmitters. The physiological and pharmacological effects of gastrin are many and varied. Principally, it stimulates gastric acid secretion but also inhibits absorption of water and electrolytes from the small intestine. Gastrin relaxes the sphincter of Oddi (Lin & Spry, 1969) and contracts the lower oesophageal sphincter. Gastrin also stimulates growth of gastric mucosa, mucosal blood flow and metabolism.

HPLC has been used by Reeve et al., (1981) in conjunction with conventional ion-exchange and gel filtration chromatography in the isolation and purification of rat gastrin. Amino acid analysis demonstrated an additional residue of arginine and proline as compared with porcine gastrin explaining the poor cross-reactivity of rat gastrin with antibodies specific for the amino terminal portion of porcine and human gastrin. Moroder et al., (1983) used preparative HPLC in the synthesis of human gastrin 1 and its leucine, norleucine and methoxinine analogues. Gastrin I is known to exhibit a high tendency to air-oxidation of its methionine-15 residue, with concomittant loss of biological activity. Production of each of the above forms was achieved in satisfactory yield combining partition chromatography and preparative HPLC. Dimaline & Reeve (1983) used reversed-phase HPLC as a novel approach to the monitoring of an enzyme reaction. Sequence determination of peptides blocked by amino-terminal pyrolidase carboxylic acid has been hampered by the lack of an efficient method for removing this moiety. Using HPLC firstly to monitor the enzyme reaction and secondly to separate the unblocked peptide from the reaction product, only a few micrograms are required for the sequencing of blocked amino-terminal peptides such as gastrin and bombesin. An example of studying metabolism using HPLC has been demonstrated by Baldwin et al., (1983) where they showed phosphorylation of gastrin-17 by epidermal growth-factor stimulated tyrosine kinase (EGFC-TK). Tyrosine phosphorylation appears to be important in the control of cellular growth; previous reports that gastrin acts as a growth factor for cells in the GI tract suggested that phosphorylation of the tyrosine at position 12 may be more important physiologically than sulphation. Using reversed-phase HPLC the authors demonstrated that Gastrin-17 is phosphorylated by EGF-TK from squamous cell membranes.

Gastrin Releasing Peptide (GRP)

Structural similarities in gastrin releasing peptide isolated from different sources have been explored by Mutt and co-workers McDonald et al. (1983) using ion-exchange and HPLC to demonstrate variant forms of GRP. Others (Roth et al., 1983) demonstrated GRP related peptides in human malignant lung carcinoid tumour by reversed-phase HPLC after immunohistochemical findings indicated the presence of GRP in normal and pathological lungs. The 27 amino acid peptide isolated from porcine gut bears considerable carboxy terminal homology with bombesin but can be separated from it by HPLC.

CCK

Cholecystokinin (CCK) was discovered in 1928 by Ivy and Oldberg when they showed that an extract of duodenum injected intravenously caused gall bladder contraction similar to that observed by fat entering the duodenum. Cross circulation experiments confirmed a circulatory mediator (CCK). Twenty years later the first CCK molecule was isolated by Jorpes and Mutt (1964) as a 33 amino acid peptide from hog intestine. This was soon shown to be the same as pancreozymin, first described by Harper & Raper in 1943. Since that time several molecular forms of CCK have been demonstrated immunochemically and isolated with 4, 8, 12, 22, 33, 39 and 58 amino acid chains. CCK in the gut has been localised immunocytochemically to the I cells of the mucosa of the upper small intestine (Polak et al., 1975). The ratio of CCK 8 to larger forms (CCK 33/39) decreases from proximal to distal small intestine (Maton et al., 1984a). CCK peptides have been demonstrated in the CNS (Dockray, 1976) in the cerebral cortex in concentrations higher than any other neuropeptide described to date (Rehfeld, 1978). CCK is thought to be responsible for stimulation of pancreatic enzyme secretion, post prandially and for contraction of the gallbladder after a fat meal; other factors are likely to play a part in the mediation of these processes e.g. innervaton and other hormones. In addition CCK potentiates secretin-induced pancreatic HCO_3^- secretion and increases intestinal motility. In the brain CCK is localised to and synthesised in neurones, and probably acts as a neurotransmitter (Rehfeld et al., 1979). Other actions of CCK, probably pharmacological in nature, include stimulation of canalicular bile flow, potentiation of gastric acid secretion and inhibition of intestinal fluid transport.

CCK is a GI hormone which lends itself well to separation by reverse-phase HPLC as the different molecular forms differ in hydrophobicity both from each other and from gastrin and can be separated with retention of bioactivity (Selden et al., 1981) and immunoactivity for subsequent analysis by radioimmunoassay (Selden et al., 1984; Maton et al., 1982, 1984b). Physiological concentrations can, therefore, be measured in this way. It is important to ensure that the methionine (Bacarese-Hamilton et al. 1985), and tryptophan residues are not oxidised by separation systems employed during the procedure and that the tyrosine residue at position 7 remains sulphated.

The largest form of CCK demonstrated to date, CCK 58, has been purified (Eysselein et al., 1982) from canine proximal small intestine. Purification consisted of sequential steps of affinity chromatography gel filtration and finally HPLC. Microsequencing of the purified peptide revealed an identical sequence to CCK 39 in addition to a previously unknown sequence. This 58 amino acid peptide shows both bio- and immuno-reactivity and is thought to be a biosynthetic precursor of the smaller CCK molecules.

HPLC fractionation of extracts from pig brain and gut demonstrated greatest concentrations of CCK immunoreactivity in the cortex of the brain and duodenal mucosa of the gut. Both brain and gut contained CCK 8, while intact CCK 33 was only found in gut.

Uncharacterised NH_2-terminal peptides found in brain begs the question as to whether differences in the pattern of CCKs between neuronal and mucosal tissues result from post-translational processing or differences in DNA templates (Eng et al., 1982). Several other authors have used HPLC in its various modes in the study of neuronal CCK (Beinfeld et al., 1980, 1981; Eskay & Beinfeld, 1982).

An interesting way of applying HPLC to the study of rat brain metabolism was used by Deschodt-Lanckman et al., (1981) when they monitored the reaction products of a crude rat synaptosomal fraction incubated with CCK forms. Looking at tryptophan fluorescence, they showed that sulphated CCK 8 was more resistant to degradation than CCK 4 (half disappearance time 52 min for CCK 8 and 3 min for CCK 4). They went on to show inhibitory effects of zinc, protease inhibitors and other agents. Sauter and Frick (1983) used electro-chemical detection (EDC) in conjunction with HPLC of CCK 4 and 8 in different regions of rat brain. ECD is not sensitive enough to detect CCKs in gut or plasma but levels in the brain are such that concentrations measured in the cortex, hippocampus, striatum and brain stem compared favourably with those found with radioimmunoassay, making it an attractive alternative to existing methods.

Turnover studies of CCK in the brain utilised HPLC and ^{35}S radiolabelled sulphate allowing the measurement of synthesis and degradation of radiolabelled CCK 8. Meek et al. (1983) showed that the turnover of CCK 8 (half-life 16 hr) is far slower than that of biogenic amines and aminoacid neurotransmitters (half lives <4 hr).

Secretin

Bayliss & Starling in 1902 demonstrated the first example of hormone action when they showed that acidification of the denervated small intestine resulted in prompt alkaline pancreatic secretion. Intravenous injections of small bowel extracts caused a similar response in the denervated pancreas. Secretin, 27 amino acids in length (Mutt & Jorpes, 1966), is one of a family of peptides which share structural similarities including glucagon, gastric inhibitory peptide (GIP) and vasoactive intestinal polypeptide (VIP) (Bodansky et al., 1973). Secretin has been localised to the S cell predominating in the duodenum but is also present in the jejunum (Polak et al., 1971). The major action of this hormone is stimulation of pancreatic bicarbonate secretion. Release of secretin into plasma is achieved by acidification of the upper small intestine and can be inhibited by somatostatin or suppression of gastric acid by H_2 receptor antagonist drugs. Other actions of secretin include inhibition of gastric acid secretion and gastrin release, inhibition of lower oesophageal sphincter pressure and stimulation of insulin release, however, each of these effects requires large doses of secretin, much greater than those released endogenously suggesting these effects are pharmacological.

Reverse phase HPLC has been used to purify biosynthetic or native secretin from intestine and brain (Carlquist et al., 1981; Beyerman et al., 1981; Voskamp & Beyerman, 1981; O'Donohue et al., 1981a; Charlton et al., 1981). Olieman used the novel approach of in situ silylation of an octadecylsilyl-silica stationary phase for the analysis of secretin (Olieman et al., 1981) using perfluoroalkanoic acids as lipophilic ion-pairing reagents in a reverse-phase system. Perfluorooctanoic acid and trifluoroacetic acid were shown to be detrimental to reverse-phase column packing, possibly causing hydrolysis of siloxane bonds. Elution of chlorodimethyl octadecylsilane in toluene through a damaged C18 column results in complete regeneration of the stationary phase.

Pancreatic Glucagon

Glucagon was discovered when Murlin et al. (1923) demonstrated that pancreatic extracts could cause hyperglycaemia in animals. Subsequently it was purified (Staub et al., 1953) and synthesised. Glucagon is a 29 amino acid polypeptide originating in the alpha cell of the islets of Langherhans and it is a circulating hormone. Pancreatic glucagon reaches the liver, via the portal blood, stimulates glycogenolysis and gluconeogenesis. Hyperglycaemia is the main stimulus for glucagon release.

Chang and co-workers have used glucagon as a polypeptide to illustrate amino acid and C-terminal sequence analysis at the picomole level using carboxypeptidase and dimethylaminoazobenzenesulphonyl chloride (DABSCl) derivatization prior to separation on reversed-phase HPLC with detection in the visible region at 436 nm. Using a Zorbax ODS column with acetonitrile/acetate buffers (pH 4.1) less than 1 pmol of an individual DABS-amino acid can be quantitatively analysed, and only 10–30 ng of the dimethylaminoazobenzenesulphonylated protein hydrolysate is required for the entire amino acid analysis (Chang, 1981; Chang et al., 1981, 1982). N-terminal analysis is also achieved with dimethylaminoazobenzene isothiocyanate (Chang, 1981).

A novel procedure for sequencing peptide mixtures by a combination of Edman degradation and field desorption mass spectrometry utilised HPLC in the measurement of phenylthiohydantoins (PTH). The field desorption ionization technique allows qualitative PTH analysis of peptides and their successively degraded fragments, but at present relies on HPLC for quantification. This technique provides a different method for confirmation of primary structure of peptides, requiring minimal pre-separation of a mixture of peptides and allowing several amino acids to be identified at one time. The procedure is illustrated with glucagon; the complete amino acid sequence can be established with two degradation cycles of tryptic and four cycles of chymotryptic peptides (Shimonishi et al., 1980).

The chemistry of glucagon has been studied by applying HPLC to competitive labelling studies allowing the chemical properties of the functional groups on glucagon to be elucidated (Cockle et al., 1982).

The time course of isolated hepatocyte U-^{14}C-palmitate utilization in the presence of glucagon has been plotted by Singh et al. using HPLC to separate the radiolabelled products from the reaction mixture (Singh et al., 1982).

Receptor probes are important tools in furthering hormone receptor studies. Rojas et al. (1983) and others (Birnbaumer et al., 1983) have used reverse phase HPLC to separate in pure form, the chemically synthesised [^{125}I-Tyr10] monoiodoglucagon from glucagon and carrier free ^{125}I using the iodogen procedure. Separation was achieved on C18 μBondapak columns using two consecutive linear gradients between methanol - 10 mM H_3PO_4 pH 3.0 (60/40) and acetonitrile - 0.1 m Tris HCl pH 9.0 (50/50). The resultant radiolabelled receptor probe was shown to be 5 to 8-fold more potent in activating adenyl cyclase in liver membranes than native glucagon, and bound specifically to sites on liver plasma membranes characteristic of glucagon receptors, indicating that monoiodoglucagon synthesised and purified in this way is a suitable probe for studying structural and functional properties of the glucagon receptor. More receptor work, by Sueiras-Diaz et al. (1984) has utilised preparative HPLC with solid phase synthesis methodology to study the N-terminal hexapeptide region of the glucagon molecule. Separation of several analogues differing in the N-terminal region and their effects in blood glucose levels in male rats indicated that whilst amino or imidazole nitrogens of the histidine residues are not essential for biological activity, an aromatic group in position 1 is important. Use of D-Phe4glucagon synthesised in the above manner to test the hypothesis that a beta-bend conformation occurs at the aromatic ring position in glucagon showed that this was the case and had important implications for receptor-recognition requirements of glucagon and its related peptides, secretin and VIP. HPLC-mass spectrometry is a technique recently developed for peptide sequencing and is well demonstrated using the octapeptide of the C-chain of glucagon (Yu et al. 1984). N-acetyl-O, and S-permethylated derivatives of the C-chain octapeptide were analysed by HPLC-MS using a moving interface with heated-gas nebuliser allowing a wide water/methanol gradient of 5–95% water at flow rates of 0.5 ml/min. This technique overcomes some of the problems encountered with certain peptides during conventional gas-chromatography mass spectrometry.

Enteroglucagon

Enteroglucagon (EG) was discovered when glucagon like immunoreactivity was demonstrated in gut using antibodies raised to pancreatic glucagon. It has recently been isolated from porcine intestine (Bataille et al., 1982) and is 69 amino acids in length containing the pancreatic glucagon sequence within it (residues 33–61). Enteroglucagon is found throughout the gut with highest concentrations in the ileum. This hormone is released after oral administration of fat and carbohydrate.

Mutt and colleagues have isolated bioactive "enteroglucagon" (oxyntomodulin) from porcine jejuno-ileum and characterised it by immunological, biological and chemical means. HPLC analysis followed by dansylation of peptide fragments after enzymatic digestion of both this peptide and glucagon suggests that the bioactive EG is glucagon plus a C-terminal octapeptide extension (lys-Arg-Asn-Lys-Asn-Asn-Ile-Ala-COOH) (Bataille et al., 1981, 1982).

Thim and Moody (1982) have isolated a fragment of pro-glucagon, a glicentin related polypeptide, from porcine pancreas using HPLC to demonstrate its purity.

Sheikh et al. 1985 used several chromatographic techniques including gel permeation and reversed-phase HPLC to provide evidence which supports the hypothesis that small amounts of glicentin are present in the pig pancreas.

Pancreatic Polypeptide

Pancreatic polypeptide (PP) was isolated as a contaminant polypeptide during purification of human proinsulin (Chance et al., 1979). It is a 36 amino acid polypeptide with a carboxyl terminal amide. Biological potentcy resides in the C-terminal hexapeptide (Leu-Thr-Arg-Pro-Arg-Try-NH_2). PP is localised in a specific cell type in the pancreas and is released in response to a fat and protein rich meal. It is a circulating hormone with inhibitory effects on the exocrine pancreas and gallbladder.

Meyers and Coy (1980) utilised reverse-phase HPLC to check the homogeneity of solid phase synthesised human pancreatic polypeptide, to characterise its fragments after cyanogen bromide cleavage, and to compare the synthesised peptide with native human, bovine and porcine pancreatic polypeptide isolated from natural sources. HPP, purified in this way was a potent inhibitor of pancreatic secretion in the dog. HPLC has also been used in the comparison of vertebrate pancreatic polypeptide with a neuropeptide in the blowfly *Calliphora vomitoria* illustrating that the primary structure of the *Calliphora* neuropeptide is very similar to mammalian pancreatic polypeptide.

Gastric Inhibitory Polypeptide (GIP)

Brown & Pederson in 1970 isolated a peptide from the upper small intestine of the pig which suppressed gastric acid secretion. Subsequently it was shown to mediate insulin release during hyperglycaemia and since this is probably its main physiological function it was renamed glucose-dependent insulinotropic polypeptide. The 42 amino acid straight chain peptide shares several sequence homologies with pancreatic glucagon, VIP and secretin. It is localised to the mucosal K cells in the duodenum and jejunum. Plasma GIP concentrations rise after a mixed meal, or after ingestion of oral glucose or fat.

Mutt & co-workers (Jornvall et al., 1981) used HPLC to demonstrate the heterogeneity of gastric inhibitory polypeptide in intestinal extracts. The same group identified GIP 1–42 and GIP 3–42 using HPLC showing that the former has both insulinotropic and somatostatinotropic activities whilst the minor component GIP 3–42 has little activity (Brown et al., 1981). More recently Mueller et al. (1983) have shown by further HPLC purification that natural purified porcine gastric inhibitory polypeptide - enterogastrone III contains a contaminant which is responsible for the observed stimulation of exocrine pancreatic secretion; the contaminating factor was found to be CCK8 co-eluting on gel chromatography but readily separated by HPLC. Bacarese-Hamilton et al. (1984) have demonstrated differences between human and porcine GIP after gel filtration revealed two major forms, 5KDa and 8KDa in size. Size exclusion chromatography showed these two forms to be similar in both species, however, isocratic reverse-phase HPLC revealed the smaller molecule to be less hydrophobic in the human than the corresponding porcine form.

Motilin

Brown et al. (1966) discovered motilin by showing that increased motor activity occurred in denervated transplanted canine stomach pouches after alkalinization of the duodenum. Motilin is a basic 22 amino acid peptide with an N-terminal phenylalanine residue and a C-terminal glutamine (Brown et al., 1973). The bioactive region of motilin appears to be its N-terminal pentapeptide. Motilin is localised primarily in endocrine cells of the duodenum and jejunum. Motilin has stimulatory effects on gastric emptying of both liquids (Christofides et al., 1981) and solids (Christofides et al., 1979) and on colonic myoelectrical activity (Rennie et al., 1979). Plasma motilin concentrations were found to be elevated in all groups of patients with diarrhoea, although the cause and mechanism of elevation is not yet known.

Solid phase synthesis of peptides, 20–40 amino acid residues long, uses standard methodologies, but major advances have been made in new and powerful purification techniques for the synthesised gastrointestinal peptides. Coy et al. (1982) have synthesised motilin and secretin with full biological activity, using analytical HPLC for final purification. The speed of this methodology resulted in no loss of activity previously reported by other workers due to rearrangement during purification. Motilin immunoreactivity has been investigated in rat brain using gel-filtration by conventional and high resolution procedures, showing the higher molecular weight form predominating in the brain whilst the small molecule is the major form in the duodenum. Species differences in the low molecular weight motilin between rat and synthetic porcine motilin were indicated by differing retention times on hydrophobic reverse phase HPLC (O'Donohue et al., 1981b).

Vasoactive Intestinal Polypeptide

Said and Mutt (1970) first isolated this polypeptide from porcine intestine and showed it to have powerful vasodilatory action in dogs hence its name, Vasoactive Intestinal Polypeptide (VIP). Subsequently it was shown to be a powerful stimulant of intestinal fluid and electrolyte secretion. Tumours of pancreatic and neural origin secreting VIP cause profound diarrhoea and electrolyte depletion. VIP is located in both gut and brain and predominates in nerves rather than cells, consistant with a primary role as a neurotransmitter rather than a hormone. VIP is a basic 28 amino acid polypeptide similar in structure to GIP, secretin and pancreatic glucagon. Its methionine (position 17) renders it easily oxidised and the two asparagine residues are susceptible to deamidation. These factors and its adsorption onto glass require great care in sample handling when using VIP in physiological studies. There are probably several molecular forms of VIP in tissues such as colonic muscle and mucosa, and plasma; tumour extracts from patients with VIP secreting tumours have shown the existence of large molecular forms.

HPLC has been applied to VIP studies primarily during isolation and purification procedures. Bovine intestinal VIP, human intestinal VIP and porcine cerebral VIP have all been isolated and characterised by Mutt & co-workers using gel-chromatography and reverse-phase HPLC (Carlquist et al., 1979). Using three modes of HPLC, reversed-phase, gel permeation and ion-exchange Murphy et al. (1984) identified VIP immunoreactivity from guinea-pig enteric nerves as chromatographically identical to the synthetic porcine VIP concluding that this material was authentic vasoactive intestinal peptide. Identification of VIP forms in tumours will be discussed below.

Bombesin

Bombesin was first isolated from the frog, *Bombina bombina* and was found to be a potent stimulant for gastrin release. The bioactive portion of the molecule, the C-terminal heptapeptide, is common to a number of different species. Bombesin has several powerful pharmacological actions in man and experimental animals, stimulating release of the whole spectrum of gastrointestinal peptide hormones and in addition, growth hormone, prolactin, TSH, TRH, catecholamines and calcitonin. Bombesin has effects on the lungs, kidneys, exocrine pancreas, and on gastrointestinal tract motility and secretion. It has been shown to be widespread in the neurones of the central and peripheral nervous systems of all mammalian species.

Three groups have used HPLC to localise bombesin-like peptides in the brain (Panula et al., 1983). Moody et al. (1981a, b) showed bombesin-like peptide concentrations to be greater in the higher regions (substantia gelatinosa, trigeminal nucleus, tractus solitarius, interpeduncular nucleus and arcuate nucleus) than lower regions. Brain extracts subjected to HPLC showed two major peaks of immunoactivity, the more hydrophobic peptide co-eluting wih synthetic bombesin, supporting the contention that bombesin may be a neuroactive peptide in the brain. Reeve et al. (1983) isolated and determined the primary structure of three bombesin-like peptides from canine intestine extracts indicating another example of mammalian neuropeptides existing in more than one biologically active form. The same group developed a simplified method for purifying gastrointestinal neuropeptides from relatively small amounts of tissue using sequential adsorption, gel-filtration, ion-exchange chromatography and HPLC and showed this method to be suitable for the preparation, with high yields, of somatostatin, bombesin, and VIP, all small basic peptides (Reeve et al., 1982). Yoshizaki et al. (1984) and Price et al. (1983) have isolated bombesin-like peptides from human fetal lung. Four different forms (I, II, III, IV) were isolated by reverse-phase HPLC using an acetonitrile gradient (14–38%) containing 0.1% trifluoroacetic acid (TFA). None of the four peptides shared identity with amphibian bombesin or porcine gastin releasing peptide. Using two different hydrophobic counter ions (TFA or heptafluorobutyric acid) the chomatrographic behaviour of peak II mimicked the C-terminal fragments of GRP, GRP18–27 or 19–27). Tryptic digestion of peak IV converted the activity to that in peak II. The authors suggest that the conversion may provide one of the possible biosynthetic steps in the formation of bombesin-like peptides in human fetal lung.

An interesting application of HPLC in public health concerns the isolation of a bombesin immunoreactive peptide from milk; (Jahnke & Lazarus, 1984) reverse-phase HPLC separated this activity from both amphibian bombesin and gastrin releasing polypeptide. The peptide was shown to have a molecular weight of 3200, to be unaffected by boiling and to be unaltered by denaturing reagents. Jahnke and Lazarus (1984) showed data to suggest that milk is an exogenous source of bombesin-like immunoreactivity, which may account for the increase of gastric acid and gastrointestinal hormone levels after milk ingestion.

Substance P

Substance P was discovered in brain and intestine of rabbits in 1931, and shown to cause smooth muscle contraction and vasodilatation (von Euler & Gaddum, 1931). It was first isolated by Chang and Leeman (1970) and is 11 amino acids in length, with a molecular weight 1,345 daltons. Within the central nervous system substance P is localised in regions of brain and spinal cord associated with pain perception. In the spinal cord it is concentrated in the dorsal horn and is probably a neurotransmitter.

Aronin et al. (1983) have studied Huntingdon's chorea, an autosomal dominant hereditary disorder characterised by early neuronal cell death, particularly in the neostratum. By HPLC and radioimmunoassay they showed levels of substance P in the caudate lobes to be markedly reduced in this condition but somatostatin levels in the same extracts to be increased. HPLC also demonstrated that brain substance P was identical to the synthetic peptide whilst somatostatin was in several forms including somatostatin 14 and 28. They concluded that neuronal elements containing somatostatin are affected differentially by the disease process from neurons containing immunoreactive substance P.

Neurotensin

Neurotensin (NT) was discovered in bovine hypothalamus and shown to cause vasodilation and transient hypotension in laboratory animals (Carraway & Leeman, 1973). It has since been isolated in the same form from bovine, rat and human intestine. It is a 13 amino acid peptide with a bioactive pentapeptide in the C-terminal region. Neurotensin may subserve a neurotransmitter or neuromodulator role in CNS (Uhl & Snyder, 1976), and infusion studies in man suggest inhibitory effects on gastric function (Blackburn et al., 1980), and stimulating effects on water and bicarbonate secretion from the pancreas (Fletcher et al., 1981).

Shulkes et al. (1984a) have explored HPLC and region specific antibody technology in turnover studies with neurotensin in man. It is known that much of the NT, released into the circulation by a meal, consists of N-terminal fragments which have no known biological activity. The sites and rate of NT metabolism are not known. Half-time clearance rates of NT1–13 were estimated in 10 normal subjects and patients with chronic renal failure (CRF). The nature of the metabolities was characterised by HPLC showing that in plasma at equilibrium only 20% of the immunoreactivity was NT1–13, most being present as NT1–8. C-terminal fragments were not demonstrated. Incubation of NT1–13, 1–8 and 8–13, in vitro, in plasma showed the N-terminal fragments to be stable in plasma whilst the C-terminal fragments were completely metabolised. In patients wih chronic renal failure they found the predominant effect was on the initial degradation of NT1–13; HPLC demonstrated 60% of the NT present as NT1–13. The in vitro degradation of NT1–13 was also slowed in CRF plasma, indicating that the metabolism of NT is influenced by the kidney, and suggesting the presence of predominantly N-terminal fragment of NT in the peripheral circulation of normal subjects to be a function of both renal and extrarenal factors.

Somatostatin

Brazeau et al. (1973) isolated a peptide from ovine hypothalamus which inhibited growth hormone release from the pituitary; this peptide was named somatostatin. It is a basic peptide which spontaneously cyclises in plasma via a disulphide bridge between cysteine residues at positions 3 and 14. It is localised in neurones throughout the brain and spinal cord and in the D cells of the gastrointestinal tract. Infusion studies have demonstrated that somatostatin has inhibiting effects on the release of many other circulating hormones and on gastric, pancreatic, biliary and intestinal exocrine secretions. It is generally thought that in normal physiology it is released locally and affects cells in the vicinity by so called "paracrine" action.

Several HPLC methods have been applied to the isolation of somatostatin in its various molecular forms from numerous organs in many species (Pradayrol et al., 1978; Lee et al., 1981; Oyama et al., 1980a,b; Pearson et al., 1980; Sagar et al., 1982; Bohlen et al., 1981a, b; Benoit et al., 1980). A good example of HPLC isolation is given by Marshak et al. (1983): the structure of somatostatin isolated from bovine retina was determined. Using an anti-somatostatin affinity column, a reverse phase octadecylsilane HPLC column, and three recycles on a reverse-phase phenyl column the peptide was purified 103 000-fold from the original extract with a yield of 14.4%. The amino acid sequence determined by Edman degradation and HPLC revealed an apparent identity of this peptide with somatostatin 28 from other mammalian tissues. These experiments indicate the phylogenetic conservation of its structure, facilitating the use of the retina as a model system for studying the neurotransmitter function of somatostatin.

Berelowitz et al. (1981) determined the diurnal variation of immunoreactive thyrotropin releasing hormone and somatostatin in the cerebrospinal fluid of the rhesus monkey combining HPLC separation and radioimmunoassay. Märki et al. (1981) used HPLC to follow the rapid conversion of somatostatin to its active metabolite [des-Ala^1]-somatostatin in human plasma. Morel et al. (1981) showed the effect of mouse hypothalamic extracts on the pro-hormone (MW 15 000 daltons) in selectively releasing the somatostatin 14. Hepatic clearance studies on different somatostatin molecules using HPLC to follow S-28 conversion to cyclic somatostatin (SRIF) during liver perfusion showed 86% of the somatostatin-like immunoreactivity to be retained with S-28 peak and only 14% with SRIF indicating no major conversion of one form to the other (Sacks et al., 1984).

Gomez et al. (1983) studied the regional distribution of three different immunoreactive forms of somatostatin (MW 15 000, 3000 and 1600 daltons) in six neural structures of rat brain. HPLC showed the two smaller forms to be somatostatin 14 and 28 as compared with bovine S-14 and S-28, whilst the larger was somatostatin precursor. The different brain structures had varying proportions of the three molecules suggesting that the intracellular processing of the high molecular weight precursor may be different in the various somatostatinergic pathways of the brain, arguing that there is a biological role for the precursor in providing distinct relative proportions of S-14 and S-28 in each rat brain region. Shoelson has used HPLC to provide further evidence for the existance of somatostatin-28 in plasma (Shoelson et al., 1982) and also showed in the dog, that levels of S-28 are consistently greater in portal venous blood than simultaneously measured peripheral venous blood (approx x2). Peters (1982) studied the mechanisms of clearance and degradation of exogenous ^3H-somatostatin in rats using reverse-phase extraction and HPLC separation methods. Three apparent consecutive plasma half lives of 1, 3 and 20 min were estimated after one injection of a pharmacological dose. The initial rapid clearance was due to uptake by large peripheral tissue masses, muscle, skin and intestine which accounted for 70% of injected radioactivity at 1 min; liver and kidney uptake was much lower (10%). Large scale degeneration occurred following uptake with release of small peptides and free amino acids into the circulation almost immediately. At later times loss of activity was slowed suggesting a return of intact peptide from extravascular storage sites to maintain plasma concentration. Penman et al. (1983) have established the gastrointestinal distribution of immunoreactive somatostatin (IRS) in human tissue. The antrum, pyloris, duodenum and pancreas showed co-chromatography with S-14 by HPLC and gel filtration of IRS whilst in the body of the stomach, jejunum, ileum and colon a large peak of activity co-eluted with synthetic S-28. Some extracts also exhibited minor peaks of 6000 and 15 000 daltons. Tissues obtained at adult surgery, rapid post mortem and human fetuses showed similar patterns. More than 90% of the IRS was localised in mucosa in endocrine-type cells, with less than 10% present in muscle localised in nerve fibres and neurones of the myenteric plexus.

Starting from a prediction from a mRNA sequence that channel catfish pancreas had a somatostatin gene product different from the somatostatin previously reported in that species (Oyama et al., 1980) Andrews and Dixon (1981) used reverse-phase HPLC to purify the gene product after observation of a cDNA reverse transcript, and showed it to be a peptide structurally identical with S-14 from other species, by tryptic mapping, amino acid composition and sequence.

Insulin

Insulin plays a major anabolic role in metabolism causing increased carbohydrate metabolism, glycogen storage, fatty acid synthesis, amino acid uptake and protein synthesis. It is a 2-chain, 51 amino acid molecule, localised in the B-cells of islets of Langerhans in the pancreas. This hormone has been available in pure form for many years, is readily analysed by HPLC, and has hence been widely used in analytical studies of insulin action, structure and function, and synthesis (Stentz et al., 1982; Schartmann et al., 1983; Parman, 1983; Damgaard & Markussen, 1979; McCaleb & Myers, 1982; Birch et al., 1984; Terabe et al., 1979; Atherton et al., 1981; Jonczyk & Gattner, 1981; Rivier & McClintock, 1983; Hsi et al., 1982; Laubli et al., 1982). Stentz et al. (1983) used HPLC to study the metabolism of A14–125 I-insulin in intact human fibroblasts, detecting and separating the early degradation products previously thought by gel chromatography and TCA precipitation to be intact insulin; HPLC separated the intact insulin from two additional peaks with decreased immunoprecipitability and receptor binding but which had the same TCA precipitability as the intact molecule. These studies showed that intact human fibroblasts can metabolise insulin within 2 minutes at 37° C.

Biosynthetic human insulin derived via recombinant DNA technology was tested exhaustively to establish its equivalence chemically, physically and immunologically with pancreatic human insulin and its biological equivalence to porcine insulin. Methods used were circular dichroism and crystallization, as well as HPLC in amino acid analysis and sequence (Chance et al., 1981).

Wetzel et al. (1981) used HPLC to monitor the expression in *E. Coli* of a chemically synthesised gene for a mini-C analog of human proinsulin. The 35 amino acid connecting peptide was replaced by a six amino acid peptide, and was shown to have high pressure liquid chromatographic and radioimmunoassay characteristics suggesting a proinsulin-like structure. Several groups have used HPLC in insulin receptor studies, both in their purification and their isolation of receptor-bound from free insulin (Frank et al., 1983; Pang & Shafer, 1983; Roth et al., 1983; Parman & Rideout, 1983; Finn et al., 1982). Roth et al. (1983) demonstrated the insulin binding of a protease degraded beta-subunit by HPLC showing it to be as good as the intact receptor, however, the digested receptor lacked the normal protein kinase moiety suggesting the beta-subunit of the receptor to be a protein kinase.

Three of the most interesting studies using HPLC in insulin research investigate some clinical disorders of insulin metabolism.

Robbins and co-workers (1984) looked at two families with familial hyperproinsulinemia, an hereditary syndrome in which individuals secrete large amounts of a proinsulin-like material, of molecular weight 9000 daltons. Previous work indicated that the proinsulin like peptide was a conversion intermediate in which, in one family, the C-peptide remained attached to the insulin B chain but in the other was attached to the A-chain. HPLC analysis of the conversion intermediates in two different buffer systems showed that the proinsulin related peptides in both families elute at a single position close to that of the normal des-Arg 31, Arg 32 proinsulin; further enzymatic treatment demonstrated the presence of Lys 64 and absence of Arg 65 in both the abnormal forms concluding that in familial hyperproinsulinaemia the intermediate proinsulin conversion is the C-peptide attached to the A chain, with Arg 65 being replaced by another amino acid residue.

Shoelson et al. (1983) used evidence from the relative HPLC retention time of abnormal insulin present in the serum of a hyperinsulinaemic diabetic patient together with genetic data to predict that the mutant insulin contained a serine for phenylalanine substitution at position B24 or B25. Preparation of Ser B24 and Ser B25 biosynthetically, and separation by reverse phase HPLC of these two semisynthetic insulins from each other and from normal insulin, allowed the characterisation of the abnormal patient insulin, showing that the patients' insulin eluted at the position of B-24 ser-insulin and co-eluted with it when chromatographed as a mixture. Further studies showed that these two mutant insulins had very low activities (B24 16% and 0.5% B25) compared with normal insulin in stimulating glucose oxidase by isolated rat adipocytes concluding that abnormal insulin with amino acid replacements at both B24 and B25 position can be associated with human diabetes.

PHI and PYY

Mutt and co-workers have isolated two new peptides using HPLC to purify and further characterize them (Tatemoto & Mutt, 1980). Both peptides were named after their C- and N-terminal amino acids since their biological actions had not been established. PHI, peptide histidine isoleucine, is 27 amino acids in length, has been isolated from porcine upper intestinal tissue and has a C-terminal isoleucine amide structure. It shows considerable sequence homology with vasoactive intestinal polypeptide (Yiangou et al., 1983, 1985), secretin, glucagon and GIP. PHI has been shown to activate adenylate cyclase, and to induce insulin release from isolated rat pancreas at both basal and raised glucose levels. PYY is a 36 amino acid peptide with an N-terminal tyrosine and an amidated tyrosine at the C-terminal. PYY has some sequence homology with both pancreatic polypeptide and neurotensin. Both peptides have been found in the intestine and in the brain.

Seino et al. (1985) characterised the molecular forms of circulating insulin in patients with hyperinsulinaemia of diverse etiology, using reverse phase HPLC and noted the differences observed in separation when various gradient conditions, and different C18 reverse-phase columns were used. A small gradient rate change from 0.17% to 0.33% prevented the abnormal patient insulin being separated from normal human insulin. Altering the solvent system from 0.2% TFA and triethylamine (TEA) with acetonitrile as the organic modifier to acetonitrile isopropanol (3:1 v/v) containing 0.1% TFA and omitting the TEA achieved the same separation as had been observed on the previous C18 column. This study serves as an example of the precise conditions of RP-HPLC required which may be critical when subtle changes in hormone variants are to be identified. In the same study six patients with benign sporadic insulinomas appeared to have insulin identical to the normal insulin. For a more general discussion of closely related molecular forms of peptides using different solvent systems see Janssen et al. (1984).

Whilst Sephadex chromatography of rat PHI-like peptides (PLPs) showed that PHI-27 and rat brain PLP had identical molecular weights, HPLC revealed not only that rat and porcine PLP were chromatographically distinct from VIP or secretin, but also that porcine brain PLP elutes at the same position of PHI-27 confirming the reported sequence (Tatemoto et al., 1983), whereas rat brain PLP immunoreactivity was consistently separated from synthetic PHI, suggesting an altered amino acid sequence or post-translational modification, (Beinfeld et al., 1984).

PHI like immunoreactivity measured in several species was chromatographically identical on Sephadex superfine gel filtration but HPLC identified several variations between species underlining the caution required in extrapolating experimental results between species (Christofides et al., 1984).

HPLC Studies in Tumours of the GI Organs

Some tumours of the gastrointestinal tract have been associated with high levels of some GI hormones e.g. VIPoma, Gastrinoma, Glucagonoma etc. Böhlen et al. (1983) isolated human growth hormone releasing factor (hGRF) and somatostatin from two pancreatic tumours. Homogeneous peptides were isolated by gel-filtration and two cycles of reverse phase HPLC. One tumour produced three growth hormone releasing factor-like peptides, one a 44 amino acid peptide synonymous with hypothalamic GRF from several species and two smaller fragments with reduced bioactivity, hGRF-40 and hGRF-37; the second tumour contained only one of the three hGRF peptides, hGRF-40. One of the tumours also contained two immunoreactive forms of somatostatin, somatostatin-14 based on chromatographic and sequence studies, and the other tentatively called somatostatin 28, based on only chromatographic behaviour. Shulkes et al. (1984b) has recently characterised VIP, neurotensin (NT) and pancreatic polypeptide (PP) from a tumour removed from a patient with watery diarrhoea, hypokalemia, hypochlorhydria and steatorrhea. Tumour levels of all three hormones were high and the ratio of these peptides was similar to that seen in plasma. Gel chromatography indicated that all the PP activity in plasma and tumor extracts co-eluted with standard human PP. Tumour extracts also subjected to HPLC however, separated two forms of VIP, the predominant fraction similar to porcine VIP and a smaller more hydrophobic form which was detected by C-terminal but not N-terminal VIP antisera; the smaller form was not demonstrated in normal ileum. HPLC showed approximately 75% of the neurotensin was present as inactive N-terminal fragments whereas normal ileum contained only NT1–13. Gamse et al. (1981) studied substance P localization and activity in pheochromocytomas and carcinoid tumour of the caecum. HPLC analysis of the pheochromocytoma tumour extracts showed the immunoreactive substance P (I-SP) to be enriched in the chromaffin granule fraction. About 40% of the I-SP was shown chemically on gel-filtration and HPLC to behave like substance P, and also in bioassays indicating that authentic substance P is present in these tumours and that in the pheochromocytoma it is located in chromaffin granules.

Unknown putative hormones

HPLC has provided a tool for the isolation and characterization of unknown peptides and putative hormones in different phyla of the animal kingdom.

Grimm-Jorgensen isolated immunoreactive somatostatin (IRSRIF) from two species of pond snails (Grimm-Jorgensen, 1983), in extracts of circumesophageal ganglia and hemolymph. In lymnoeastagnalis the IRSRIF roughly paralleled that of the perikanya and neurochemical regions of the neurosecretory neurons involved in the stimulus of growth. Older snails had lower concentrations of IRSRIF in hemolymph, but removal of the shell edge, a procedure stimulating growth, raised hemolymph levels. Immunoreactive somatostatin was present in acid extracts of various parts of the GI tract of physa, but IRSRIF from neither species co-chromatographed with synthetic somatostatin during reverse-phase TLC indicating that the immunoreactive material may be a growth hormone and cross react with somatostatin antibodies but is not somatostatin.

Preparative reverse-phase HPLC of porcine anterior pituitary gland extracts purified a polypeptide of 21 000 daltons of approximately 173 amino acid residues with an isoelectric point of 4.9 (Hsi et al., 1982). This peptide, termed APPG, is not a glycoprotein and has sequence homologies with a number of peptides including duck proinsulin, rous sarcoma virus transforming protein TVFV60 and pig secretin. More recently Vandermeers et al., (1984) have purified a novel pancreatic secretory factor (PSF) from the venom of the Gila Monster *(Heloderma suspectum)* (Vandermeers et al., 1984). Using three different HPLC techniques, purification to homogeneity of two bioactive components was achieved. A protein of 17.5 kDA had a potent secretory effect on dispersed rat pancreatic acini, termed PSF; a second peptide, of molecular weight 5.9 kDa, termed helodermin, was able to stimulate adenylate cyclase in rat pancreatic membranes.Neither one peptide could exert the actions of the other.

In this chapter we have attempted to illustrate the many applications of high pressure liquid chromatography in its various modes to peptide research. The gastrointestinal hormones provide a group of peptides of considerable diversity where knowledge of these hormones has been greatly increased with the development of this highly sensitive, reproducible, rapid technique.

References

Andrews PC, Dixon JE (1981) Isolation and structure of a peptide hormone predicted from a mRNA sequence. A second somatostatin from the catfish pancreas. J Biol Chem 256:8267–8270

Aronin N, Cooper PE, Lorenz LJ, Bird ED, Sagar SM, Leeman SE, Martin JB (1983) Somatostatin is increased in the basal ganglia in Huntington disease. Ann Neurol 13:519–526

Atherton E, Hubscher W, Sheppard RC, Woolley V (1981) Synthesis of a 21-residue fragment of human proinsulin by the polyamide solid phase method. Hoppe Seylers Z Physiol Chem 362:833–839

Bacarese-Hamilton AJ, Adrian TE, Bloom SR (1984) Human and porcine immunoreactive gastric inhibitory polypeptides (IR-GIP) are not identical. FEBS Lett 168:125–128

Bacarese-Hamilton AJ, Adrian TE, Chohan P, Antony T, Bloom SR (1985) Oxidation/reduction of methionine residues in CCK: a study by radioimmunoassay and isocratic reverse phase high pressure liquid chromatography. Peptides – (Fayetteville) 6.1:17–22

Baldwin GS, Knesel J, Monckton JM (1983) Phosphorylation of gastrin-17 by epidermal growth factorstimulated tyrosine kinase. Nature 301:435–437

Bataille D, Tatemoto K, Coudray AM, Rosselin G, Mutt V (1981) Bioactive enteroglucagon (oxyntomodulin): evidence for a C-terminal extension of the glucagon molecule. C R Seances Acad Sci 293:323–328

Bataille D, Tatemoto K, Gespach C, Jornvall H, Rosselin G, Mutt V (1982) Isolation of glucagon-37 (bioactive enteroglucagon/oxyntomodulin) from porcine jejuno-ileum. Characterization of the peptide. FEBS Lett 146:79–86

Bayliss WM, Starling EH (1902) The mechanism of pancreatic secretion. J Physiol London 28:325–335

Beinfeld MC, Meyer DK, Brownstein MJ (1980) Cholecystokinin octapeptide in the rat hypothalamoneurohypophysial system. Nature 288:376–378

Beinfeld MC, Meyer DK, Brownstein MJ (1981) Cholecystokinin in the central nervous system. Peptides 2:77–79

Beinfeld MC, Korchak DM, Roth BL, O'Donohue TL (1984) The distribution and chromatographic characterization of PHI-27-like peptides in rat and porcine brain. J Neuroscience. 4.11:2681–2688

Benoit R, Böhlen P, Brazeau P, Ling N, Guillemin R (1980) Isolation and characterization of rat pancreatic somatostatin. Endocrinology 107:2127–2129

Berelowitz M, Perlow MJ, Hoffman HJ, Frohman LA (1981) The diurnal variation of immunoreactive thyrotropinreleasing hormone and somatostatin in the cerebrospinal fluid of the rhesus monkey. Endocrinology 109:2102–2109

Beyerman HC, Kranenburg P, Schaaper WM, Voskamp D (1981) Improvements in the synthesis of secretin by the repetitive excess mixed anhydride (REMA) method. Int J Pept Protein Res 18:276–283

Birch NP, Christie DL, Renwick AG (1984) Immunoreactive insulin from mouse brain cells in culture and whole rat brain. Biochem J 218:19–27

Birnbaumer L, Swartz TL, Rojas FJ, Garber AJ (1983) Synthesis, purification by HPLC and characterization of monoiodoglucagon labeled with carrier free 125I: use as glucagon receptor probe. J Recept Res 3:3–20

Blackburn AM, Fletcher DR, Bloom SR, Christofides ND, Long RG, Fitzpatrick ML, Baron JH (1980) Effect of neurotensin on gastric function. Lancet i:987–989

Bloom SR, Long RE (1982) Radioimmunoassay of Gut Regulatory Peptides. W B Saunders: London

Bodansky M, Klavsner YS, Said SI (1973) Biological activities of synthetic peptides corresponding to fragments of and the entire sequence of the vasoactive intestinal polypeptide. Proc Natl Acad Sci 70:382–384

Böhlen P, Benoit R, Ling N, Guillemin R, Brazeau P (1981a) Isolation and characterization of rat hypothalamic somatostatin-14. Endocrinology 108:2008–2010

Böhlen P, Brazeau P, Esch F, Ling N, Guillemin R (1981b) Isolation and chemical characterization of somatostatin-28 from rat hypothalamus. Regul Pept 2:359–369

Böhlen P, Brazeau P, Esch F, Ling N, Wehrenberg WB, Guillemin R (1983) Human growth hormone releasing factor and somatostatin from two pancreatic tumors: isolation and characterization. Regul Pept 6:343–353

Brazeau P, Vale W, Burgus R, Ling N, Butcher M, Rivier J, Guillemin R (1973) Hypothalamic polypeptide that inhibits the secretion of immunoreactive pituitary growth hormone. Science 179:77–79

Brown JC, Cook MA, Dryburgh JR (1973) Motilin, a gastric motor activity stimulating polypeptide: the complete amino acid sequence. Can J Biochem 51:533–537

Brown JC, Dahl M, Kwauk S, McIntosh CH, Otte SC, Pederson RA (1981) Actions of GIP. Peptides 2:241–245

Brown JC, Johnson LP, Magee DF (1966) Effect of duodenal alkalinization on gastric motility. Gastroenterology 50:333–339

Brown JC, Pederson RA (1970) A multiparameter study on the action of preparations containing cholecystokininpancreozymin. Scand J Gastroenterol 5:537–541

Carlquist M, Jornvall H, Mutt V (1981) Isolation and amino acid sequence of bovine secretin. FEBS Lett 127:71–74

Carlquist M, Mutt V, Jornvall H (1979) Isolation and characterization of bovine vasoactive intestinal peptide (VIP). FEBS Lett 108:457–460

Carraway R, Leeman SE (1973) The isolation of a new hypotensive peptide, neurotensin, from bovine hypothalami. J Biol Chem 248:6854–6861

Chance RE, Kroeff EP, Hoffmann JA, Frank BH (1981) Chemical, physical, and biologic properties of biosynthetic human insulin. Diabetes Care 4:147–154

Chance RE, Moon NE, Johnson MG (1979) Human pancreatic polypeptide (HPP) and bovine pancreatic polypeptide (BPP) In: Jaffe BM, Behrman HR (eds) Methods of hormone radioimmunoassay. Academic Press, New York, pp 657–672

Chang JY (1981) Isolation and characterization of polypeptide at the picomole level. Pre-column formation of peptide derivatives with dimethylaminoazobenzene isothiocyanate. Biochem J 1999:537–545

Chang JY, Knecht R, Braun DG (1981) Amino acid analysis at the picomole level. Application to the C-terminal sequence analysis of polypeptides. Biochem J 199:547–555

Chang JY, Knecht R, Braun DG (1982) Isolation and sequence determination or polypeptides at the picomole level. In: Elzinga M (ed) Methods in protein sequence analysis. Humana Press, New Jersey, p 113

Chang MM, Leeman SE (1970) Isolation of a sialogogic peptide from bovine hypothalamic tissue and its characterisation as substance P. J Biol Chem 245:4784–4790

Charlton CG, O'Donohue TL, Miller RL, Jacobowitz DM (1981) Secretin immunoreactivity in rat and pig brain. Peptides 2:45–49

Christofides ND, Long RG, Fitzpatrick ML, McGregor GP, Bloom SR (1981) Effect of motilin on the gastric emptying of glucose and fat in humans. Gastroenterology 80:456–460

Christofides ND, Modlin IM, Fitzpatrick ML, Bloom SR (1979) Effect of motilin on the rate of gastric emptying and gut hormone release during breakfast. Gastroenterology 76:903–907

Christofides ND, Yiangou Y, Tatemoto K, Bloom SR (1984) Characterization of peptide histidine isoleucine-like immunoreactivity in rat, human, guinea-pig and cat gastrointestinal tracts - evidence of species differences. Digestion 30:165–170
Cockle SA, Kaplan H, Hefford MA, Young NM (1982) Application of high-performance liquid chromatography to competitive labeling studies: the chemical properties of functional groups of glucagon. Anal Biochem 125:210–216
Coy DH, Coy EJ, Lee KY, Chey WY (1982) Solidphase synthesis and biological activities of gastrointestinal hormones: secretin and motilin. Peptides 3:137–141
Damgaard U, Markussen J (1979) Analysis of insulins and related compounds by HPLC. Horm Metab Res 11:580–581
Deschodt-Lanckman M, Bui ND, Noyer M, Christophe J (1981) Degradation of cholecystokinin-like peptides by a crude rat brain synaptosomal fraction: a study by high pressure liquid chromatography. Regul Pept 2:15–30
Dimaline R, Reeve Jr JR (1983) Reversed-phase highperformance liquid chromatography used to monitor enzymatic cleavage of pyrrolidone carboxylic acid from regulatory peptides. J Chromatogr 257:355–360
Dockray GJ (1976) Immunochemical evidence of cholecystokinin-like peptides in man. Nature 264:568–570
Eng J, Shiina Y, Straus E, Yalow RS (1982) Posttranslational processing of cholecystokinin in pig brain and gut. Proc Natl Acad Sci USA 79:6060–6064
Eskay RL, Beinfeld MC (1982) HPLC and RIA of cholecystokinin peptides in the vertebrate neural retina. Brain Res 246:315–318
von Euler USO, Gaddum JH (1931) An unidentified depressor substance in certain tissue extracts. J Physiol Lond 72:74–87
Eysselein VE, Reeve Jr JR, Shively JE, Hawke D, Walsh JH (1982) Partial structure of a large canine cholecystokinin (CCK58): amino acid sequence. Peptides 3:687–691
Finn FM, Titus G, Nemoto H, Noji T, Hofmann K (1982) Insulin receptor isolation studies. Metabolism 31:691–698
Fletcher DR, Blackburn AM, Adrian TE, Chadwick VS, Bloom SR (1981) Effect of neurotensin on pancreatic function in man. Life Sci 29:2157–2161
Frank BH, Peavy DE, Hooker CS, Duckworth WC (1983) Receptor binding properties of monoiodotyrosyl insulin isomers purified by high performance liquid chromatography. Diabetes 32:705–711
Gamse R, Saria A, Bucsics A, Lembeck F (1981) Substance P in tumors: pheochromocytoma and carcinoid. Peptides 2:275–280
Gomez S, Morel A, Nicolas P, Cohen P (1983) Regional distribution of the Mr 15 000 somatostatin precursor, somatostatin-28 and somatostatin-14 in the rat brain suggests a differential intracellular processing of the high molecular weight species. Biochem Biophys Res Commun 112:297–305
Grimm-Jorgensen Y (1983) Immunoreactive somatostatin in two pulmonate gastropods. Gen Comp Endocrinol 49:108–114
Harper AA, Raper HS (1943) Pancreozymin. A stimulant of the secretion of pancreatic enzymes in extracts of the small intestine. J Physiol 102:115–125
Hsi KL, Seidah NG, De Serres G, Crietien M (1982) Isolation and NH2-terminal sequence of a novel porcine anterior pituitary polypeptide. Homology to proinsulin, secretin and Rous sarcoma virus transforming protein TVFV60. FEBS Lett 147:261–266
Ivy AC, Oldberg E (1928) A hormone mechanism for gallbladder contraction and evacuation. Amer J Physiol 86:599–613
Jahnke GD, Lazarus LH (1984) A bombesin immunoreactive peptide in milk. Proc Natl Acad Sci USA 81:578–582
Janssen PSL, Van Nispen JW, Hamelinck RLAE, Melge PATA, Gorede BC (1984) Application of reversed phase HPLC in some critical peptide separations. J Chromatogr Sci 22:234–238
Jonczyk A, Gattner HG (1981) A new semisynthesis of human insulin. Tryptic transpeptidation of porcine insulin with L-threonine tert-butyl ester. Hoppe Seylers Z Physiol Chem 362:1591–1598
Jornvall H, Carlquist M, Kwauk S, Otte SC, McIntosh CH, Brown JC, Mutt V (1981) Amino acid sequence and heterogeneity of gastric inhibitory polypeptide (GIP). FEBS Lett 123:205–210

Jorpes JE, Mutt V, Toczko K (1964) Further purification of cholecystokinin and pancreozymin. Acta Clin Scand 18:2408–2410

Knip M (1984) Analysis of pancreatic polypeptide hormones by reversed-phase high performance liquid chromatography. Horm Metab Res 16:487–491

Laubli UK, Baier W, Binz H, Celio MR, Humbel RE (1982) Monoclonal antibodies directed to human insulin-like growth factor I (IGF I). Use for radioimmunoassay and immunopurification of IGF. FEBS Lett 149:109–112

Lee CM, Emson PC, Iversen LL (1981) Chromatographic behaviour and post-mortem stability of somatostatin in the rat and mouse brain. Brain Res 220:159–166

Lin TM, Spry GF (1969) Effect of pentagastrin, cholecystokinin, caerulein and glucagon on the choledochal resistance and bile flow of conscious dog. Gastroenterology. 56:1178

Märki F, Schenkel L, Petrack B, Czernik AJ, Ansell J, Allen M, Brundish DE, Martin JR, McMartin C, Peters GE, Wade R (1981) Rapid conversion of somatostatin to active metabolite in human plasma. FEBS Lett 127:22–24

Marshak DW, Reeve JR, Shively JE, Hawke D, Takami MS, Yamada T (1983) Structure of somatostatin isolated from bovine retina. J Neurochem 41:601–606

Maton PN, Selden AC, Chadwick VS (1984a) Differential distribution of molecular forms of cholecystokinin in human and porcine small intestinal mucosa. Regulat Peptides 8:9–19

Maton PN, Selden AC, Chadwick VS (1982) Large and small forms of cholecystokinin in human plasma: measurement using high pressure liquid chromatography and radioimmunoassay. Regul Pept 4:251–260

Maton PN, Selden AC, FitzPatrick ML, Chadwick VS (1984b) Infusion of cholecystokinin octapeptide in man: relation between plasma cholecystokinin concentrations and gallbladder emptying rates. Eur J Clin Invest 14:37–41

McCaleb ML, Myers RD (1982) 2-Deoxy-D-glucose and insulin modify release of norepinephrine from rat hypothalamus. Am J Physiol 242:R596–R603

McDonald TJ, Jornvall H, Tatemoto K, Mutt V (1983) Identification and characterization of variant forms of the gastrin-releasing peptide (GRP). FEBS Lett 156:349–356

Meek JL, Iadarola MJ, Giorgi O (1983) Cholecystokinin turnover in brain. Brain Res 276:375–378

Meyers CA, Coy DH (1980) Solid-phase synthesis of human pancreatic polypeptide. Int J Pept Protein Res 16:248–253

Mikes O (1979) Rapid separation of proteins and their higher-molecular fragments by means of Spheron ionexchanges. Int J Pept Protein Res 14:393–401

Moody TW, O'Donohue TL, Jacobowitz DM (1981a) Biochemical localization and characterization of bombesin-like peptides in discrete regions of rat brain. Peptides 2:75–79

Moody TW, Thoa NB, O'Donohue TL, Jacobowitz DM (1981b) Bombesin-like peptides in rat spinal cord: biochemical characterization, localization and mechanism of release. Life Sci 29:2273–2279

Morel A, Lauber M, Cohen P (1981) Selective processing of the 15 000 Mr prosomatostatin by mouse hypothalamic extracts releases the tetradecapeptide. FEBS Lett 136:316–318

Moroder L, Gohring W, Nyfeler R, Scharf R, Thamm P, Wendlberger G (1983) Syntheses of human little gastrin-I and its leucine-15, norleucine-15 and methoxinine-15 analogs. Hoppe Seylers Z Physiol Chem 364:157–171

Mueller MK, Demol P, Goebell H, Fladrich G, Brown JC (1983) Natural purified porcine gastric inhibitory polypeptide (GIP) stimulates exocrine pancreas secretion due to contamination with a cholecystokinin-like substance. Regul Pept 5:307–315

Murlin JR, Clough HD, Gibbs CBF, Stokes AM (1923) Aqueous extract of pancreas. I. Influence on the carbohydrate metabolism of depancreatized animals. J Bio Chem 56:253–296

Murphy R, Furness JB, Costa M (1984) Measurement and chromatographic characterization of vasoactive intestinal peptide from guinea-pig enteric nerves. J Chromatogr 336.1:41–50

Mutt V, Jorpes JE (1966) Secretin: isolation and determination of structure. In: Proceedings, International Symposium on the Chemistry of Natural Products. Stockholm. 26th June: Section 2C-3

O'Donohue TL, Beinfeld MC, Chey WY, Chang TM, Nilaver G, Zimmerman EA, Yajima H, Adachi H, Poth M, McDevitt RP, Jacobowitz DM (1981a) Identification, characterization and

distribution of motilin immunoreactivity in the rat central nervous system. Peptides 2:467–477
O'Donohue TL, Charlton CG, Miller RL, Boden G, Jacobowitz DM (1981b) Identification, characterization, and distribution of secretin immunoreactivity in rat and pig brain. Proc Natl Acad Sci USA 78:5221–5224
Olieman C, Sedlick E, Voskamp D (1981) In situ silylation of an octadecylsilyl-silica stationary phase applied to the analysis of peptides, such as secretin and glucagon. J Chromatogr 207:421–424
Oyama H, Bradshaw RA, Bates OJ, Permutt A (1980a) Amino acid sequence of catfish pancreatic somatostatin. J Biol Chem 255:2251–2254
Oyama H, Hirsch HJ, Gabbay KH, Permutt A (1980b) Isolation and characterization of immunoreactive somatostatin from fish pancreatic islets. J Clin Invest 65:993–1002
Pang DT, Shafer JA (1983) Stoichiometry for the binding of insulin to insulin receptors in adiopocyte membranes. J Biol Chem 258:2514–2518
Panula P, Hadjiconstantinou M, Yang HY, Costa E (1983) Immunohistochemical localization of bombesin/gastrinreleasing peptide and substance P in primary sensory neurons. J Neurosci 3:2021–2029
Parman AU (1983) Comparison of carboxymethyl-cellulose cation-exchange chromatography and high-performance liquid chromatography in the purification of guinea-pig insulin from pancreatic extracts. J Chromatogr 256:293–301
Parman AU, Rideout JM (1983) Purification of porcine proinsulin by high-performance liquid chromatography. J Chromatogr 256:283–291
Paul S, Wood K, Said SI (1984) Purification of (125I)-vasoactive intestinal polypeptide by reverse phase HPLC. Peptides (Fayetteville) 5.6:1085–1087
Pearson D, Shively JE, Clark BR, Geschwind II, Barkley M, Nishioka RS, Bern HA (1980) Urotensin II. a somatostatin-like peptide in the caudal neurosecretory system of fishes. Proc Natl Acad Sci USA 77:5021–5024
Penman E, Wass JA, Butler MG, Penny ES, Price J, Wu PRLH (1983) Distribution and characterization of immunoreactive somatostatin in human gastrointestinal tract. Regul Pept 7:53–65
Peters GE (1982) Distribution and metabolism of exogenous somatostatin in the rat. Regul Pept 3:361–369
Pingoud V, Trautschold I (1984) High performance liquid chromatography of iodine-labelled insulin and glucagon derivatives with on-line gamma detection. Anal Biochem 140:305–314
Polak JM, Coulling I, Bloom SR, Pearse AGE (1971) Immunofluorescent localization of secretin and enteroglucagon in human intestinal mucosa. Scand J Gastroenterol 6:739–744
Polak JM, Pearse AGE, Bloom SR, Buchan AM, Rayford PL, Thompson JC (1975) Identification of cholecystokinin-secreting cells. Lancet ii:1016–1018
Pradayrol L, Chayvialle JA, Carlquist M, Mutt V (1978) Isolation of a porcine intestinal peptide with C-terminal somatostatin. Biochem Biophys Res Commun 85:701–708
Price J, Penman E, Bourne GL, Rees LH (1983) Characterization of bombesin-like immunoreactivity in human fetal lung. Regul Pept 7:315–322
Reeve Jr J, Yamada T, Chew P, Walsh JH, Dimaline R (1982) Rapid high-yield purification of neuropeptides from canine intestinal muscle. J Chromatogr 229:57–65
Reeve Jr JR, Dimaline R, Shively JE, Hawke D, Chew P, Walsh JH (1981) Unique amino terminal structure of rat little gastrin. Peptides 2:453–458
Reeve Jr JR, Walsh JH, Chew P, Clark B, Hawke D, Shively JE (1983) Amino acid sequences of three bombesin-like peptides from canine intestine extracts. J Biol Chem 258:5582–5588
Rehfeld JF (1978) Immunochemical studies on cholecystokinin II. Distribution and molecular heterogeneity in the central nervous system and small intestine of man and dog. J Biol Chem 253:4022–4030
Rehfeld JF, Goltermann N, Larsson L-I, Emson PM, Lee CM (1979) Gastrin and cholecystokinin in central and peripheral neurons. Fed Proc 38:2325–2329
Rennie JA, Christofides ND, Bloom SR, Johnson AG (1979) Stimulation of human colonic activity by motilin. Gut 20:A912
Rivier J, McClintock R (1983) Reversed-phase high-performance liquid chromatography of insulins from different species. J Chromatogr 268:112–119

Robbins DC, Shoelson SE, Rubenstein AH, Tager HS (1984) Familial hyperproinsulinemia. Two cohorts secreting indistinguishable type II intermediates of proinsulin conversion. J Clin Invest 73:714–719
Rojas FJ, Swartz TL, Iyengar R, Garber AJ, Birnbaumer L (1983) Monoiodoglucagon: synthesis, purification by high pressure liquid chromatography, and characteristics as a receptor probe. Endocrinology 113:711–719
Roth KA, Evans CJ, Weber E, Barchas JD, Bostwick DG, Bensch KG (1983) Gastrin-releasing peptide-related peptides in a human malignant lung carcinoid tumor. Cancer Res 43:5411–5415
Roth RA, Mesirow ML, Cassell DJ (1983) Preferential degradation of the beta subunit of purified insulin receptor. Effect on insulin binding and protein kinase activities of the receptor. J Biol Chem 258:14456–14460
Sacks HS, Terry LC, Wright RK, Stentz FB (1984) Somatostatin metabolism: differences in clearance of N-terminal and central portions of molecule during perfusion of rat liver. Am J Physiol 246:G226–G234
Sagar SM, Rorstad OP, Landis DM, Arnold MA, Martin JB (1982) Somatostatin-like immunoreactive material in the rabbit retina. Brain Res 244:91–99
Said SI, Mutt V (1970) Polypeptide with broad biological activity: isolation from small intestine. Science 169:1217–1218
Sauter A, Frick W (1983) Determination of cholecystokinin tetrapeptide and cholecystokinin octapeptide sulfate in different rat brain regions by high-pressure liquid chromatography with electrochemical detection. Anal Biochem 133:307–313
Schartmann B, Gattner HG, Danho W, Zahn H (1983) Increased yields of insulin by recycling of non-combined insulin chains. Hoppe Seylers Z Physiol Chem 364:179–186
Seino S, Funakoshi A, Fu ZZ, Vinik A (1985) Identification of insulin variants in patients with hyperinsulinaemia by reversed-phase, high performance liquid chromatography. Diabetes 34:1–7
Selden AC, Maton PN, Chadwick VS (1981) Analysis of cholecystokinin peptides by HPLC and bioassay. Horm Metab Res 13:363
Selden C, Maton PN, Chadwick VS (1984) Reversed-phase high-performance liquid chromatography and separation of peptide hormones in human tissues and plasma with reference to cholecystokinins. J Chromatogr 305:43–52
Sheikh SP, Baldissera FG, Karlsen FO, Holst JJ (1985) Glicentin is present in the pig pancreas. FEBS Lett 179.1:1–6
Shimonishi Y, Hong YM, Kitagishi T, Matsuo T, Matsuda H, Katakuse I (1980) Sequencing of peptide mixtures by Edman degradation and field-desorption mass spectrometry. Eur J Biochem 112:251–264
Shoelson S, Fickova M, Haneda M, Nahum A, Musso G, Kaiser ET, Rubenstein AH, Tager H (1983) Identification of a mutant human insulin predicted to contain a serine-for-phenylalanine substitution. Proc Natl Acad Sci USA 80:7390–7394
Shoelson S, Polonsky K, Docherty H, Jaspan J, Rubenstein A (1982) Evidence for the presence of somatostatin 28 in plasma. Diabetes 31:474–477
Shulkes A, Bijaphala S, Dawborn JK, Fletcher DR, Hardy KJ (1984a) Metabolism of neurotensin and pancreatic polypeptides in man: role of the kidney and plasma factors. J Clin Endocrinol Metab 58:873–879
Shulkes A, Boden R, Cook I, Gallagher N, Furness JB (1984b) Characterization of a pancreatic tumor containing vasoactive intestinal peptide, neurotensin, and pancreatic polypeptide. J Clin Endocrinol Metab 58:41–48
Singh B, Osmundsen H, Borrebaek B (1982) The time course of glucagon action on the utilization of [U-14C]palmitate by isolated hepatocytes. Arch Biochem Biophys 217:244–250
Staub A, Sinn L, Behrens OK (1953) Purification and crystallization of hyperglycaemic glycogenolytic factor (HGF). Science 117:628–629
Stentz FB, Harris HL, Kitabchi AE (1983) Early detection of degraded A14-125I-insulin in human fibroblasts by the use of high performance liquid chromatography. Diabetes 32:474–477
Stentz FB, Wright RK, Kitabchi AE (1982) A rapid means of separating A14-125I-insulin from heterogeneously labeled insulin molecules for biologic studies. Diabetes 31:1128–1131

Sueiras-Diaz J, Lance VA, Murphy WA, Coy DH (1984) Structure-activity studies on the N-terminal region of glucagon. J Med Chem 27:310-315

Tatemoto K, Mutt M (1980) Isolation of two novel candidate hormones using a chemical method for finding naturally occurring polypeptides. Nature 285:417-418

Tatemoto K, Carlquist M, McDonald T, Mutt V (1983) Isolation of a brain peptide identical to the intestinal PHI (peptide HI) FEBS Letts 153:248-252

Tempst P, Hunkapiller MW, Hood LE (1984) Separation of peptides by reverse phase high performance liquid chromatography using propyl- and cyanopropylsilyl supports. Anal Biochem 137:188-195

Terabe S, Konaka R, Inouye K (1979) Separation of some polypeptide hormones by high-performance liquid chromatography. J Chromatogr 172:163-177

Thim L, Moody AJ (1982) Purification and chemical characterization of a glicentin-related pancreatic peptide (proglucagon fragment) from porcine pancreas. Biochem Biophys Acta 703:134-141

Uhl GR, Snyder SH (1976) Regional and subcellular distributions of brain neurotensin. Life SCI 19:1827-1832

Vanderhaeghen JJ, Signeau JC, Gepts W (1975) A new peptide in the invertebrate central nervous system reacting with antigastrin antibodies. Nature 257:604-606

Vandermeers A, Mandermeers-Piret MC, Robberecht P, Waelbroeck M, Dehaye JP, Winand J, Christophe J (1984) Purification of a novel pancreatic secretory factor (PSF) and a novel peptide with VIP- and secretin-like properties (helodermin) from Gila monster venom. FEBS Lett 166:273-276

Voskamp D, Beyerman HC (1981) Biological potency and chromatographic properties of [aspartoyl3]-secretin, [beta-Asp3]-secretin, and secretin (4-27). Products of intramolecular reactions in secretin. Int J Pept Protein Res 18:284-288

Watson KJR, Shulkes A, Smallwood RA, Douglas MC, Hurley R, Kalnins R, Moran L (1985) Watery diarrhoea - Hypokalemia - Achlorhydria Syndrome and carcinoma of the oesophagus. Gastroenterol 88:798-803

Wetzel R, Kleid DG, Crea P, Heyneker HL, Yansura DG, Hirose T, Kraszewksi A, Riggs AD, Itakura K, Goeddel DV (1981) Expression in Escherichia coli of a chemically synthesized gene for a mini-C analog of human proinsulin. Gene 16:63-71

Yalow RS, Berson S (1970) Size and change distinctions between endogenous human plasma gastrin in peripheral blood and heptadecapeptide gastrins. Gastroenterology 58:609-615

Yiangou Y, Christofides ND, Evans JE, Bloom SR (1983) The presence of peptide histidine isoleucine in human, dog, guinea-pig and rat pancreas. Diabetologia 25:125-127

Yiangou Y, Christofides ND, Blank MA, Yanaihara N, Tatemoto K, Bishop AE, Polak JM, Bloom SR (1985) Molecular forms of peptide histidine isoleucine-like immunoreactivity in the gastrointestinal tract. Gastroenterol 89:516-524

Yoshizaki K, de Bock V, Solomon S (1984) Origin of bombesin-like peptides in human fetal lung. Life Sci 34:835-843

Yu TJ, Schwarz HA, Cohen SA, Vouros P, Karger BL (1984) Sequence analysis of derivatised peptides by high performance liquid chromatography-mass spectrometry. J Chromatogr 301.2:425-440

High-Performance Liquid Chromatography of Steroid Hormones

H. L. J. MAKIN, and E. HEFTMANN

Introduction

Steroid hormones originate in man from the C_{27} steroid, cholesterol. In the steroidogenic tissues (adrenal cortex, gonads, placenta, etc.), the cholesterol side chain is removed and C_{21} progestagens, C_{21} corticosteroids, C_{19} androgens and C_{18} oestrogens are successively formed. The active steroid hormones are secreted into the blood and are carried to their site or sites of action, bound either to a specific binding globulin (e.g. sex hormone binding globulin, cortisol-binding globulin) or non-specifically to albumin. In the peripheral tissues they are further metabolised, as a result of which they are inactivated. In some tissues, metabolism of the secreted hormone is required to transform the steroid to an active metabolite (i.e. testosterone reduced to 5α-dihydrotestosterone in prostate or testosterone aromatised to oestradiol in pituitary), after which further metabolism is required for inactivation. Inactive metabolites are then lost, predominantly in the urine, but also in the faeces. Some metabolism may also occur during entero-hepatic circulation.

A single secreted steroid can be metabolised to a large number of metabolites and thus the problem of analysing the multiplicity of steroid hormones present in plasma and their metabolites in urine is a considerable one. Until the advent of suitable chromatographic methods for adequate separation became available, these metabolites were estimated by group procedures which measured those steroids with a common chemical characteristic (e.g. 17-oxosteroids, 17-oxogenic-steroids, Porter-Silber chromogens, etc.). A brief account of these group estimations is given by Makin (1984a).

Early methods of separation involved paper chromatographic procedures, partition chromatography on Celite columns, ion-exchange chromatography, adsorption chromatography on silica gel (silicic acid) and alumina (both on thin-layer plates and in columns), and gel fitration through LH-20 and Lipidex 5000. Many of these techniques are still in widespread use today and are reviewed by Makin (1984a) and Heftmann (1976) where references to the original publications can be found. "High-pressure" or rather "high-performance" liquid chromatography (HPLC) developed from the early conventional column chromatographic systems by careful optimisation of the parameters which determined speed and efficiency of separation (but see Kissinger, 1986). High performance is achieved with columns with ultrafine particles of sorbents, and the use of pressure for driving eluents through the columns is incidental. Thus, the use of microparticulate silica, reverse phase bonded silica, and ion-exchange material in small cartridges (Sep-Pak) or syringes (Bond-Elut) is certainly an example of high-performance chromatography, albeit not at high pressure. These simple systems have been very successful in the preliminary extraction of steroids from urine and plasma (Shackleton & Whitney, 1980; Heikkinen, Fotsis & Adlercreutz, 1981; Axelson & Sahlberg, 1981, Cannell et al., 1982, Risk, Holland & Brown, 1985; Dawson, Kontur & Monjan, 1985), and for the fractionation of vitamin D metabolites (Redhwi, Anderson & Smith, 1981; Adams, Clemens & Holick, 1981; Rhodes et al., 1983) and bile acid conjugates (Street, Trafford & Makin, 1985). Despite the value of these techniques and similar procedures, such as high-performance thin-layer chromatography (HPTLC) (e.g. Grassini-Strazza & Nicoletti, 1985), this chapter will only deal with HPLC of steroid hormones in conventional columns, and will not deal with HPLC of synthetic steroids of pharmaceutical interest, which has been reviewed by Smith (1981) and is discussed by Heftmann & Lin (1982).

HPLC was first applied to the separation of steroid hormones by Siggia & Dishman (1970) who evaluated a number of different sorbents for the separation of corticosteroids, androgens and oestrogens. Early developments in HPLC, as applied to the separation of steroid hormones, are reviewed by Heftman (1976), Fitzpatrick (1978) and Heftmann & Hunter (1979). The use of HPLC for the fractionation of steroids has increased greatly over the years, as evidenced by the number of references quoted by Fitzpatrick in 1978 (40), Heftmann & Hunter in 1979 (185), Heftmann & Lin in 1982 (430). It is clearly not possible to provide in this chapter a comprehensive and detailed review of the multiplicity of published HPLC methods for the separation of steroid hormones. The more recent reviews of Kautsky (1981), Heftmann & Lin (1982) and Heftmann (1983) cover the literature up to 1981. In this chapter most attention will be focussed on HPLC methods described since 1981 and applications of these methods to steroid metabolic studies or to the measurement of steroids in human body fluids.

While it is probably true to say that the particular difficulties of separating the large number of naturally occurring steroid metabolites has been one of the incentives towards the development of modern methods of HPLC, it is still not widely utilised by those involved in steroid endocrinology. Capillary gas chromatography without (Shackleton, Taylor & Honour, 1980) and with (Dehennin et al., 1984) mass spectrometry and with modern techniques of soft ionisation (Shackleton, Mattox & Honour, 1983), thin-layer chromatography (TLC), conventional liquid chromatography, and even paper chromatography (Schoneshofer, 1977; Witzgall & Hassan-Ali, 1980; Takada et al., 1984) are still utilised instead of HPLC by many workers (see Table 1). The development of high-affinity high specificity antisera has also contributed towards an increasing reliance on immunoassay for the direct measurement of individual steroids and their conjugates in a variety of body fluids, such as saliva (Few, Chaudry & James, 1984), serum (Perry, Al-Dujaili & Edwards, 1982), in vitro incubation medium (Racz et al., 1984) or urine (Shah et al., 1984). It is interesting to note that in the sections on steroids in "Hormones in Blood" (Gray & James, 1983), HPLC is not mentioned once, except with regard to the estimation of vitamin D and metabolites.

Table 1. A survey of papers published in the Journal of Clinical Endocrinology & Metabolism (JCEM 1984, vol. 58 and 1985, vol. 61) and the Journal of Steroid Biochemistry (JSB 1984, vol. 20 and 1985, vols 22 & 23) with regard to methods of separation and/or quantitation of steroid hormones

Journal	Number of papers surveyed	Method of separation used*					
		HPLC	RIA/CPB	TLC	PC	CC	GLC
JSB (1984)	68	9	30	22	7	25	9
JSB (1985)	115	24	47	40	11	24	13
JCEM (1984)	41	3	32	4	3	12	1
JCEM (1985)	31	2	28	2	2	11	1

* Some papers use more than one method. If so, all are included. RIA: radioimmunoassay, CPB: competitive protein binding, TLC: thin-layer chromatography, PC: paper chromatography, CC: conventional column chromatography, GLC: gas-liquid chromatography

While many systems for the separation of the various steroid hormone groups have been described and the literature on HPLC separation of steroids is growing, much of this work has been directed towards the development of separation systems for the various steroid hormones, but the application of these systems to problems in-steroid endocrinology has been limited. Although HPLC is non-destructive and thus the sample is recoverable, claims that it does not require the formation of derivatives are not always true, since many underivatised steroids cannot be detected with sufficient sensitivity by available HPLC detectors. Practical problems with HPLC systems are frequently encountered. Frits become blocked or very high back pressures develop, particular with sorbents of 5 μm or less, due to extraneous matter in the sample. Careful attention to sample preparation and "clarification" or filtration of sample prior to injection is often required. Pre-packed columns are very expensive and packing material is not always readily available; thus experimentation with differing column material is not encouraged. Many commercial HPLC systems are over-elaborate, involving unnecessary microprocessors, and are thus very expensive. These are perhaps some of the reasons why many workers in the steroid field do not use HPLC, despite the impressive separations of steroids by HPLC which have been published.

Typical HPLC columns have efficiencies of approximately 5 000–10 000 theoretical plates in comparison with conventional packed GC columns at around 6 000 theoretical plates. Capillary GC columns, on the other hand, may reach efficiencies as high as 100 000 theoretical plates. Micro-bore HPLC (Novotny, Alasandro & Konishi, 1983) can, however, give efficiencies comparable to capillary GC, although there may be problems in loading samples onto these columns and detecting them. Tsuji & Binns (1982) have used microbore (1 mm ID) straight-phase and reversed-phase columns and claimed a 16-fold increase in sensitivity when compared to conventional columns, but this system was not applied to the analysis of biological samples.

Biochemistry of Steroid Hormones

Detailed consideration of the pathways of biosynthesis and subsequent catabolism of steroid hormones can be found in a number of reviews (e.g. Hobkirk, 1979; Bondy & Rosenberg, 1980; Makin, 1984b) and thus only a brief outline will be given here.

Side-chain Cleavage of Cholesterol

All steroid hormones are synthesised from cholesterol by side chain cleavage, which occurs inside the mitochondria. This process is the site of action of the pituitary trophic hormones, which control steroid secretion. Hydroxylation at C20 and C22 is followed by a cleavage of the C20–C22 bond, giving rise to the C_{21} steroid, pregnenolone. These reactions are catalysed by a mixed-function oxidase enzyme system, which utilises the specific cytochrome P450 designated Cytochrome $P450_{scc}$ (Fig. 1).

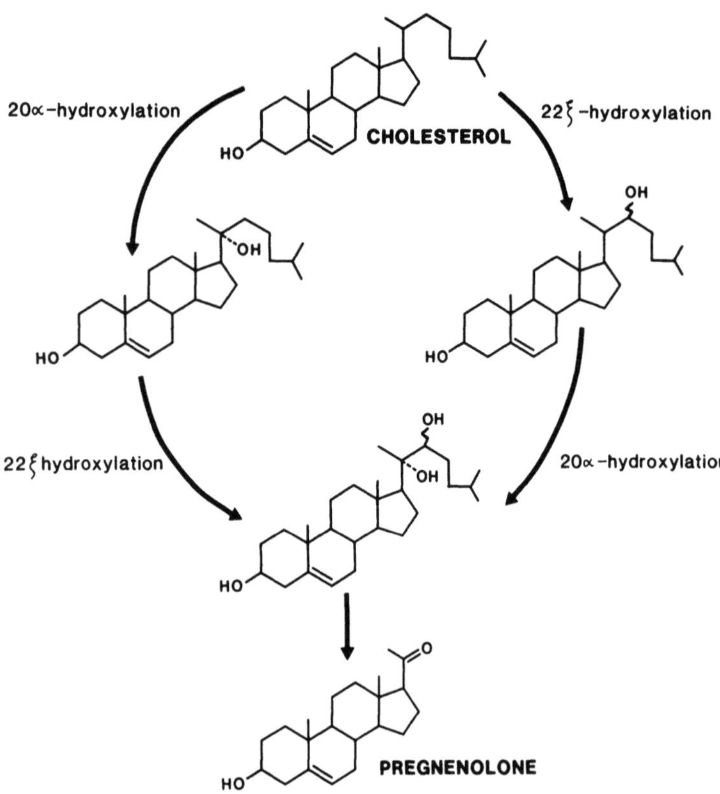

Fig. 1. An outline of the biosynthesis of pregnenolone

Synthesis of Progestagens and Corticosteroids

Pregnenolone can be transformed either inot progesterone, which is secreted by the placenta and corpus luteum, or into 17-hydroxypregnenolone, depending upon the particular steroidogenic tissue. In the human adrenal cortex, 17-hydroxypregnenolone is the precursor of the main glucocorticoid, cortisol, and progesterone is the precursor of the mineralocorticoid, aldosterone, both of which are secreted under trophic stimuli (ACTH for cortisol and Angiotensin II for aldosterone). The secretion rate of cortisol by the adrenal cortex (approx. 20 mg/day) is roughly 100x that of aldosterone. Cortisol is synthesised by zona fasciculata/reticularis zone of the adrenal, whereas aldosterone synthesis occurs in the zona glomerulosa. The pathways illustrated in Figures 2 and 3 outline the major routes of synthesis, but other, probably less important pathways have been described.

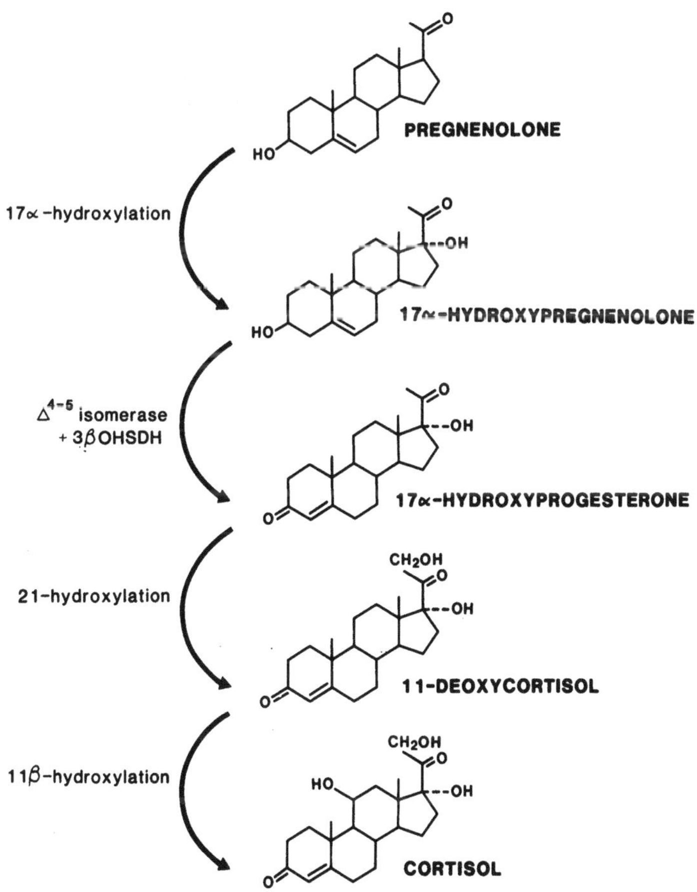

Fig. 2. Biosynthesis of cortisol from pregnenolone in the zona fasciculata/reticularis of the adrenal cortex

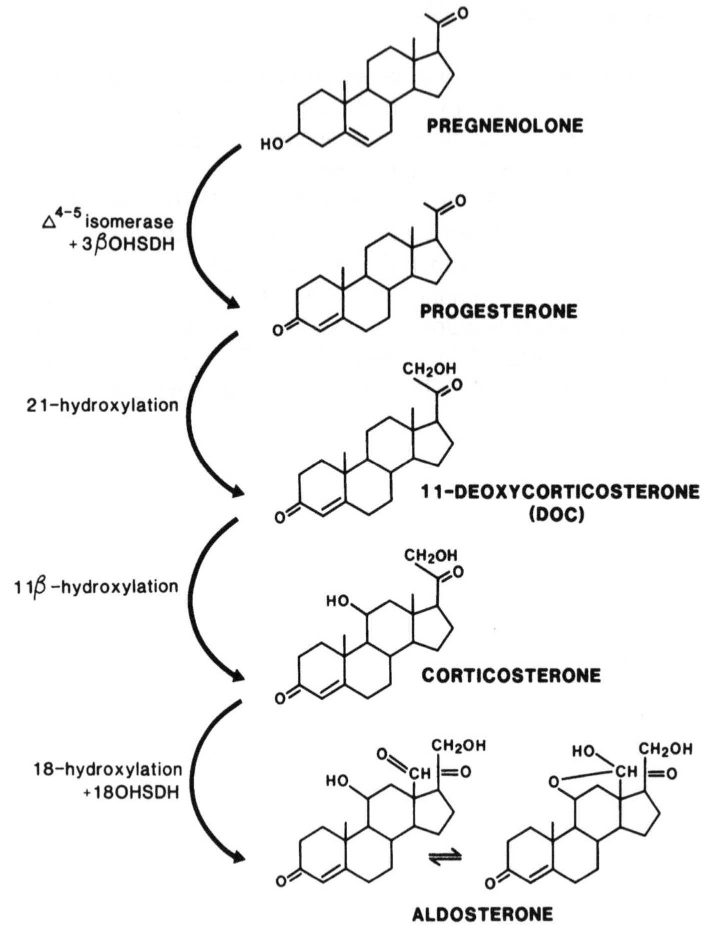

Fig. 3. Biosynthesis of progesterone (placenta & corpus luteum) and aldosterone (zona glomerulosa of adrenal cortex) from pregnenolone

Synthesis of Androgens and Oestrogens

Androgens and oestrogens are primarily synthesised in the gonads, although the adrenal cortex also secretes these steroids, from 17-hydroxypregnenolone and 17-hydroxyprogesterone, as illustrated in Figure 4. With the exception of the oestrogens, which have an aromatic A ring, all active secreted steroid hormones contain a Δ^4-3-oxo grouping in the A ring, the reduction of which normally leads to inactivation.

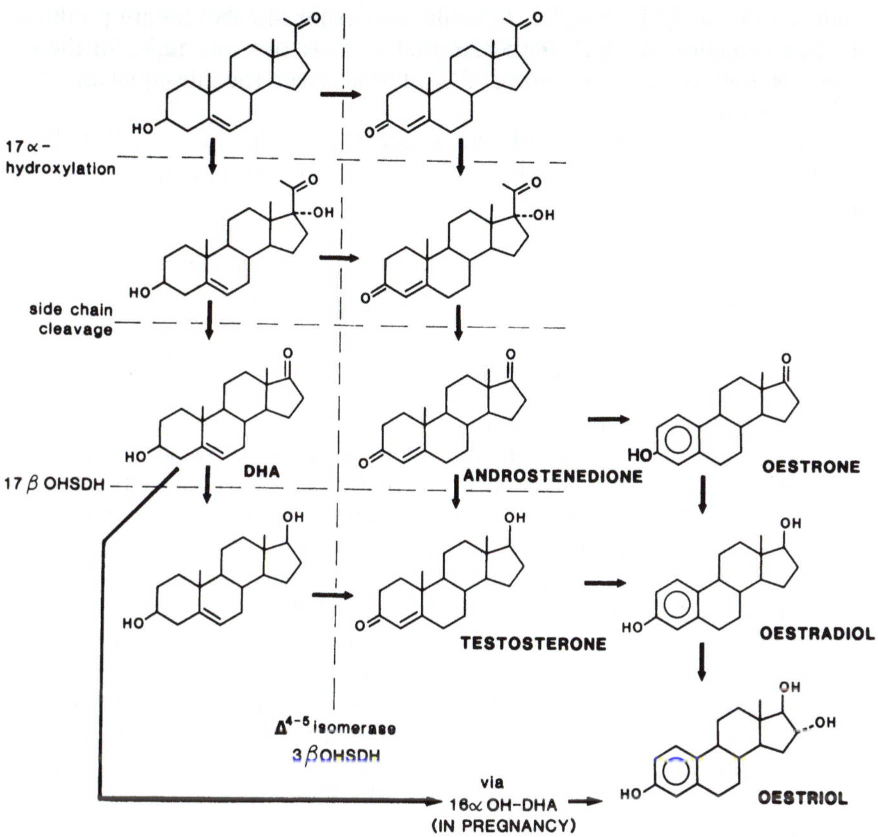

Fig. 4. Biosynthesis of the androgens and oestrogens

Catabolism in the Peripheral Tissues

The secreted steroid hormones are metabolised in their target cells and in the peripheral tissues. In the main, these catabolic reactions render the steroid molecule more water-soluble and include conjugation on hydroxyl groups with sulphuric and glucuronic acids. Mixed glucuronides and sulphates can occur, particularly with the oestrogens (Levitz & Young, 1977). There are four major catabolic reactions of the C_{21} and C_{19} steroids (see Gower, 1984).

1. Oxidation of the 11β-hydroxyl to an oxo group (e.g. cortisol - cortisone) by the enzyme 11β-hydroxysteroid dehydrogenase. Oxidation of the 11β-hydroxyl renders the molecule inactive.
2. Reduction of the 20-oxo group to give either 20α- or 20β-hydroxyls. There are two enzymes which catalyse this reaction, 20α- and 20β-hydroxysteroid dehydrogenase.
3. Tetrahydroreduction of the A ring. Four isomers can be produced when the Δ^4 double bond is reduced (5α-H & 5β-H) and when the 3-oxo group is

reduced (3α- & 3β-hydroxyls). Actually, two major metabolites are produced by this reduction of which the $5\beta,3\alpha$ predominates over the $5\alpha,3\alpha$ for the C_{21} steroids. Reduction of the A ring of C_{19} steroids given roughly equal amounts of the two isomers.
4. Side-chain cleavage of the 17-hydroxylated C_{21} steroids by the C17,20-lyase enzyme system. Approximately 5–10% of secreted cortisol is metabolised in this way.

These four catabolic pathways are not mutually exclusive and thus it can be seen that there are a large number of potential metabolites from a single secreted steroid. In addition, a small proportion of the C_{21} steroids are converted to C_{21} carboxylic acids (Monder & Bradlow, 1980). Testosterone can also be metabolised by hydroxylation at various sites (van der Hoeven, 1981), to produce metabolites with less androgenic activity.

The catabolism of the oestrogens does not follow the pathways given above. Instead, more complex reactions occur, involving further hydroxylations at C2 and C4 and leading to the so-called "catechol" oestrogens (reviewed by MacLusky et al., 1981). There are a multiplicity of different reactions which can occur and it is beyond the scope of this chapter to discuss this aspect of steroid catabolism, which has been reviewed by Fotherby (1984a). Oestrogens, like other steroids, undergo entero-hepatic recirculation and can thus be further modified by gastrointestinal flora before being reabsorbed. This may be an important process in the catabolism of oestrogens but is of less quantitative importance in the catabolism of C_{21} and C_{19} steroids (reviewed by Honour, 1984). Further complexities in steroid synthesis and catabolism occur during pregnancy (Fotherby, 1984b), in the foetus and neonate (Shackleton, 1984), and in pathological situations where the pattern of secretion is disturbed.

Steroid Analysis by HPLC

HPLC in the analysis of steroids is largely confined to the use of micro-particulate silica (3–10 μm), eluted with organic solvents of low polarity (so-called straight-phase systems), and reversed-phase systems, using polar solvents (including water) and packing materials (silica or spherical bead supports) to which bonded surface coats have been applied. A common reversed-phase material is bonded C_{18} (ODS, octadecasilane). There are a considerable number of different supports and reversed-phase coatings which can be utilised, and 200 of these have been summarised in an excellent review by Majors (1980).

A variety of different columns have found favour with different investigators. Schoneshofer & Dulce (1979) examined a number of systems and decided on the slightly polar DIOL bonded phase, which has been used in all subsequent studies (e.g. Schoneshofer, Fenner & Dulce,, 1981; Schoneshofer & Weber, 1983), except in those on the excretion of 20α- and 20β-dihydrocortisol (Schoneshofer & Weber, 1982; and Schoneshofer, Weber & Nigam, 1983). In this case since the DIOL column did not resolve the 20α- and 20β-isomers, a reversed-phase ODS column was required. Reversed-phase systems are the most widely used, and the silica straight-phase systems which were utilised in the early years are now losing popularity. Although ODS is the most common bonded phase, other similar phases of differing chain lengths (C1–C22) have been used. While, in general, increasing the chain length increases retention times (Majors, 1980), the selectivity of steroid separations seems to be relatively unaffected (O'Hare & Nice, 1981). Other types of bonded phases have also been introduced for particular steroid separations, such as -NO_2 (Lantto, 1982), -NH_2 (McGinley, Braughler & Hall, 1982), PRP-1 (a polystyrene divinylbenzene, Oliphant & Bowers, 1982), -CN (Tan, Navennec & Gevry, 1984). Metal-impregnated matrices (with silver nitrate, called argentation chromatography) have been used for oestrogen separation (e.g. Tscherne & Capitano, 1977). Surprisingly, -CN bonded columns, which show selective retention of steroids with oxo groups and have been successfully applied in the vitamin D field (Jones, 1983), have not been widely applied in the steroid field (but see Nakamura & Yakata, 1982 and 1983; and Ando et al., 1986). The mechanism of steroid separations on HPLC varies from adsorption in straight-phase systems to partition in reversed-phase systems, although in individual systems the division is probably not as simple as this and a complex interplay of mechanisms is involved. Straight-phase eluents containing significant quantities of water are clearly introducing a considerable element of partition.

Columns can be prepared in the laboratory or bought directly from the manufacturer. It would be expected that columns packed with similar materials would show similar retention characteristics. That this is not so, has been clearly shown by Nice & O'Hare (1978) in a study of three reversed-phase ODS columns (Fig. 5). O'Hare, Nice & Capp (1980) and O'Hare & Nice (1981) also demonstrated that the degree of capping (coverage of active polar groups of the silica support by the bonded phase, or by subsequent silanisation) varied widely between packing materials supplied by different manufacturers and that inadequately capped supports were not suitable with some 18-hydroxylated corticosteroids (see Fig. 6). Helboa (1986) studied eight straight-phase silica and six reverse-phase C_{18} coated silica systems and concluded that straight-phase silica, dynamically modified by cetyltrimethyl ammonium bromide in the eluent was more reproducible for the separation of cortisol, corticosterone acetate and a number of synthetic steroids.

O'Hare & Nice (1981) suggest that the best separation is achieved with columns at elevated temperature (45°) and noted that, when their separations were carried out at room temperature, reproducibility decreased, making identification of peaks from their retention times more difficult. Although there are reports of HPLC at elevated temperatures (30°; Kobayashi et al., 1984, 35°, Kawasaki, Maeda & Tsuji, 1979, 37°, Imaizumi et al., 1984, 40°; Cavina et al., 1984, 45°; Schoneshofer & Dulce, 1979; Watanabe & Yoshizawa, 1983a and b; Bucala, Fishman & Cerami, 1984, 50°; Gotelli et al., 1981; Dawson, Kontur & Monjan, 1985, 52°; Patterson et al., 1984), there is little agreement about the best temperature for steroid HPLC, and for the majority of separations the temperature is not specified and is thus assumed to be ambient. Interestingly, Patterson et al. (1984), who carried out HPLC at the highest temperature, 52°, felt the need to interpolate a short "solvent-conditioning" column of silica between pump and injector to saturate the solvent with silica and thus protect the μBondapak C_{18} analytical

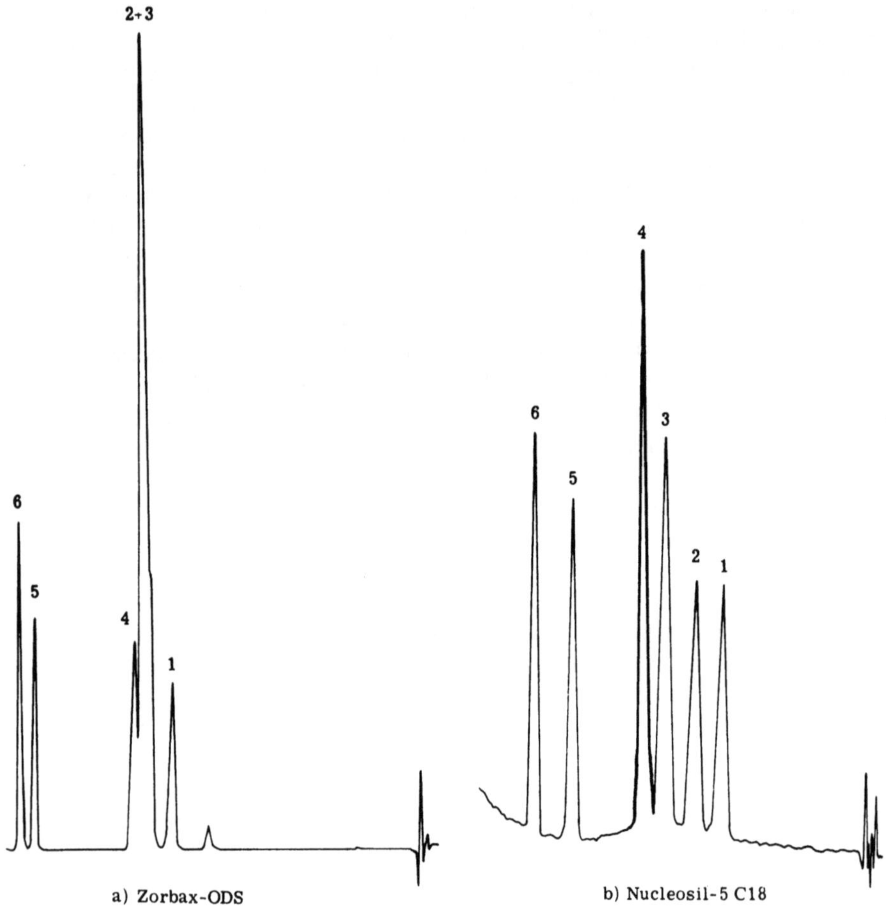

Fig. 5

column. There have been few reports of HPLC carried out at low temperatures, although there is some evidence that temperatures of 0–4° improve selectivity of microparticulate silica (Rhodes et al., 1983). In an investigation of the biosynthesis of 18-oxocortisol by frog interrenal using a reversed-phase μBondapak C_{18} column Ulick, Chu & Land, (1983) commented that enhanced resolution was observed at 0°. In contrast, Schoneshofer & Dulce (1979) showed improved separation of isomeric steroids by raising the column temperature to 40°.

In addition to temperature and column material, solvent flow rate and composition are important considerations when devising a separation system for a steroid mixture. Viscosity of common solvents may impose restrictions on solvent flow rates, which are usually in the 0.5–2 ml/min range. Probably solvent composition has the most profound effect on steroid separations on a given column, and apparently minor changes in solvent can substantially alter the separations achieved. O'Hare & Nice (1981) give a very good example of the effect

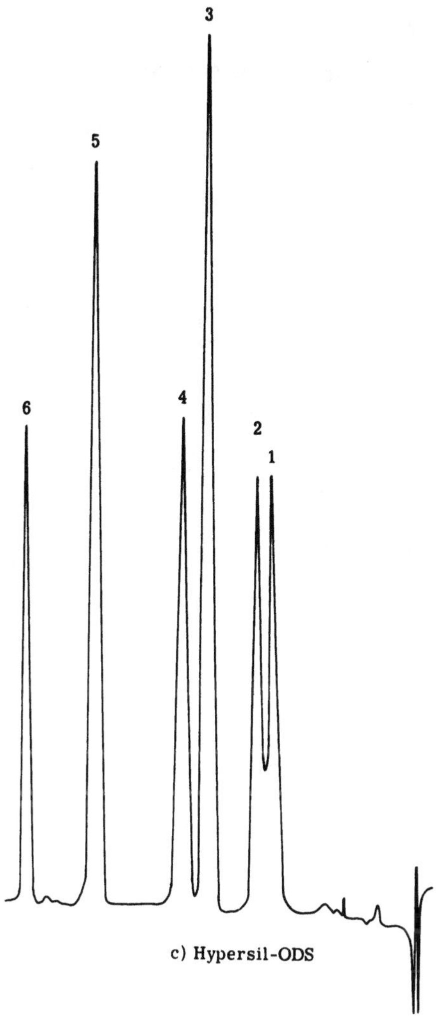

Fig. 5. Selective effects of reversed-phase column packings on the separation of steroid hormones by HPLC. Chromatograms were obtained under identical conditions of gradient elution (32–100% acetonitrile over 50 min). Wide differences in effective separations are seen, although the column packings are ostensibly the same. 1 = 17α-hydroxy-20α-dihydroprogesterone; 2 = testosterone; 3 = androstenedione; 4 = 17α-hydroxyprogesterone; 5 = 20α-dihydroprogesterone; 6 = progesterone. (Nice & O'Hare, 1978 with permission of authors and publisher)

on the separation of a number of steroids of altering from methanol/water to acetonitrile/water during reversed-phase HPLC. The selection of solvents for HPLC is restricted by the need to ensure, if reasonable sensitivity is required with UV detection, that the solvent does not have significant absorbance at the wavelength selected for detection. Solvent systems and columns for the separation of the steroid hormones are considered in the appropriate section below.

A further limitation of HPLC for steroid analysis in biological samples has been the difficulty of detection with sufficient sensitivity. The commonest detector attached to HPLC systems is the UV detector with either fixed (usually 254 nm) or variable wavelength. Although oestrogens and Δ^4-3-oxo steroids do

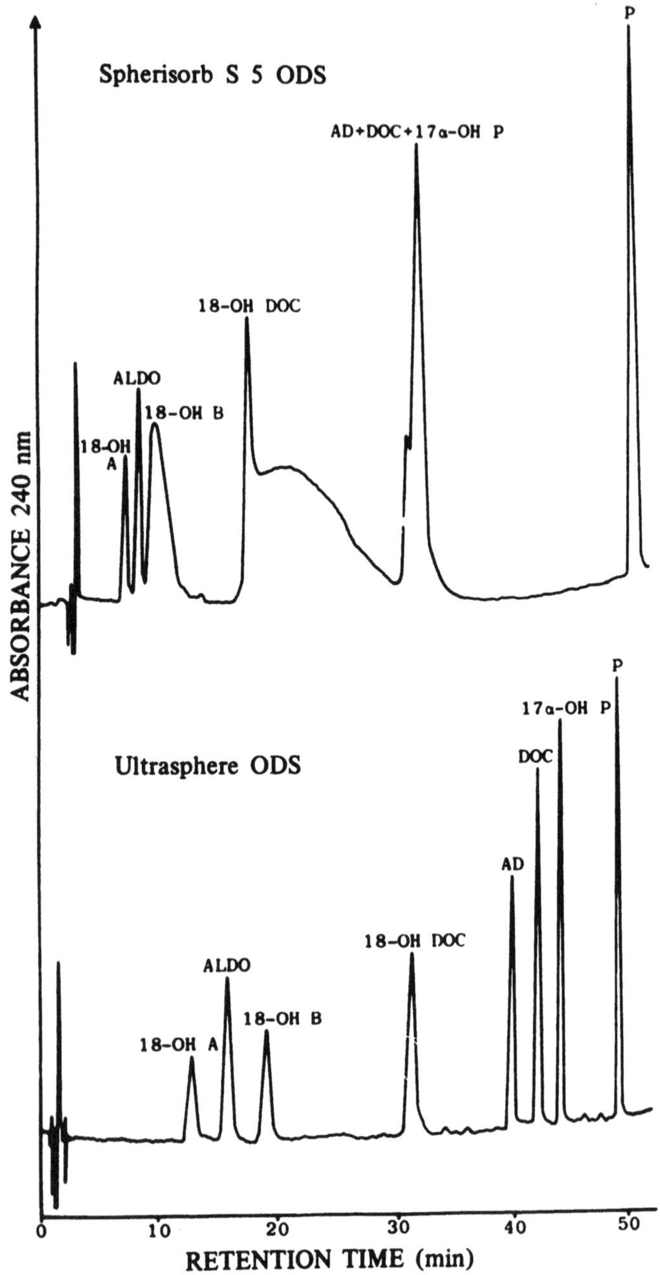

Fig. 6. The behaviour of 18-hydroxylated and some other adrenal steroids during HPLC using methanol/water solvent gradients on partially capped (Spherisorb S5 ODS) and fully covered (Ultrasphere ODS) columns. 18-OHA: 18-hydroxy-11-dehydrocorticosterone, ALDO: aldosterone, 18-OHB: 18-hydroxycorticosterone, 18-OHDOC: 18-hydroxy-11-deoxycorticosterone, AD: 4-androstenedione, DOC: 11-deoxycorticosterone, 17α-OHP: 17-hydroxyprogesterone, P: progestone. (From O'Hare & Nice, 1981, with permission of authors and publisher)

have absorbance maxima in the UV, many steroids do not. Sensitivity (i.e. minimum detectable limit = x2 background) of UV detection is usually around 5 ng, but this may be affected by the composition of the eluent. Refractive index detectors (e.g. Sunde & Lundmo, 1982) are relatively insensitive and cannot be used in gradient elution. A variety of approaches have been adopted to increase sensitivity of detection and these will be discussed in the appropriate section below. One interesting approach for the detection of Δ^5-3β-hydroxysteroids used immobilised 3β,17β-hydroxysteroid dehydrogenase and measured the reduction of NAD by fluorescence of NADH (Wu et al., 1986) in an on-line post-column system.

HPLC of C_{21} Steroids

The corticosteroid which, in normal circumstances, circulates in human plasma in highest concentration, is cortisol. Because of its relatively high concentration and ease of estimation, measurement of plasma cortisol level as a means of assessing adrenocortical function has been used for some years. Many methods for its estimation in plasma and other body fluids have been described using HPLC methodology. These are summarised in Table 2. An examination of these methods reveals that they fall into two categories:

- straight-phase silica columns and elution with a chlorinated hydrocarbon (usually dichloromethane), ethanol (or methanol) and water mixtures. Occasionally, other solvents are added, such as tetrahydrofuran, with which Canalis, Reardon & Caldarella (1982) improved the separation of interfering substances from cortisol;
- reversed-phase systems (primarily C_{18} ODS bonded phases), which are more popular and use methanol/water or acetonitrile/water for elution. For detection and quantitation, UV absorbance at 254 nm is common and thus it should be possible to detect cortisol in extracts from as little as 0.1 ml in serum or plasma (5-20 ng). However, because of the problem of non-specific background in biological samples, it is usual to analyse substantially more sample (around 1 ml) and it is not obvious why the method of Matsuzawa, Sugimoto & Ishiguro (1981) is so sensitive.

While it is clear that the interpolation of an HPLC step prior to radioimmunoassay of plasma or urinary free cortisol greatly improves specificity (Schoneshofer et al., 1980; Schoneshofer, Fenner & Dulce, 1980; Schoneshofer, Skobolo & Dulce, 1981; Schoneshofer, Weber & Nigam, 1983), it is arguable that similar improvements could be obtained with other simpler techniques such as differential solvent extraction or use of small Celite partition columns. From an analytical point of view, the greater selectivity/resolving power of HPLC does not seem appropriate for the routine estimation of a single steroid, although for isolation and identification of unknown metabolites it is clearly extremely valuable. For the separation and subsequent analysis of a number of different steroids from a single sample (i.e. "profiling" so successfully carried out with capillary GC - e.g. Shackleton, Taylor & Honour, 1980), HPLC has found application (see Shackle-

Table 2. Some HPLC methods for the measurement of cortisol in blood and urine

Reference	Column	Solvent	Detection method	Volume of plasma (P), serum (S) or urine (U) used (ml)
Hesse, Pietrizik & Hoetzel (1974)	Silica gel	Dichloromethane/ ethanol/water	UV 240 nm	1 ml (P)
Trefz, Byrd & Kochen (1975)	Zorbax SIL	Dichloromethane/ ethanol/water	UV 240 nm	1 ml (P)
Van den Berg et al. (1977)[1]	Nucleosil-NO_2	Dichloromethane/ 2-propanol/water	UV 240 nm	1 ml (P)
Scott & Dixon (1979)	Hypersil-ODS	Methanol/water	UV 239 nm	1 ml (P)
Reardon, Caldarella & Canalis (1979) Canalis, Caldarella & Reardon (1979)	μBondapak C18	Methanol/water	UV 254 nm	1 ml (S)
Kabra, Tsai & Marton (1979)	μBondapak C18	Acetonitrile/ phosphate buffer	UV 254 nm	1 ml (P)
Smith (1979)	Spherisorb C18	Acetonitrile/ wate/acetic acid	UV 254 nm	0.5 ml (S)
Kawasaki, Maeda & Tsuji (1979)	Hitachi gel 3042	Dichloromethane/ ethanol/water	Fluorescence of dansyl derivatives	0.5 ml (P) 1 ml (S)
Goehl, Sunderesan & Vandlament (1979)	Lichrosorb Si60	Dichloroethane/ butanol/water	Fluorescence of dansyl hydrazones	1 ml (P)
Frey, Frey & Benet (1979)	5 μm Silica	Dichloromethane/ methanol/water/ tetrahydrofuran/ acetic acid	UV	1 ml (P)
Rose & Jusko (1979)	Zorbax SIL	Dichloromethane/ methanol	UV 254 & 280 nm	1 ml (P, U & saliva)
Cavina et al. (1979)	Lichrosorb 100	Chloroform/water/ methanol or chloroform/ 2-propanol	UV 254 nm	0.25– 1 ml (P)
Schoneshofer et al. (1980)	Merck DIOL	Hexane/ 2-propanol	RIA	(U)
Scott, Chakraborty & Marks (1980)	Partisil	Dichloromethane/ ethanol/water	UV 239 nm	1 ml (P)
Petersen, Nation & Ashley (1980)	RP-8	Methanol/water	UV 254 & 240 nm	1 ml (P)
De Vries et al. (1980)	10 μm Silica	Chloroform/iso- octane/methanol/ water	UV 254 nm	0.6 ml (P)

Table 2. (Continued)

Reference	Column	Solvent	Detection method	Volume of plasma (P), serum (S) or urine (U) used (ml)
Matsuzawa, Sugimoto & Ishiguro (1981)	μPorasil	Dichloromethane/ ethanol/water	UV 254 nm	0.1 ml (S)
Schoneshofer, Skobolo & Dulce (1981)	Merck DIOL	Hexane/ 2-propanol	UV 254 nm & RIA	1 ml (S)
Gotelli et al. (1981)	μBondapak C18	Tetrahydrofuran/ acetonitrile/ water/phosphate buffer, pH 4.4	Fluorescence of sulphuric acid/ethanol fluorophores	0.05 ml (S)
Schoeneshofer, Fenner & Dulce (1981)	Merck DIOL	Hexane/2-propanol	RIA	1 ml (S)
Nakamura & Yakata (1982)[2]	Zorbax SIL Zorbax CN	Dichloromethane/ ethanol/water. Methanol/water	UV 254 nm	1 ml (U)
Shihabi, Andrews & Scaro (1982)	5 μm ODS	Acetonitrile/ phosphate buffer	Fluorescence of sulphuric acid induced fluorophores	0.5 ml (U)
Ost et al. (1982)	Nucleosil-NO_2	Dichloromethane/ ethanol	UV 254 nm	1 ml (S)
Lantto (1982)	Nucleosil-NO_2	Dichloromethane/ ethanol	UV 254 nm	2 ml (U)
Caldarella, Reardon & Canalis (1982)	μBondapak C18	Methanol/water	UV 242 & 254 nm	1 ml (S)
Canalis, Reardon & & Caldarella (1982)	μBondapak C18	Methanol/water/ tetrahydrofuran	UV 254 nm	2 ml (U)
Lewbart & Elverson (1982)	Zorbax ODS	Methanol/water	UV 254 nm	1 ml (U)
Lambert et al. (1983)	RSIL C18 HL	Methanol/water	UV 254 mn	0.2 ml (S)
Schoneshofer & Weber (1983)	DIOL column	Hexane/2-propanol	RIA	2 ml (U)
Goto et al. (1983)	Cosmosil 5 SL	Hexane/ethyl acetate	Fluorescence as 9-anthroyl nitriles	0.1 ml (S)
Hofreiter et al. (1983)	μBondapak C18	Methanol/water	UV 254 nm	1 ml (P)

Table 2. (Continued)

Reference	Column	Solvent	Detection method	Volume of plasma (P), serum (S) or urine (U) used (ml)
Nilsson (1983)	Affinity column (antiserum linked to Lichrosorb Si 60 and then onto Lichrosorb RP18)	Water Methanol/water	UV 254 nm	2 ml
Farwell et al. (1983)	Ultrasphere C8	Methanol/water	UV 242 nm	3 ml (P)
Ebling, Szefler & Jusko (1984); Carson & Jusko (1984)	Zorbax SIL	Dichloromethane/ hexane/ethanol/	UV 254 nm	1 ml (P)
Loche et al. (1984)	Waters Radial Compression C18	Methanol/water	UV 254 nm	0.5 ml (S)
Dawson, Kontur & Monjan (1984)	Octyl C8	Tetrahydrofuran/ acetonitrile/ water	UV 239 nm	1.0 ml (P)
Schoneshofer et al. (1985)	ODS-Hypersil	Acetonitrile/ water	UV 254 nm	0.2 ml (U)
Gutenberger et al. (1985)	Ultrasphere C8	Methanol/ water	UV 242 nm	2.5 ml (P)
Stoner et al. (1986)	Novapak C18	Methanol/ water	UV 254 nm	0.5–1.0 ml (S)

[1] Comparison of this method with six other (Thijssen et al., 1980)
[2] Clinical application of this method (Nakamura & Yakata, 1983) and extension to measure 6β-hydroxycortisol in addition to cortisol (Nakamura & Yakata, 1985) using 2 ml urine.

ton, 1986). Schoneshofer, Fenner & Dulce (1981) have separated eleven adrenal steroids in plasma prior to quantitation by RIA and, using the same HPLC system (Fig. 7), fifteen unmetabolised "free" steroids were measured in urine (Schoneshofer & Weber, 1983) and in serum (Eibs & Schoneshofer, 1984). The selectivity of HPLC has also been exploited for the simultaneous measurement of a number of adrenal and testicular steroids in in vitro experiments, O'Hare et al. (1976) reported the resolution of 43 naturally occurring steroids using three different solvent systems. Resolution of 25 different 18-hydroxylated steroids using a single reversed-phase ODS column and concave gradient elution with methanol/water and dioxane/water has been reported (Figure 8 – O'Hare, Nice & Capp, 1980), and O'Hare & Nice (1981) have investigated 13 different reversed-

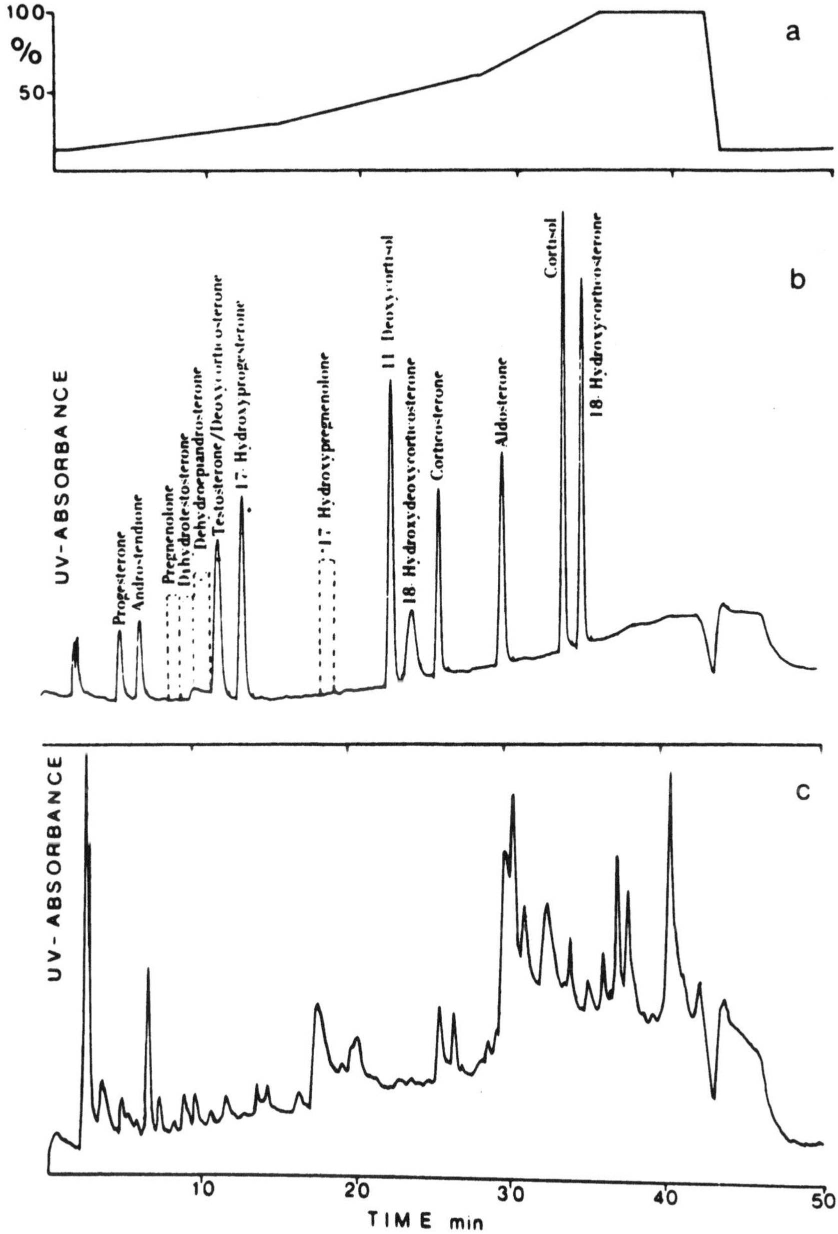

Fig. 7. UV chromatogram of a mixture of steroid standards (b) and of the ether extract from a normal urine specimen (c). Amount of each steroid injected was 500 ng (b). Upper part (a) represent gradient profile applied and indicates % of solvent B (n-hexane:2-propanol::3:1) in a mixture of solvent A (n-hexane) and solvent B. A DIOL column was used at 40° (from Schoneshofer & Weber, 1983 with permission of authors and publisher)

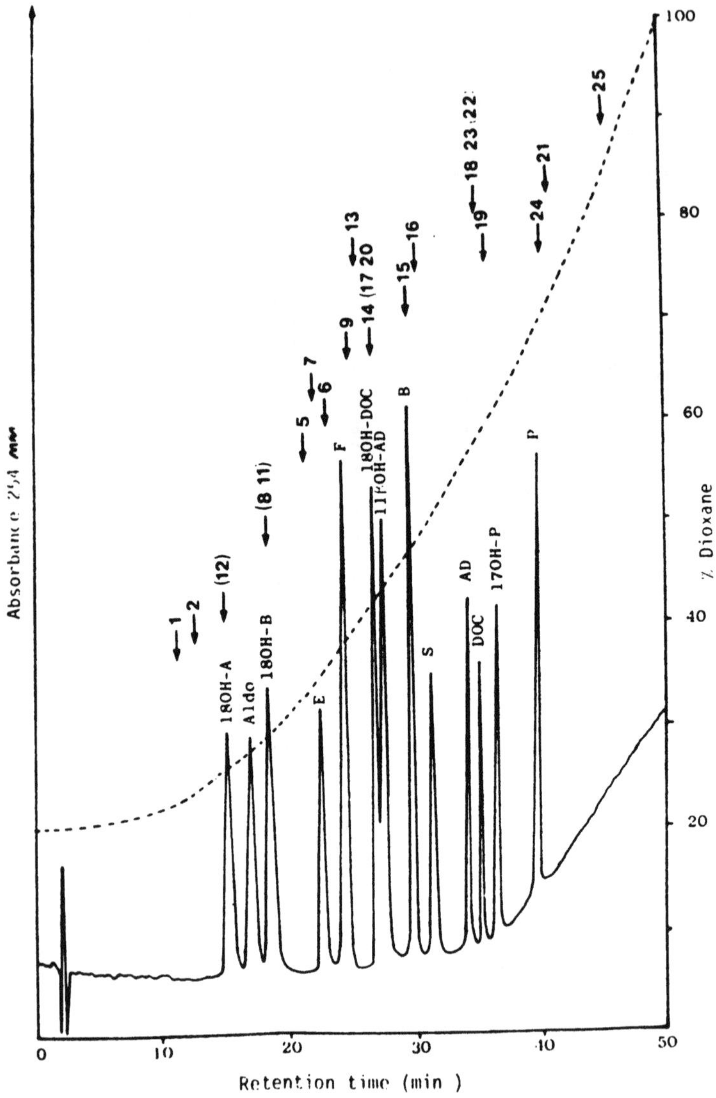

Fig. 8. Retention times of a number of steroid hormones during HPLC on a 150 × 4.6 mm ODS-Hypersil column with dioxane-water gradient elution 8a and with methanol-water gradient elution 8b. E: cortisone, F: cortisol, 11βOH-AD: 11β-hydroxy-4-androstenedione, B: corticosterone, S: 11-deoxycortisol. Other abbreviations are given in subscript to Figure 6. The numbers indicate the positions of standard steroids, whose identities can be found by reference to original publication (from O'Hare, Nice & Capp, 1980 with permission of authors and publisher)

phase columns for the separation of 23 different steroids. Cavina et al. (1979) adopted a similar approach but used a straightphase silica column with the result that the number of steroids completely separated was less than that achieved with reversed-phase systems. However, Cavina et al. (1984) have carried out a

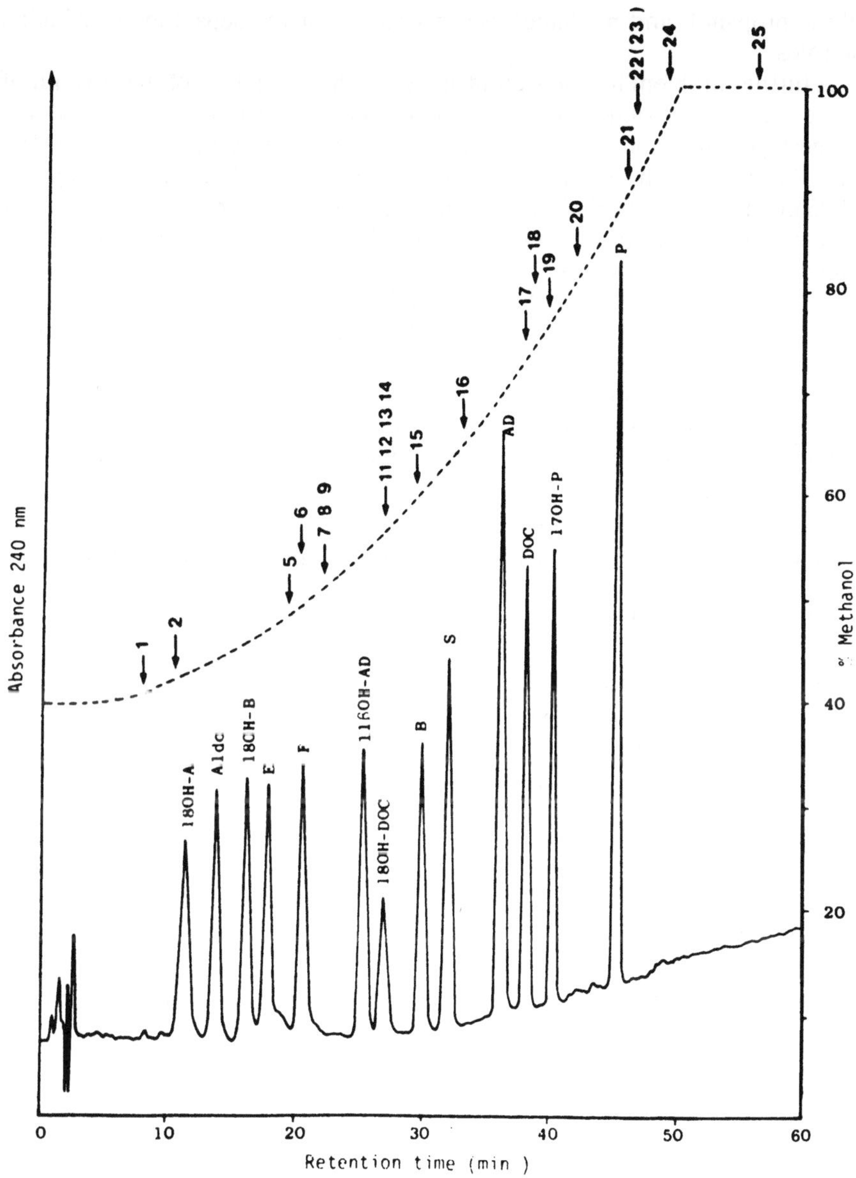

Fig. 8

comprehensive study of the separation of 19 different C_{19} and C_{21} steroids using two different straight-phase and four reversed-phase columns, a number of different solvent systems with isocratic, linear- and concave-gradient elution at room temperature and 45°, recording capacity factors for each system. Capp & Simonian (1985) have described a multi-step gradient elution using reversed-phase propanol- and methanol- water systems for the separation of 25 adrenal steroids.

A further problem for the steroid analyst is the separation of the large number of reduced metabolites of pregnenolone and progesterone. The analysis of progestagen metabolites by HPLC has been reviewed by Purdy et al. (1981) and Lin, Heftmann & Hunter (1980) have undertaken a careful study of the separation of 25 pregnane diones, diols and monohydroxymonoketones, using straight- and reversed-phase columns. Francis and Kinsella (1985) studied the separation of C_{21} metabolites of progesterone by reversed-phase HPLC with similar results to those reported for C_{19} steroids by Lin & Heftmann (1982). Allenmark & Boren (1981) studied the separation of chemically reduced progesterone and pregnenolone isomers and applied these studies to an investigation of pregnenolone and progesterone metabolism in vitro (Allenmark et al., 1981a). A number of HPLC systems have been described for the analysis of corticosteroids other than cortisol and some of those not already mentioned above are listed in Table 3.

Table 3. Some HPLC methods for the measurement of various corticosteroids in blood or urine

Steroid	Reference	Column	Solvent	Detection
11-deoxycortisol in plasma	Reardon, Caldarella & Canalis (1979), Canalis, Caldarella & Reardon (1979)	μBondapak C18	Methanol/water	UV 254 nm
Cortisone, aldosterone, corticosterone, 11-dehydro-corticosterone in plasma	Cavina et al. (1979)	Zorbax SIL Lichrosorb Si 1000	Chloroform/ methanol/water	UV 254 nm
Cortisone in plasma	Frey, Frey & Benet (1979)	Silica 5 μm	Dichloromethane/ methanol/acetic acid/tetrahydro-furan	UV 254 nm
11-deoxycortisol in plasma	De Vries, Lomecky-Janousek & Popp-Snijders (1980)	Silica	Chloroform/iso-octane/methanol/ water	UV 254 nm
6β-hydroxysteroids in urine	Kraan, Derks & Drayer (1980)	μPorasil	Chloroform/ methanol/water	UV 280 nm
6β-hydroxy-cortisol in urine	Geber-Taras, Park & Ohnhaus (1981)	Lichrosorb Si 60	Dichloromethane/ n-hexane/ethanol/ water	UV 240 nm
19-nor-11-deoxycortico-sterone in urine	Dale, Holbrook Melby (1981)	Zorbax ODS	Methanol/water	UV 254 nm

Table 3. (Continued)

Steroid	Reference	Column	Solvent	Detection
9 cortico-steroids in plasma	Schoneshofer, Fenner & Dulce (1981)	Merck DIOL	Hexane/2-propanol	RIA
Pregnenolone & 17-hydroxy-pregnenolone in serum	Schoneshofer, Maxeimer & Fenner (1981)	DIOL	Hexane/2-propanol	RIA
17-hydroxy-corticosteroids in urine	Kawasaki, Maeda & Tsuji (1982a)	Zorbax SIL	Dichloromethane/ethanol/water	Fluorescence of dansyl derivatives (pre-column)
	Seki & Yamaguchi (1984)	Finepak C18	Methanol/water	Fluorescence of benzamidine derivatives (post-column)
Aldosterone & 18-hydroxy-corticosterone in urine	Schoneshofer, Weber & Dulce (1982) and Schoneshofer & Weber (1983)	Merck DIOL	Hexane/2-propanol	UV 254 & RIA
Corticosterone in rat plasma	Hofreiter et al. (1982)	μBondapak C18	Methanol/water	UV 254 nm
20α-reduced cortisol in urine*	Schoneshofer & Weber (1982), Schoneshofer, Weber & Nigam (1983)	Hypersil-ODS	Methanol/water	UV 254 nm
Cortisone & 6β-hydroxy-cortisol in urine	Weber & Schoneshofer (1982)	Hypersil-ODS	Methanol/water	UV 254 nm
Cortisone in urine	Lewbart & Elverson (1982)	Zorbax ODS	Methanol/water	UV 254 nm
Cortisone in serum	Goto, Shamsa & Nambara (1983)	Cosmosil 5SL	Hexane/ethyl acetate	Fluorescence of 9-anthroyl nitrile derivatives
6β-hydroxy-cortisol in urine	Goto et al. (1983)	Cosmosil 5SL Lichrosorb RP18	Hexane/ethyl acetate	Fluorescence of 9-anthroyl nitrile derivatives
15 steroids in urine	Schoneshofer & Weber (1983)	Merck DIOL	Hexane/2-propanol	RIA
Corticosterone in rat & mouse plasma	Shimizu, Amagaya & Ogihara (1983)	Jasco FINE SIL C_8-5	Acetonitrile/sulphuric acid	UV 254 nm
15 steroids in serum	Eibs & Schoneshofer (1984)	Merck DIOL	Hexane/2-propanol	RIA

Table 3. (Continued)

Steroid	Reference	Column	Solvent	Detection
Corticosteroids in urine	Seki & Yamaguchi (1983)	Finepak C18	Methanol/water	Fluorescence of glycinamide derivatives (post-column)
18-hydroxy-corticosterone in plasma	Imaizumi et al. (1984)	Finepak SIL C18-5	Methanol/HCl	UV 245 nm
11-deoxycortisol & 17-hydroxy-progesterone in serum	Weisman et al. (1984)	μBondapak C18	Acetonitrile (or methanol)/ water	UV 254 nm
6β-hydroxy-cortisol in urine	Dumont, Sclavons & Desager (1984)	Rsil C18 HL	Acetonitrile/ KH_2PO4 and trichloracetic acid	UV 243 nm
Cortisol and 6β-hydroxy-cortisol in urine	Nakamura & Yakata (1985)	Zorbax SIL & Zorbax CN	Dichloromethane/ ethanol/water methanol/water	UV 242 nm
Progesterone in serum	Lagana et al. (1986)	C18 reverse phase-5μ	Acetonitrile/ water	UV 242 nm
17OHprogesterone, 11-deoxycortisol and cortisol in serum	Stoner et al. (1986)	Novapak C18	Methanol/ water	UV 254 nm
20-reduced cortisol and cortisone in urine	Schoneshofer et al. (1986a)	Hypersil-ODS(5μ)	Acetonitrile/ water	UV 254 nm

* Clinical application in diagnosis of chronic hypercorticoidism (Schoneshofer et al. 1986b)

Although the majority of methods of detection have utilised UV absorbance at 240 or 254 nm, these methods are only applicable to those naturally occurring corticosteroids containing the Δ^4-3-oxo grouping (absorbance maximum at 240 nm). Attempts have been made to utilise other detection methods, apart from RIA. Two methods which utilise the pre-column formation of sulphuric acid-induced fluorophores have been used for the estimation of cortisol in plasma (Gotelli et al., 1981) and urine (Shihabi, Andrews & Scaro, 1982). This allowed the formation of single peaks and reduced the volume of plasma required for assay (0.05 ml; Gotelli et al., 1981) to levels comparable to those required in routine RIAs. Similar improvement in sensitivity was achieved using 9-anthroyl nitrile derivatives (Goto et al., 1983a). Other techniques of improving sensitivity, and presumably, specificity of detection have also been used: post-column derivatisation with glycinamide (Seki & Yamaguchi, 1983) or benzamidine (Seki & Yamaguchi, 1984), and pre-column formation of dansyl hydrazones (Kawasaki, Maeda & Tsuji, 1982a). Goehl, Sundaresan & Prasna (1979) had previously reported a method for the measurement of plasma cortisol using dansyl hydrazone derivative and fluorescence detection, although they found that two derivatives (syn and anti isomers) were formed. Horikawa, Tanimura & Tamura (1979) had used isonicotinic hydrazine which, in the presence of methanolic aluminium chloride, gave fluorescence derivatives. No increase in sensitivity over that achieved with UV absorbance was found, although an improvement in selectivity was claimed. A number of other detectors have been reported and are reviewed by Heftmann & Lin (1982). An interesting detection system, using 2-t-butylanthroquinone in the elution solvent and photoreduction fluorescence, has been reported, but its sensitivity is similar to that of UV absorbance (Gandelman & Birks, 1983). The formation of benzoyl derivatives, originally described by Fitzpatrick & Siggia (1973), has been exploited, together with micro-column HPLC (1 m $\times$ 0.22 mm ID packed with 3 μm "bonded spherical particles" and stepwise elution with increasing concentrations of acetonitrile in water) to produce urinary steroid profiles comparable to those achieved with capillary GLC (Novotny, Alasandro & Konishi, 1983 - see Figure 9). Novotny and his colleagues have further extended micro-column liquid chromatography to improve resolution and sensitivity, describing separations of C_{19} and C_{21} steroids as dansyl- and 7-(chlorocarbonylmethyl)-4-methylcoumarin derivatives and their glucuronide conjugates as dansyl derivatives using normal (Novotny et al., 1984) and laser-induced (Gluckman, Shelly & Novotny, 1984) fluorescence detection.

Considering the number of variables involved in setting up a suitable HPLC system for any particular separation, it can obviously be very difficult to achieve the maximum efficiency of separation of the components of a complex mixture on empirical grounds. There seem to have been very few systematic approaches to the optimisation of parameters for steroid separations by HPLC. One such approach has been "simplex optimisation" (Fast, Culbreth & Sampson, 1982) to optimise the separation of cortisol during the evaluation of a candidate definitive method for its estimation in plasma using gas chromatography-mass spectrometry (Patterson et al., 1984). Solvent systems for the resolution of polar adrenal steroids by reversed-phase HPLC have been optimised using computer-predicted retention times (D'Agostino et al., 1984), and Antle (1982) optimised solvent selectivity for six C_{21} steroids on Zorbax-NH_2 using the so-called "selectivity triangle" and computerised mapping techniques. Computer aided solvent optimisation in HPLC has recently been reviewed (D'Agostino et al., 1985).

Apart from the method of Nilsson (1983), who packed the injection loop with immobilised antiserum (linked to Lichrosorb Si) to purify cortisol before separation and measurement on a reversed-phase C_{18} column, there have been few attempts to incorporate purification or pre-concentration of C_{21} steroid analytes as part of the HPLC system. However an apparently cumbersome but comprehensive on line, extraction, evaporation, and injection (LEEI) system for corticosteroids in plasma has been described by Oka et al. (1984), although it is not clear just how practical this system would be. Nakamura & Yakata (1982) have utilised a combination of straight-phase Zorbax-SIL (for prepurification) and reversed-phase (for analysis) in the measurement of urinary free cortisol and

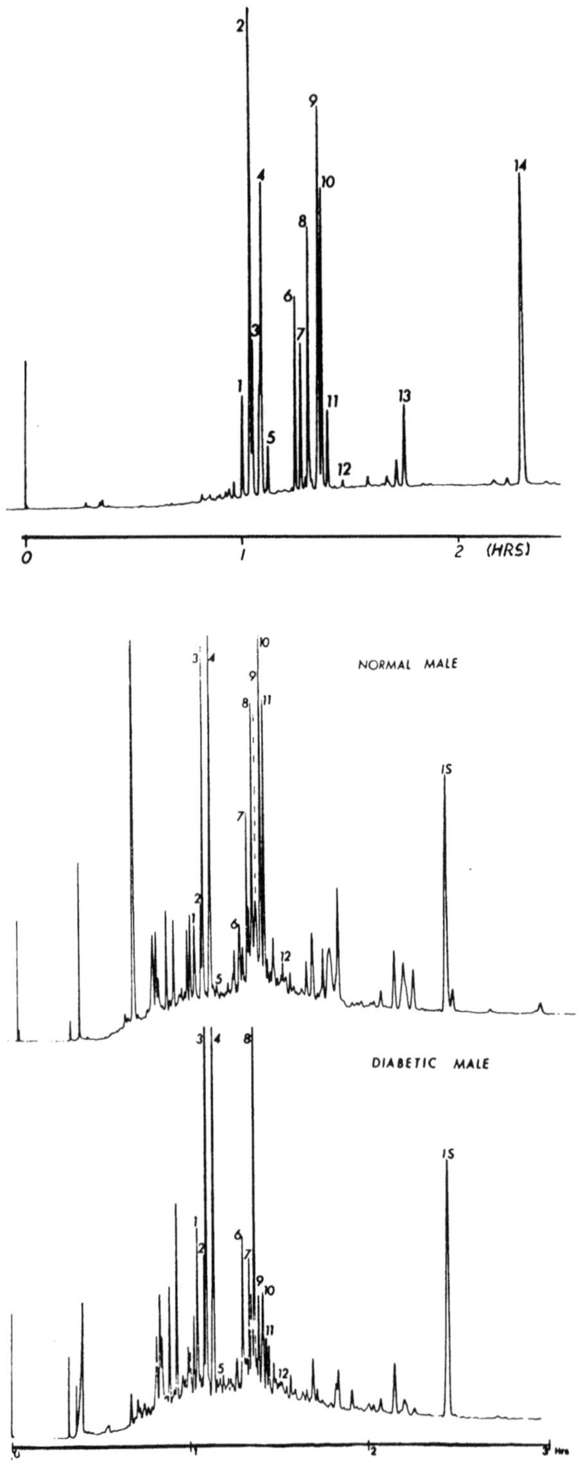

Fig. 9. Separation of 14 benzoylated steroid standards 9a and representative chromatographic metabolic profiles of urinary steroids from a normal 9b and diabetic human male 9c. Stepwise gradient conditions: 80% acetonitrile (ACN)/H_2O (15 min); 85% ACN/H_2O (14 min); 90% ACN/H_2O (15 min); 95% ACN/H_2O (18 min); 100% ACN (∞). Column was a fused silica capillary (1 m × 0.24 mm ID) packed with 3 μm bonded spherical particles. Detection by UV absorbance at 230 nm. (1) 11-hydroxyandrosterone; (2) 11-hydroxyaetiocholanolane; (3) allotetrahydrocortisol; (4) tetrahydrocortisol; (5) tetrahydrocortisone; (6) β-cortolone; (7) β-cortol; (8) α-cortolone; (9) α-cortol; (10) aetiocholanolone; (11) androsterone; (12) dihydroepiandrosterone; (13) pregnanetriol; (14) androstanediol; (15) internal standard (benzoylated 5α-androstan-3β,17βdiol) (from Novotny, Alasandro & Konishi, 1983 with permission of authors and publisher)

obtained much cleaner tracings after Zorbax-SIL prepurification. As an example of the possibilities of using HPLC systems to concentrate a steroid analyte present in low concentration prior to analysis by HPLC, Schoneshofer, Kage & Weber (1983) described a microprocessor-controlled automated HPLC system for the measurement of triamcinolone in urine (Fig. 10). This system has now been adapted for the automated analysis of urine free cortisol (Schoneshofer et al., 1985) and for 20-reduced metabolites of cortisol and cortisone in urine (Schoneshofer et al., 1986a). Considerable improvement over other pre-column purification procedures was claimed. Pre-purification HPLC systems have been utilised with other steroids, particularly with the oestrogens. These will be discussed later.

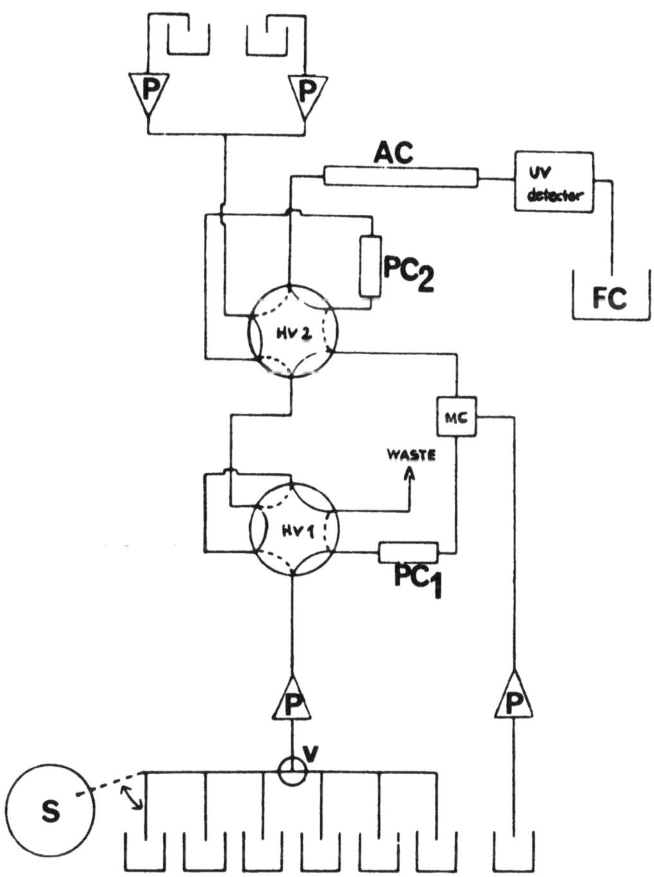

Fig. 10. Flow diagram of the fully automated HPLC system for the assay of triamcinolone in urine. The sample is initially pumped to PC1 (PRP-1, see Oliphant & Bowers, 1982) where the lipophilic material is retained on the reversed-phase matrix. The triamcinolone fraction is eluted from PC1 and delivered to PC2 (Silica-ODS). Finally the appropriate fraction is pumped to the analytical column (AC - ODS-Hypersil). P: pump, V & HV: valves, MC: mixing chamber, S: sampler, FC: fraction collector, PC: pre-column. AC: analytical column (from Schoneshofer, Kage & Weber, 1983 with permission of authors and publisher)

The application of HPLC to the investigation of C_{21} steroid metabolism has been limited for the reasons discussed above. However, the use of HPLC has been described, for example, in the diagnosis of congenital adrenal hyperplasia (Weisman et al., 1984; and Stoner et al., 1986), the identification of 19-oxygenated metabolites of deoxycorticosterone produced by rat adrenals (Gomez-Sanchez et al., 1982), studies on the isolation and purification of polar metabolites of aldosterone produced in the liver (Morris et al, 1983; McDermott, Latif & Morris, 1983), and the identification of 18-oxocortisol as a metabolite produced in the bullfrog interrenal (Ulick, Chu & Land, 1983). The HPLC systems in all these investigations were reversed-phase systems similar to those listed in Tables 2 and 3. Studies on the oxidation of corticosteroids to 20-hydroxy-21-oic acids (Monder et al., 1982; Han, Marandian & Monder, 1983) utilised the reversed-phase HPLC system for cortoic acids described by Farhi & Monder (1978) and p-bromophenacyl derivates of the carboxylic acids.

Substantial quantities of highly polar steroids are found in the urine of neonates. These steroids are not extracted efficiently by conventional solvent extraction procedures and thus they had been undetected until Derks & Drayer (1978a) used Amberlite XAD2 for their solid-phase extraction. Subsequently, 6α-hydroxylated steroids were isolated and quantified in neonatal urine (Derks & Drayer, 1978b) and in urine from pregnant and non-pregnant women (Kraan, Derks & Drayer, 1980) by HPLC and gas chromatography-mass spectrometry. This is a good example of the value of simple straight-phase HPLC systems (μPorasil, eluted with chloroform/methanol/water); 6β-hydroxysteroids were completely separated from 6α-hydroxysteroids, which were more polar.

Table 4 gives a selection of some further applications of HPLC in studying the endocrinology of the C_{21} corticosteroids. The HPLC separation of aldosterone metabolites has been reviewed by Morris & Tsai (1981).

Table 4. Some applications of HPLC in studies of C21 steroid metabolism

Application	Reference	Column	Solvent system
18-hydroxy-11-deoxycorticosterone and corticosterone in rat incubation media	Chan et al. (1977)	Bondapak phenyl/ Porasil B	Acetonitrile/ water
18-hydroxysteroids in adrenal tissue extracts	O'Hare, Nice & Capp (1980)	Various reversed phase systems	Methanol-, acetonitrile-dioxane-water systems
Δ^5-3β-hydroxy-dehydrogenase activity	Suzuki, Kadowaki & Tamaoki (1980)	μBondapak CN	Acetonitrile/ water
19-nor-DOC in regenerating rat adrenals	Dale, Holbrook & Melby (1980)	Zorbax ODS	Methanol/water

Table 4. (Continued)

Application	Reference	Column	Solvent system
Progesterone metabolites in bovine liver, kidney & milk	Purdy et al. (1980)	Chromegabond DIOL	2-Propanol/ heptane
Metabolism of aldosterone to more polar metabolites	Morris & Tsai (1981) Morris et al. (1983)	Zorbax C8	Methanol/water
Separation of pregnenolone and progesterone metabolites from human teratoma cell line	O'Hare et al. (1981)	Ultrasphere ODS & Lichrosorb RP18	Methanol/ & Acetonitrile/ water
Oxidation of corticosteroids to 21-oic acids	Purkaystha et al. (1982)	Zorbax ODS	Methanol/water
A-ring reduction of progesterone & formation of fatty acid esters	Albert, Prasad & Lieberman (1982)	C18 Corasil	Acetonitrile/ water
Isolation & identification of 18,19-dihydroxy-corticosterone	Okamoto et al. (1982) Fujii et al. (1984)	TSK-GEL LS310 ChemcosorbODS ChemcosorbODS	Dichloromethane/ ethanol/water Ethanol/water Acetonitrile/ water
Regulation of 21-hydroxylation in vitro	Hornsby (1982)	Regis C18	Methanol/ & acetonitrile/ water
Steroids in adrenal cell cultures	Ramirez, Millot & Maume (1982)	Micropak C18	Acetonitrile/ water
Corticosteroids in adrenal tissues and tumours	Saito et al. (1983)	Zorbax SIL Zorbax CN	Dichloromethane/ cyclohexane/ethanol Methanol/water
19-hydroxylation of 18-hydroxy-DOC by cytochrome P-450	Mormoi et al. (1983)	TSK-GEL LS310	Dichloromethane/ ethanol/water
Regulation of 18-hydroxy-corticosterone formation in bovine adrenal mitochondria	Kim et al. (1983)	TSK-310	Dichloromethane/ methanol/water
Effects of ACTH on monolayer adrenal cell culture	Kramer et al. (1983)	μBondapak C18	Methanol/water

Table 4. (Continued)

Application	Reference	Column	Solvent system
Identification of 19-nor-DOC in normal rat serum	Kobayashi et al. (1984)	Chemcosorb7ODS	Methanol/water
Corticosterone metabolism by reconstituted Cyt P-450	Wada et al. (1984)	ChemcosorbODS	Methanol/water
Metabolism of 16α-hydroxy-progesterone by intestinal bacteria	Glass & Burley (1984)	C18 column	Methanol/water
Assay of 3β-hydroxysteroid dehydrogenase/isomerase	Marston et al. (1985)	μBondapak C18	Methanol/water
Isolation of aldosterone dimer from rat adrenals	Cozza, Lantos & Burton (1985)	Alltech R-SIL C18	Methanol/water
Aldosterone metabolism in rat kidney	McDermott et al. (1985)	Zorbax C8	Methanol/water/acetonitrile/THF/acetic acid solvent mixtures

HPLC of C_{19} Steroids

Metabolites of the main secreted androgen, testosterone, are either A-ring reduction products (Lundmo & Sunde 1984), including a number of isomeric androstane- or androstenediols, or products of hydroxylation at various positions on the steroid nucleus. Other naturally occurring C_{19} steroids arise by side chain cleavage of the secreted 17-hydroxylated corticosteroids and are either derived from cortisol ($C_{19}O_3$ steroids) or cortisol precursors ($C_{19}O_2$ steroids). In contrast to its value in studying C_{21} steroid metabolism, HPLC has not found such wide application in the study of androgens (see Heftmann & Hunter, 1979; Heftmann & Lin, 1982). Many studies of HPLC separations have not been confined to the C_{19} steroids but have included them in a more general study (e.g. O'Hare et al., 1976; Allenmark et al., 1981; Kessler, 1982a). There have been several studies of the metabolism of testosterone in a variety of tissues (e.g. Wood et al., 1983). HPLC systems have been described for 11 major C_{19} steroids, based on a straight-phase adsorption system (Hunter, Walden & Heftmann, 1979), and for hydroxylated androgen metabolites, using a reversed-phase system (van der Hoeven, 1980). A comprehensive study of 69 androstane derivatives on straight- and reversed-phase systems has been undertaken by Lin & Heftmann (1982), examining $C_{19}O_2$, $C_{19}O_3$, and hydroxylated metabolites of testosterone at carbons

4,5,6,16,18 & 19. Five separate HPLC systems were described with Zorbax-SIL and Zorbax-ODS columns. Reversed-phase HPLC was most useful for the separation of the hydroxylated metabolites, and judicious combination of both columns enabled complete separation of all the androstane derivatives studied.

In studies of gonadal steroid synthesis, androgens are not the exclusive steroid products and thus some effort has been expended in developing HPLC systems which cater for the androstane precursors, 17-hydroxyprogesterone and 17-hydroxypregnenolone, as well as the oestrogen products of A-ring aromatisation. Walters, Foster & Cottrell (1981) have described a reversed-phase system (Partisil ODS), using a gradient elution solvent (tetrahydrofuran/water) to separate progesterone, 17-hydroxyprogesterone, androstenedione and testosterone. Kessler (1982a) employed a similar reversed-phase system to separate 7α-, 11α-, 11β-, 16β-, 19-hydroxytestosterones, androstenedione, testosterone, 17-hydroxyprogesterone, progesterone, and pregnenolone. Kessler (1982b) extended this work and demonstrated the separation of 18 different gonadal steroids using a C_{18} reverse-phase column. Subsequent use of a C_8 radial compression column allowed the resolution of three further steroids which had not been separated on the C_{18} column. This system was applied to an investigation of the steroid content of a virilising adrenal tumour, excised from a 2-year-old girl (Richards et al., 1983).

For the resolution of most C_{19} steroid mixtures, reversed-phase systems are common. In a study of the synthesis of 6-hydroperoxyandrostenedione and a demonstration of its properties as an aromatase inhibitor, a bonded CN column was used (Tan, Hrycay & Matsumoto, 1983; Tan, Navennec & Gevry, 1984). Hausknecht et al. (1982) separated androgen metabolites from human endometrium with a straight-phase μPorasil column, although they found it necessary to include a reversed-phase system to separate the 3β,17β-diol and 3α,17β-diol metabolites, which is surprising in view of the excellent separation achieved by Lin & Heftmann (1982) using a very similar straight-phase system.

Application of HPLC for the measurement of androgens in human body fluids has been minimal because of the shift of interest to measurements in plasma or serum, where, because of their low concentrations, they are generally undetectable by conventional HPLC detectors and also on account of the availability of specific antisera for the major physiologically important androgens, testosterone and 5α-dihydrotestosterone. Table 5 lists some further applications of HPLC in studies of C_{19} steroid metabolism.

Apart from testosterone and androstenedione and their hydroxylated metabolites, which absorb at 240 nm, most C_{19} steroids do not show strong UV absorbance. In investigations of suitable HPLC systems, UV detection at 205 nm (Lin & Heftmann, 1982), 206 nm (Kling & diZerega, 1982), and 280 nm (Heftmann & Lin, 1982) or the refractive index detector (Lin & Heftmann, 1982; Sunde & Lundmo, 1982), have been used in addition to the specific UV absorbance at 240 nm. Various attempts have been made to find suitable derivatives which would increase the specificity and sensitivity of detection so that HPLC could be applied to the direct measurement of C_{19} steroids in human plasma. Kawasaki, Maeda & Tsuji (1981) described one such method for the measurement of the major 17-oxosteroids in human plasma (androsterone, etiocholanolone and DHA) and in urine (+ 11-oxoetiocholanolone, 11-oxoandrosterone & 16α-

hydroxyDHA) from patients with ovarian cystoma, adrenogenital syndrome, and hypertension. 17-Dansylhydrazones were formed, prior to straight-phase HPLC on Zorbax-SIL or Hitachi Gel 3042, and eluted with a dichloromethane/ethanol/water solvent at 35°. Fluorescence detection gave a minimum detectable limit for DHA of 0.6 μg/100 ml with 0.1 ml of serum. A careful study of the conditions for the formation of the hydrazones was made (see also Weinberger, Koziol & Millington, 1984). This method has also been applied to the determina-

Table 5. Recent applications of HPLC in studies of C_{19} steroid metabolism

Application	Reference	Column	Solvent
Lipoproteins stimulate androgen biosynthesis in cultured rat testis cells	Schreiber, Weinstein & Hsueh (1982)	Analtec RP-18	Acetonitrile/ water
Nuclear metabolites of testosterone in monkey brain	Bonsall et et al. (1983)	Spherisorb 5-ODS	Acetonitrile
Metabolism of 5α-DHT and testosterone by human leukocytes	Clair et al. (1984)	Spherosil XOA 600	Dichloromethane/ ethanol/hexane
Effect of Angiotensin II on release of 19-OH- & 6-OH-androstenedione	Sekihara (1984)	μBondapak C18	Acetonitrile/water
Helix pomatia-induced oxidation of 3β-OH-steroids	Messeri et al. (1984)	Perkin-Elmer C8	Methanol/water
In vitro adrenal androgen synthesis in tissue culture	Hornsby & Aldern (1984)	Regis C18	Methanol/ & acetonitrile/water
Androstane-3α,17β-diol metabolism by rat ventral prostate	Lundmo & Sunde (1984)	Supelcosil LC18	Methanol/ acetonitrile/water
Microsomal Metabolites of testosterone	Mancilla & Gil (1984)	Lichrosorb DIOL	Hexane/ 2-propanol
Aromatisation of 16α-hydroxylated androgens	Numazawa et al. (1985)	Radial-Pak C18	Methanol/ phosphate pH 3.0
Metabolism of testosterone by rat lung microsomes	Halperin-Walega & Greene (1985)	Zorbax ODS	Acetonitrile/ methanol/water
Metabolism of DHA and androstenedione by human alveolar cells	Milewich & Cain (1986)	μBondapak C18	Methanol/water

tion of 17-oxosteroid glucuronide and sulphate conjugates in plasma (Kawasaki, Maeda & Tsuji, 1982b). Dansylhydrazine as a pre-column derivatisation reagent seems to be the only successful use of HPLC for the direct measurement of androgens in plasma or urine, although Darney, Wing & Ewing (1983) measured testosterone, androstenedione, 17-hydroxyprogesterone & progesterone in venous effluent from perfused rat testis using reversed-phase HPLC and UV absorbance at 240 nm. This was a very similar procedure to that described previously (Cochran, Darney & Ewing, 1979) but using a ternary solvent system. Yamada & Aizawa (1983) have described a method for measuring testosterone in rat testis & urine and human urine, using a straight-phase silica column and UV absorbance at 240 nm. The testosterone peak was eluted 2 minutes after injection, the solvent peak being eluted at around 1.7 minutes. It was thus so close to the solvent front that the specificity of this method must be suspect, particularly since no preliminary purification was carried out prior to HPLC. HPLC has, however, proved extremely valuable as a separation procedure prior to immunoassay. Cook, Rawlings & Kennedy (1982) separated testosterone, androstenedione, 5α-dihydrotestosterone, androsterone, 5α-androstane-3α,17β-diol, and 5α-androstane-3β,17β-diol from 0.5 ml serum from adult rams with reverse-phase Ultrasphere-ODS (methanol/water) prior to radioimmunoassay. Stocks & Benraad (1983) utilised HPLC with automated fraction collection prior to radioimmunoassay to measure plasma androstenedione, testosterone, 17-hydroxyprogesterone, and progesterone. Together with a number of other C_{21} steroids, Schoneshofer & Weber (1983) measured androstenedione, 5α-dihydrotestosterone, DHA, and testosterone in human urine by radioimmunoassay after HPLC on a DIOL column (hexane/2-propanol, at 40°).

HPLC of C_{18} Steroids

HPLC has been widely used for the separation of the oestrogens. Lin & Heftmann (1981) have published a detailed examination of both straight- (Zorbax SIL) and reversed-phase (Zorbax ODS) columns, using isocratic elution for the separation of 24 naturally occuring oestrogens. By a judicious combination of both HPLC systems, it was possible to separate all the oestrogens of interest, a similar conclusion to that reached for the androgens (Lin & Heftmann, 1982).

Catechol oestrogens (those with hydroxyl groups at C2 or C4) are particularly prone to oxidation and it is therefore usual to include an antioxidant in the eluent such as ascorbic acid (Gelbke & Knuppen, 1972; Aten et al., 1982) or butylated hydroxytoluene (Clancey et al., 1983). Similar separations to those achieved by Lin & Heftmann (1981), but including 2- and 4-methoxy oestrogens, have been described by Aten et al. (1982) using a Lichrosorb DIOL column and gradient elution (see Figure 11). Roos (1980) examined a number of systems for separating oestrogens, ranging from straight-phase SIL to reversed-phase C_{18} columns, but also including -CN and -NH_2 bonded phases. Slikker, Lipe & Newport (1981) evaluated six different systems for the separation of 25 free and conjugated oestrogens and suggested a scheme which resolved the oestrogens present in monkey urine and plasma. Figure 12 shows the separation of 15 oestro-

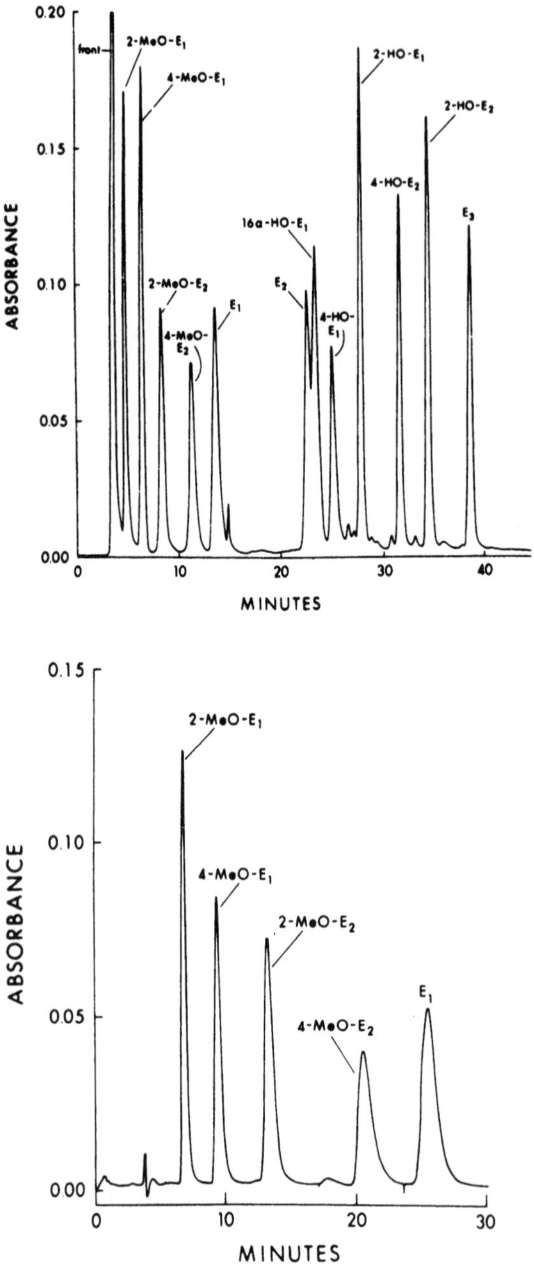

Fig. 11. HPLC of steroidal oestrogens on a Lichrosorb Diol Column; 11a: Gradient elution. Initial mobile phase, methylene chloride:iso-octane (7:3); Two minutes after injection, a concave gradient (Waters No. 7) was run for an additional 28 min to 100% of 2-propanol:dichloromethane:iso-octane (10:63:27); Flow rate was 1 ml/min and detection was by absorbance at 280 nm. 11b: Isocratic elution. Mobile phase, dichloromethane:iso-octane (1:1); Flow rate 1 ml/min; Detection at 280 nm. MeO: methoxy, HO: hydroxy, E_1:oestrone, E_2: oestradiol-17β, E_3: oestriol (from Aten et al., 1982, with permission of authors & publisher)

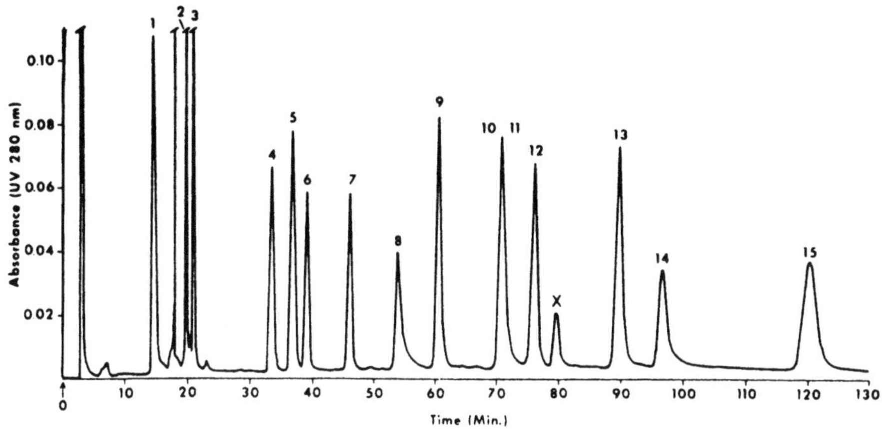

Fig. 12. HPLC profile of estrogen and metabolite standards. Chromegabond Diol, 10 μm, 300 × 4.6 mm normal-phase column; mobile phase, 100% hexane to hexane-2-propanol (80:20), linear gradient program (Whatman No. 6) set for 100 min; flow-rate, 1.5 ml/min at 300 p.s.i.; chart speed = 1 in/10 min; standards: (1) e_1-3-Met, 30 μg; (2) E_2-3-Met, 10 μg; (3) 2-MeO-E_1, 10 μg; (4) E_1, 10 μg; (5) 2-MeO-E_2, 10 μg; (6) E_3-3-Met, 10 μg; (7) E_2, 10 μg; (8) 2-OH-E_1, 20 μg; (9) 16-Epi-E_3, 30 μg; (10) 2-OH-E_2, 15 μg; (11) E_3, 10 μg; (12) 6α-OH-E_2, 10 μg; (x) unknown degradation product; (13) 15α-OH-E_3, 25 μg; (14) 2-OH-E_3, 30 μg; (15) αE_2-17Glyc, 30 μg. The unlabeled UV spike between standard (1) and (2) is the 2-propanol solvent front. Met: methyl, αE_2-17Glyc: internal standard (oestradiol-17α-glucuronide), other abbreviations are given in subscript to Figure 11 (from Slikker, Lipe & Newport, 1981 with permission of authors and publisher)

gens achieved on a Chromegabond DIOL column. They found that the addition of ammonium acetate to their reversed-phase eluent decreased tailing of peaks. Some of the other HPLC systems for the separation of oestrogens are listed in Table 6. One interesting development for the future is the development of silica-immobilised benzeneboronic acid (Hansson, Glad & Hansson, 1983), which can complex with compounds containing hydroxyl groups in close proximity, such as catechol oestrogens. Such a system might find application as a pre-column for the selection of catechol oestrogens, which could then be analysed by further HPLC on an analytical reversed-phase column.

Catechol oestrogens can be detected by electrochemical detection and all oestrogens show an absorbance maximum at 280 nm. Pre-column derivatisation has, however, been used in the form of dansyl derivatives (Schmidt, Vandemark & Slavin, 1978; Roos & Medwick, 1980). Oestrogens can also be detected from their natural fluorescence (Taylor, Knotts & Schmidt, 1980) which improves both sensitivity and specificity. Kling & DiZerega (1982) utilised the different absorbance of oestrogens (at 206 nm) and androgens (at 254 nm) to measure each in a reversed-phase system where androstenedione and oestrone, and testosterone and oestradiol did not separate. Some applications of these HPLC systems to studies of oestrogen metabolism are given in Table 7.

Table 6. Some recent HPLC separation systems used for the analysis of oestrogens

Oestrogens examined	Reference	Column	Solvent	Detection system
Oestriol in urine	Taylor, Knotts & Schmidt (1980)	Whatman PXS-ODS	Acetonitrile/ water	Fluorescence
Oestrone, oestradiol, oestriol in urine	Mourey & Siggia (1980)	Lichrosorb Si60	Methanol/ water	UV 280 nm
2-OH-oestrone, 2-OH-oestradiol, 4-OH-oestrone, 4-OH-oestradiol	Shimada et al. (1981) Shimada, Tanaka & Nambara (1981)	Hitachi Gel 3053 μBondapak C18	Methanol/ phosphate buffer, pH 2.5 Aectonitrile/ phosphate, pH 3	Electro-chemical
Serum oestriol	Kaplan & Hohnadel (1982 & 1983)	μBondapak C18	Acetonitrile/ phosphate buffer, ph 6.5/EDTA	Electro-chemical
Plasma oestradiol	Tetsuo, Eriksson & Sjovall (1982)	Lichrosorb DIOL	Dichloro-methane/hexane/ 2-propanol	GC-MS
Urinary oestriol	Cartoni & Coccioli (1983)	Lichrosorb RP18	Acetonitrile/ water	UV 280 nm
Oestrone, oestradiol, 2-OH-oestrone, 4-OH-oestradiol, 2-OH-oestradiol	Clancey et al. (1983)	Ultrasphere Si	Chloroform/ heptane/acetic acid	UV 280 nm & radio-activity
Oestradiol-17α & -17β in plasma, milk & urine	Batra & Saumande (1984)	S1-10 Silica	Chloroform/ acetone	RIA
Oestriol in plasma	Andreolini et al. (1985)	Supelco C18 packing	Acetonitrile/ phosphate buffer, pH 6.2	Fluorescence
Oestriol in urine	Gunasingham, Tay & Ang (1985)	Lichrosorb Si-60	Hexane/ ethanol	Electro-chemical
Automated system for oestriol in urine	Schoneshofer, Dhar & Ioanides (1986)	Hypersil-ODS	Acetonitrile/ water	UV 238 nm
Catechol oestrogen in rat liver & brain	Pfeiffer et al. (1986)	Lichrosorb-DIOL	Isoocatane/ methylene chloride/ isopropanol	UV 280 nm

Oestrogen metabolites may be conjugated with sulphuric and/or glucuronic acids and, usually, these conjugates are hydrolysed prior to analysis. HPLC of intact cansteroid conjugates is discussed in a following section. Bowers & Johnson (1981a & b) used a column packed with β-glucuronidase, immobilised on glutaraldehyde-activated controlled-pore glass, which was placed in the injection

Table 7. Some recent applications of HPLC in studies of oestrogen metabolism

Application	Reference	Column	Solvent
Oestrogen 2-hydroxylase activity	Allenmark et al. (1981b)	Nucleosil C_{18}	Methanol/acetate-citrate buffer, pH 5.8
Resolution of oestradiol-17α & -17β	Clark et al. (1982)	μPorasil	Chloroform/hexane/methanol
2-OH- & 4-OH-oestradiol in brain incubations	Prescott, Boyd & Seaton (1982)	HP C_{18} column	Methanol/water/acetic acid
Fatty acid esters of oestradiol in normal serum	Mellon-Nussbaum et al. (1982), Janocko & Hochberg (1983)	Lichrosorb DIOL	Chloroform/isooctane Dichloromethane
Oestrogen secretion by perfused testis	Thomson, Ewing & Lasley (1983)	Partisil ODS	Acetonitrile/water
Reaction of 16α-OH-oestrone with erythrocytes	Bucala, Fishman & Cerami (1984)	HP RP C_8 column	Methanol/phosphate buffer, pH 7.0
Assay of oestradiol-17-sulphate 4-hydroxylase activity	Watanabe & Yoshizawa (1985)	TSK-Gel ODS-120A	Methanol/phosphate buffer pH 3.0

loop position (see Figure 13). Oestrogen glucuronides were injected into the immobilised enzyme column in phosphate buffer (pH 6.5) and, after 5 minutes, the free oestrogen was eluted into the analytical column by increasing concentrations of methanol in the eluent. The oestrogen is so strongly bound to the immobilised enzyme as not to be eluted until methanol is added to the phosphate buffer. Oestriol in pregnancy urine was measured by this technique, using the natural fluorescence for detection and quantitation. This system has been extended to provide a routine analytical procedure for the analysis of oestriol in urine at a rate of 9 samples per hour, using the principle of "box-car" HPLC (Johnson & Bowers, 1982). This is an interesting approach to overcoming the objection to HPLC for routine analysis, namely that, unlike TLC, samples must be analysed sequentially. It is usually necessary to ensure that all the material injected into the column is eluted before the next sample injection, and thus the analysis time is increased. "Box-car" chromatography utilises a pre-column which carries out a preliminary separation of the compound of interest and directs only a small fraction containing the analyte to the analytical column. A number of sequential injections can therefore be made before the first separation is complete. The analysis time is therefore decreased. In the procedure of Johnson & Bowers (1982), the pre-column purification is carried out on the immobilised enzyme column. HPLC of oestrogens has been reviewed by Heft-

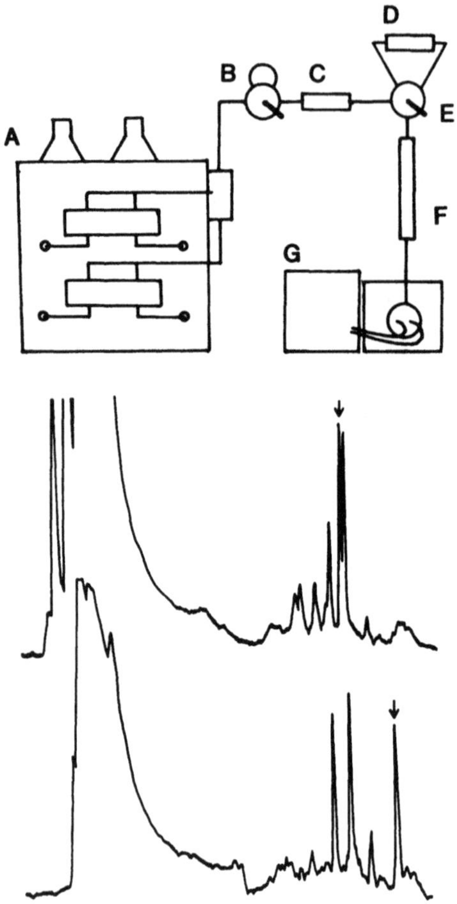

Fig. 13. A: Schematic diagram of the combined immobilised enzyme reactor/reversed-phase HPLC system. The immobilised β-glucuronidase is packed into the injection loop (D). After 5 minutes, the hydrolysed oestrogen is directed to the analytical column (F). B: Chromatographic analysis of pregnancy urine with and without the use of the immobilised enzyme. The upper and lower tracings were obtained in the absence and presence of the enzyme reactor, respectively. The arrow in the upper tracing indicates the retention time of oestriol glucuronide; this peak is replaced by a peak corresponding to oestriol (see arrow) in the lower tracing. y-axis, arbitrary fluorescence intensity units (from Bowers & Johnson, 1981a with permission of authors and publisher)

mann & Hunter (1979), Schmidt (1981), Heftmann & Lin (1982), and Heftmann (1983). Serum oestradiol and oestriol have been measured after pre-concentration on co-polymer beads in an automatic HPLC system (Dohji et al., 1984).

HPLC of Steroid Conjugates

The HPLC systems previously discussed have been primarily concerned with the separation and measurement of free steroids. However, the majority of steroids are excreted as water-soluble simple and mixed conjugates with sulphuric and/or glucuronic acids and, thus, hydrolysis by enzymic or chemical means has been undertaken prior to analysis. However, this procedure obscures information about the precise nature of the conjugates which may be of considerable importance. Efforts have therefore been made to devise methods of separating intact conjugates. In contrast to gas chromatography, the non-destructive nature of HPLC would appear to offer an ideal system for such separations. Two reviews on this subject have recently been published. Van der Wal & Huber (1982) reviewed the HPLC of a wide variety of conjugates, while Ager & Oliver (1984)

Table 8. HPLC systems used for the separation of steroid conjugates

Steroid conjugate	Reference	Column	Solvent
Oestriol 16- & 3-glucuronide	Oliphant & Bowers (1982)	PRP-1 bonded to Lichrosorb Si60	Acetonitrile/acetate buffer, pH 7.5
17-oxosteroid glucuronides in urine & serum	Kawasaki, Maeda & Tsuji (1982)	μBondapak C_{18}	Sodium acetate in methanol/water/ acetic acid
Oestriol 16- and 17-glucuronides in bile & urine	Shimada, Xie & Nambara (1982)	TSK-GEL LS410 ODS-SIL	Tetrahydrofuran/ phosphate buffer pH 3.0
DHA sulphate in urine	Causon, Collins & Fry (1982)	Hypersil SAS	Pentanol/borate buffer, pH 7.0
2-Hydroxylation of oestradiol 17β-sulphate and -glucuronide	Watanabe & Yoshizawa (1983a, 1983b)	TSK-GEL LS410 ODS-SIL	Methanol/acetate buffer, pH 5.0
17-oxosteroid sulphates in serum	Shimada, Tanaka & Nambara (1984)	μBondapak C_{18}	Methanol/phosphate buffer
Oestrogen sulphates in endometrial tumour in culture	Grosso & Larsen (1984)	RP C_{18} Alltech RP-600	Acetonitrile/tri-ethylamine buffer, pH 4.0
In vitro formation of sulphates and glucuronides of catechol oestrogens	Shimada et al. (1984a & b)	TSK-GEL LS-410 & LS-420 ODS-SIL	Ammonium phosphate/ tetrahydrofuran/ acetonotrile-methanol
Steroid 3-sulphates	Simonian & Capp (1984)	Apex-ODS or Spherisorb ODS	Methanol/ammonium sulphate
17-Oxosteroid glucuronides in urine	Novotny et al. (1984)	5 μm Spherisorb C_{18} (microcolumn)	Methanol/water Acetic acid containing 0.01 M sodium acetate
Aldosterone & tetrahydro aldosterone in urine	Risk, Holland & Brown (1985)	DIOL column	Acetonitrile/acetate buffer, pH 4.3
Steroid 3-sulphates	Dibbelt & Kuss (1985)	Radial-Pak C_{18}	Methanol/triethyl-ammonium sulphate
4-Hydroxy monosulphates & glucuronides	Shimada, Xie & Nambara (1986)	Develosil-ODS-5	0.52% NaAcetate (pH 3.2): Acetonitrile and/or THF
17oxosteroid sulphates in urine (separated as hydrazones)	Nishikaze & Iwata (1986)	Zorbax SIL	Dichloromethane/ Fluorescence of hydrazones/ethanol/water

confined their attention to oestrogen conjugates and four conjugates in particular (oestriol 16-glucuronide, 3-glucuronide, 3-sulphate 16-glucuronide, and 3-sulphate). Analysis of oestrogen conjugates has also been reviewed by Schmidt (1981).

Although a number of different HPLC systems have been introduced for the analysis of steroid conjugates (see Schmidt, 1981; Van der Wal & Huber, 1982), almost all recent systems have utilised reversed-phase columns. A summary of some of the more recent publications on the separation of steroid conjugates by HPLC is given in Table 8, from which it will be seen that only two groups did not use ODS-bonded phases. Causon, Collins & Fry (1982) used Hypersil-SAS for the measurement of DHA sulphate in plasma and Oliphant & Bowers (1982) used a microparticulate polystyrene divinylbenzene (PRP-1) phase.

Simonian & Capp (1984) made the interesting observation that incorporation of ammonium sulphate into the aqueous portion of the eluent increased the retention times of steroid sulphates on ODS columns while having no effect on the retention of free steroids. Similarly Dibbelt & Kuss (1985) incorporated triethylammonium sulphate in their elution solvent. The addition of ammonium ions in this way is an example of ion pairing, the ammonium counter ion pairing with the ionised sulphate. Detection systems for conjugates are usually based upon the methods for the detection of the unconjugated steroids, since neither glucuronic nor sulphuric acids have any particular characteristic which makes them easier to detect. UV absorbance at 190 nm for DHA sulphate (Causon, Collins & Fry, 1982), 280 nm for oestrogen conjugates (Grosso & Larsen, 1984), electrochemical detection of catechol oestrogen conjugates (Shimada, Xie & Nambara, 1982, Shimada et al., 1984a & b), electrochemical detection of p-nitrophenylhydrazones of 17-oxosteroid sulphates (Shimada, Tanaka & Nambara, 1984 and Shimada, Xie & Nambara, 1986), and fluorescence of dansyl hydrazones of 17-oxosteroid sulphates and glucuronides (Kawasaki, Maeda & Tsuji, 1982b), have been used.

HPLC of Steroid Receptors

Although the HPLC of cytoplasmic and nuclear receptors is beyond the scope of this chapter, it is perhaps worth noting that HPLC is increasingly being applied to the separation and quantitation of steroid receptors. Since steroid hormones are thought to exert their effects by interaction with the nuclear chromatin, knowledge of the nature of cytoplasmic and nuclear receptors increases our understanding of the mechanism of action of the steroid hormones. In addition the demonstration of the presence of steroid receptors in some hormone-responsive tumours has increased the awareness of their importance. Table 9 lists a few of the more recent publications where steroid receptors have been separated and/or quantitated by HPLC. The various HPLC systems available for steroid receptor studies are reviewed by Wittliff (1985).

Table 9. Some HPLC systems used for the separation and quantitation of cytoplasmic and nuclear steroid receptor proteins

Application	Reference
Oestrogen receptors; separation and quantitation	Hofmann et al. (1981)
Practical aspects of HPLC for progesterone and oestrogen receptors	Pavlik et al. (1982a & b)
Micromethod for measurement of oestrogen receptors	Kirdani et al. (1982)
Oestrogen receptors and clinical correlation with prostatic disease	Pontes et al. (1982)
Characterization of oestrogen receptor	Landsdorfer, Clements & Witliff (1983)
Micromeasurement of steroid receptors in breast cancer	Milano et al. (1983)
Nuclear oestrogen receptors	Madhok & Leung (1983)
HPLC chromatofocussing for analysis of oestrogen receptor heterogeneity	Hutchens et al. (1983)
Oestrogen and androgen receptors in prostatic disease	Kirdani et al. (1984)
Cytoplasmic and nuclear oestrogen receptors	Braunsberg (1984)
Practical considerations of HPLC in steroid receptor analysis	Pavlik et al. (1985)
Characterization of androgen, oestrogen and progesterone receptors	Brinkmann et al. (1985)
Receptor binding of nandrolone and testosterone	Bergink et al. (1985)
Isolation of cytosol oestrogen receptors	Dunaway, Hutchens & Besch, (1985) & Hyder et al. (1985)
Oestrogen receptor heterogeneity in clinical samples	Boyle et al. (1985)
Analysis of rat uterine receptors	Madhok & Leung (1986)

HPLC of Steroid Metabolising Enzymes

Being proteins, steroid metabolising enzymes are also susceptible to purification by similar HPLC procedures to those used for receptor protein studies. For example, 3β-hydroxysteroid sulfotransferase from hamster epididymis has been separated by HP exclusion chromatography on a Waters I-250 column (Bouthillier et al., 1985).

Summary

Although a considerable amount of work has been carried out in the last ten years in developing methods for the separation of steroids by HPLC, it is still not widespread for the reasons discussed above. There is however no doubt that further developments in HPLC technology, in increasing sensitivity and/or specificity of detection systems, perhaps with microbore columns, may lead to an increase in the use of this powerful analytical procedure as an additional separation method to improve specificity of assay. Solution of the problem of simple interfacing of HPLC systems with mass spectrometers (discussed in another chapter by Games) should further increase the application of HPLC. HPLC is of particular value in providing a means of separating unstable compounds prior to assay by relatively nonspecific quantitation methods. Most steroids do not fall into this category, but the steroid vitamin D and its metabolites do and HPLC has proved in this area to be invaluable (see chapter by Jones & DeLuca). There are a multiplicity of different HPLC systems for the separation of steroids, varying in column type (and manufacturer), solvent composition and method of elution, temperature of elution, etc., and only a few attempts have been made to rationalise these data. It would therefore seem that a fruitful area of future study would be the investigation of computerised systems for the selection and optimisation of HPLC systems for particular steroid separations.

References

Adams JS, Clemens TL, Holick MF (1981) Silica Sep-Pak preparative chromatography for vitamin D and its metabolites. J Chromatogr 266:198–201

Ager RP, Oliver RWA (1984) Separation of oestrogen conjugates in urine and synthetic mixtures by high-performance liquid chromatographic methods. J Chromatogr 309:1–15

Albert DH, Prasad VVK, Lieberman S (1982) The conversion of progesterone into 5α-pregnane-3,20-dione, 3β-hydroxy-5α-pregnan-20-one, and its fatty acid esters by preparations of bovine corpora lutea. Endocrinology 111:17–23

Allenmark S, Boren H (1981) Isolation by means of preparative reversed-phase liquid chromatography of epimeric alcohols formed upon reduction of pregnenolone and progesterone. J Liquid Chromatogr 4:1797–1805

Allenmark S, Son Berg AA, Hammar M, Lindstrom E (1981a) Combined paper and reversed-phase high-performance liquid chromatography method for the study of pregnenolone and progesterone metabolites. J Chromatogr 224:399–405

Allenmark S, Hjelm E, Hammar M, Son Berg AA (1981b) A new method for the determination of estrogen 2-hydroxylase activity. J Biochem Biophys Methods 4:209–214

Ando T, Koshika S, Komura K, Nakayama Y, Hara S (1986) Characterization of packing columns for the liquid chromatographic determination of corticosteroids in human plasma. J Liquid Chromatogr 9:2601–2608

Andreolini F, Borra C, DiCorcia A, Sampero R, Raponi G (1985) Total oestriol in maternal serum or plasma as measured by liquid chromatography. J Chromatogr 337:379–383

Antle PE (1982) Solvent optimization in normal-phase liquid chromatography of selected steroids. Chromatographia 15:277–281

Aten RF, Eisenfeld AJ, MacLusky NJ, Hochberg RB (1982) Separation of steroidal oestrogens and their major unconjugated metabolites by high-performance liquid chromatography. J Steroid Biochem 16:447–449

Axelson M, Sahlberg BL (1981) Solid extraction of steroid conjugates from plasma and milk. Analyt Lett 14:771–782

Batra SK, Saumande J (1984) High-performance liquid chromatographic separation of oestradiol-17α and -17β in biological fluids: application to plasma, milk and urine of cows. J Liquid Chromatogr 7:2431–2446

Bergink EW, Janssen PSL, Turpijn EW, Van der Vies J (1985) Comparison of the receptor binding properties of nandrolone and testosterone under in vitro and in vivo conditions. J Steroid Biochem 22:831–836

Bondy PK, Rosenberg LE (eds) (1980) In: Metabolic Control and Disease, 8th edn. WB Saunders, Philadelphia London Toronto

Bonsall RW, Rees HD, Michael RP (1983) Characterization by high-performance liquid chromatography of nuclear metabolites of testosterone in the brains of male rhesus monkeys. Life Sciences 33:655–663

Bouthillier M, Bleau G, Chapdelaine A, Roberts KD (1985) The purification of 3β-hydroxysteroid sulfotransferase of the hampster epididymis. J Steroid Biochem 22:733–738

Bowers LD, Johnson PR (1981a) On-line cleavage of urinary estriol conjugates with immobilized β-glucuronidase before liquidchromatographic analysis. Clin Chem 27:1554–1557

Bowers LD, Johnson PR (1981b) Immobilized β-glucuronidase as an on-line pre-column modification reagent for high-performance liquid chromatography. Analyt Biochem 116:111–115

Boyle DM, Wiehle RD, Shahabi NA, Wittliff JL (1985) Rapid, high resolution procedure for assessment of oestrogen-receptor heterogeneity in clinical samples. J Chromatogr 327:369–376

Braunsberg H (1984) Application of high performance liquid chromatography to the analysis of cytoplasmic and nuclear oestrogen receptors. J Steroid Biochem 20:417–420

Brinkmann AO, Boet-de Vries J, De Boer W, Lindh LM, Mulder E, Van Der Molen HJ (1985) Characterization of steroid hormone receptors with ion-exchange fast protein liquid chromatography. J Steroid Biochem 22:85–90

Bucala R, Fishman J, Cerami A (1984) The reaction of 16α-hydroxyestrone with erythrocytes in vitro and in vivo. Eur J Biochem 140:593–598

Caldarella AM, Reardon GE, Canalis E (1982) Analysis for cortisol in serum by liquid chromatography. Clin Chem 28:538–543

Canalis E, Caldarella AM, Reardon GE (1979) Serum cortisol and 11- deoxycortisol by liquid chromatography: clinical studies and comparison with radioimmunoassay. Clin Chem 25:1700–1703

Canalis E, Reardon GE, Caldarella AM (1982) A more specific, liquid-chromatographic method for free cortisol in urine. Clin Chem 28:2418–2420

Cannell GR, Galligan JP, Mortimer RH, Thomas MJ (1982) Extraction of steroids from plasma with reversed-phase C_{18} cartridges. Clin Chim Acta 122:419

Capp MW, Simonian MH (1985) Separation of the major adrenal steroids by reversed-phase high-performance liquid chromatography. Analyt Biochem 147:374–381

Carson SW, Jusko WJ (1984) Simultaneous analysis of cortexolone and cortisol by high-performance liquid chromatography for use in the metyrapone test. J Chromatogr 306:345–350

Cartoni GP, Coccioli F (1983) High-performance liquid chromatographic determination of estrogen in human urine. J Chromatogr 278:144–150

Causon RC, Collins SL, Fry DE (1982) Determination of urinary dehydroepiandrostone sulphate by combined high-performance liquid chromatography and radioimmunoassay. J Chromatogr 227:485–491

Cavina G, Moretti G, Alimenti R, Gallinella B (1979) Analysis of natural corticosteroids in adrenal extracts and in biological fluids by high-performance liquid chromatography. J Chromatogr 175:125–140

Cavina G, Gallinella B, Porra R, Valvo L (1984) Identification and determination of corticosteroids in adrenal gland extracts for pharmaceutical use by HPLC. Chromatographia 19:206–215

Chan TH, Moreland M, Hum WT, Birmingham MK (1977) Quantitative determination of 18-hydroxydeoxycorticosterone and corticosterone by high pressure liquid chromatography. J Steroid Biochem 8:243–245

Clair P, Patricot MC, Mathian B, Revol A (1984) Androgen metabolism in vitro by human leukocytes. Variations with age and sex. J Steroid Biochem 20:377–381

Clark JH, Williams M, Upchurch S, Eriksson H, Helton E, Markaverich BM (1982) Effects of estradiol-17α on nuclear occupancy of the estrogen receptor, stimulation of nuclear type II sites and uterine growth. J Steroid Biochem 16:323–328

Clancey CJ, Owen-Fett C, Farley DB, Van Orden DE (1983) High-performance liquid chromatographic separation of catechol estrogens: use as a screening procedure for evaluation of in vitro metabolism of [^{3}H]estradiol. J Chromatogr 256:335–340

Cochran RC, Darney KJ, Ewing LL (1979) Measurement of testosterone with a high-performance liquid chromatograph equipped with a flow-through ultraviolet spectrophotometer. J Chromatogr 173:349–355

Cook SJ; Rawlings NC, Kennedy RI (1982) Quantitation of six androgens by combined high-performance liquid chromatography and radioimmunoassay. Steroids 40:369–380

Cozza EN, Lantos CP, Burton G (1985) A highly lpophilic form of aldosterone. Isolation and characterization of an aldosterone dimer. J Steroid Biochem 23:511–516

D'Agostino G, Mitchell F, Castagnetta L, O'Hare MJ (1984) Solvent optimization of reversed-phase high-performance liquid chromatography for polar adrenal steroids using computer predicted retentions. J Chromatogr 305:13–26

D'Agostino G, Castagnetta L, Mitchell F, O'Hare MJ (1985) Computer aided mobile-phase optimization and chromatogram simulation in HPLC: A review. J Chromatogr 338:1–23

Dale SL, Holbrook MM, Melby JC (1981) 19-nor-deoxycorticosterone in the neutral fraction of human urine. Steroids 37:103–109

Darney KJ, Wing T-Y, Ewing LL (1983) Simultaneous measurement of four testicular Δ^4-3-ketosteroids by isocratic high-performance liquid chromatography with on-line ultraviolet absorbance detection. J Chromatogr 257:81–90

Dawson R, Kontur P, Monjan A (1984) High-performance liquid chromatographic (HPLC) separation and quantitation of endogenous glucocorticoids after solid-phase extraction from plasma. Hormone Res 20:89–94

Dehennin L, Blacker C, Reiffsteck A, Scholler R (1984) Estrogen 2-, 4-, 6- or 16-hydroxylation by human follicles by gas chromatography-mass spectrometry associated with stable isotope dilution. J Steroid Biochem 20:465–471

Derks HJGM, Drayer NM (1978a) Improved methods for isolating cortisol metabolites from neonatal urine. Clin Chem 24:1158–1162

Derks HJGM, Drayer NM (1978b) The identification and quantification of three new 6α-hydroxylated corticosteroids in human neonatal urine. Steroids 31:289–305

DeVries CP, Lomecky-Janousek M, Popp-Snijders C (1980) Rapid quantitative assay of plasma 11-deoxycortisol and cortisol by high-performance liquid chromatography for use in the metyrapone test. J Chromatogr 183:87–91

Dibbelt L, Kuss E (1985) Isocratic reversed-phase high-performance liquid chromatography of free and sulphoconjugated steroids. J Chromatogr 322:371–375

Dohji T, Fushimi M, Kawabe T, Kamiyama F, Mori M, Sugita N, Tanizawa O (1984) Rapid measurement of oestradiol and oestriol by high-performance liquid chromatography after automatic pretreatment. J Chromatogr 311:249–255

Dumont E, Sclavons M, Desager JP, J Lig Chromatogr (1984); 7:2051–2057. Use of an internal standard to assay 6-beta-hydroxycortisol

Dunaway HE, Hutchens TW, Besch PK (1985) Isolation of unliganded steroid receptor proteins by high-performance size-exclusion chromatography. An investigation of steroid-dependent structural alterations. J Chromatogr 327:221–235

Ebling WF, Szefler SH, Jusko WJ (1984) Analysis of cortisol, methylprednisolone, and methylprednisolone hemisuccinate. Absence of effects of troleandromycin on ester hydrolysis. J Chromatogr 305:271–280

Eibs G, Schoneshofer M (1984) Simultaneous determination of fifteen steroid hormones from a single serum sample by high-performance liquid chromatography. J Chromatogr 310:386–389

Farhi RL, Monder C (1978) Analysis of steroidal carboxylic acids by high-pressure liquid chromatography. Analyt Biochem 90:58–68
Farwell SO, Kagel RA, Gutenberger SK, Olson DP (1983) Weak-calf syndrome and the determination of cortisol: adapting literature methods to real-life problems. Analyt Chem 55:985A–988A
Fast DM, Culbreth PH, Sampson EJ (1982) Multivariate and univariate optimization studies of liquid-chromatographic separation of steroid mixtures. Clin Chem 28:444–448
Few JD, Chaudry S, James VHT (1984) The direct determination of aldosterone in human saliva. J Steroid Biochem 21:87–92
Fitzpatrick R, Siggia S (1973) High resolution liquid chromatography of derivatized non-ultraviolet absorbing hydroxy steroids. Analyt Chem 45:2310–2314
Fitzpatrick FA (1978) High-performance liquid chromatography of the steroid hormones. Advances in Chromatography 16:37–73
Fotherby K (1984a) Biosynthesis of the oestrogens. In: Makin HLJ (ed) Biochemistry of steroid hormones, 2nd edn. Blackwells, Oxford, p 207–292
Fotherby K (1984b) Endocrinology of the menstrual cycle and pregnancy. In: Makin HLJ (ed) Biochemistry of steroid hormones, 2nd edn. Blackwells, Oxford, p 409–440
Francis FE, Kinsella RA (1984) Reversed-phase high-performance liquid chromatography of C_{21} metabolites of progesterone. J Chromatogr 336:361–367
Frey FJ, Frey BM, Benet LZ (1979) Liquid chromatographic measurement of endogenous and exogenous glucocorticoids in plasma. Clin Chem 25:1944–1947
Fujii S, Momoi K, Okamoto M, Yamano T, Okada T, Terasawa T (1984) 18,19-dihydroxydeoxycorticosterone, a new metabolite produced by cytochrome $P\text{-}450_{11\beta}$. Chemical synthesis and structural analysis by ^1H NMR. Biochemistry 23:2558–2564
Gandelman MS, Birks JW (1983) Liquid chromatographic detection of cardiac glycosides, saccharides and hydrocortisone based on the photoreduction of 2-tert-butylanthraquinone. Analyt Chim Acta 155:159–171
Gelbke HP, Knuppen R (1972) A new method for preventing oxidative decomposition of catechol estrogens during chromatography. J Chromatogr 71:465–471
Gerber-Taras E, Park BK, Ohnhaus EE (1981) The estimation of 6β-hydroxycortisol in urine – a comparison of two methods: high-performance liquid chromatography and radioimmunoassay. J Clin Chem Clin Biochem 19:525–527
Glass TL, Burley CZ (1984) Biotransformation of 16-dehydroprogestone by the intestinal anaerobic bacterium, Eurbacterium sp. 144. J Steroid Biochem 21:65–72
Gluckman J, Shelly D, Novotny M (1984) Laser fluorimetry for capillary liquid chromatography:high sensitivity of derivatized biological compounds. J Chromatogr 317:443–453
Goehl TJ, Sundaresan GM, Prasad V (1979) Fluorimetric high-pressure liquid chromatographic determination of hydrocortisone in human plasma. J Pharm Sci 68:1374–1376
Gomez-Sanchez CE, Gomez-Sanchez EP, Shackleton CHL, Milewich L (1982) Identification of 19-hydroxydeoxycorticosterone, and 19-oic-deoxycorticosterone as products of deoxycorticosterone metabolism by rat adrenals. Endocrinology 110:384–389
Gotelli GR, Wall JH, Kabra PM, Marton LJ (1981) Fluorometric liquid-chromatographic determination of serum cortisol. Clin Chem 27:441–443
Goto J, Shamsa F, Goto N, Nambara T (1983) The simultaneous determination of cortisol and cortisone by high-performance liquid chromatography with fluorimetric detection. J Pharmaceut Biomed Anal 1:83–88
Goto J, Shamsa F, Nambara T (1983) Studies on steroids. CLXXXII Determination of 6β-hydroxy cortisol in urine by high-performance liquid chromatography with fluorescence detection. J Liq Chromatogr 6:1977–1985
Gower DB (1984) Steroid catabolism and urinary excretion. In: Makin HLJ (ed) Biochemistry of steroid hormones, 2nd edn. Blackwells, Oxford, p 349–382
Grassini-Strazza G, Nicoletti I (1985) High-performance thin-layer chromatography on amino-bonded silical gel: application to barbiturates and steroids. J Chromatogr 322:149–158
Gray CH, James VHT (eds) (1983) In: Hormones in Blood, 3rd edn. Vol 4, Academic Press, London
Grosso DS, Larsen BR (1984) Estrogen sulfoconjugation by human endometrial cancer cells (RL 95-2) in culture. J Clin Endocr Metab 58:368–373

Gunasingham H, Tay BT, Ang KP (1985) Determination of oestriol in pregnancy urine by normal-phase high-performance liquid chromatography with electrochemical detection. J Chromatogr 341:271–278

Gutenberger SK, Olson DP, Kagel RA (1985) Comparison of reversed-phase high-performance liquid chromatography and competitive protein-binding assay in the quantitation of cortisol in bovine plasma. J Liquid Chromatogr 8:107–124

Han A, Maradici A, Monder C (1983) Metabolism of corticosterone in the mouse. Identification of 11β,20α-dihydroxy-3-oxo-4-pregnen-21-oic acid as a major metabolite. J Biol Chem 258:13703–13707

Hansson L, Glad M, Hansson C (1983) Boronic acid-silica: a new tool for the purification of catecholic compounds on-line with reversed-phase high-performance liquid chromatography. J Chromatogr 265:37–44

Hausknecht V, Lopez de la Osa E, Gurpide E (1982) In vitro metabolism of C_{19} steroids in human endometrium. J Steroid Biochem 17:621–629

Heftmann E (1976) Chromatography of Steroids, Elsevier, Amsterdam Oxford New York

Heftmann E, Hunter IR (1979) High-pressure liquid chromatography of steroids. J Chromatogr 165:283–299

Heftmann E, Lin JT (1982) Steroid analysis by high-performance liquid chromatography. J Liq Chromatogr 5 (Suppl No 1):121–173

Heftmann E (1983) Steroids: In: Heftmann E (ed) Chromatography, Elsevier, Amsterdam Oxford New York, Journal of Chromatography Library Series 22B:191–222

Heikkinen R, Fotsis T, Adlercreutz H (1981) Reverse-phase C_{18} cartridge for extraction of estrogens from urine and plasma. Clin Chem 27:1186–1189

Helboa P (1986) Separation of corticosteroids by high-performance liquid chromatography on dynamically modified silica. J Chromatogr 366:191–196

Hesse C, Pietrzik K, Hotzel D (1974) Spezifische Bestimmung von Corticosteron und Cortisol im Nanogramm-Bereich. Z Klin Chem Klin Biochem 12:193–197

Hobkirk R (ed) (1979) Steroid Biochemistry, CRC Press, Florida

Hofmann GE, Fuchs A, Wiehle RD, Wheeler Feldhoff PA, Wittliff JL (1981) Separation and analysis of estrogen receptors by high-pressure liquid chromatography using [^{3}H]estradiol-17β and [16α-^{125}I]estradiol as ligands. Clin Chem 27:1078

Hofreiter BT, Allen JP, Mizera AC, Powers CD, Masi AM (1982) High-performance liquid chromatography and radioimmunoassay of rat plasma corticosterone. Steroids 39:547–555

Hofreiter BT, Mizera AC, Allen JP, Masi AM, Hicok WC (1983) Solid state extraction of cortisol from plasma or serum for liquid chromatography. Clin Chem 29:1808–1809

Honour JW (1984) Biliary excretion and enterohepatic circulation. In: Makin HLJ (ed) Biochemistry of Steroid Hormones, 2nd edn. Blackwells, Oxford, p 382–408

Horikawa R, Tanimura T, Tamura Z (1979) Method for fluorescence detection in the high-performance liquid chromatography of Δ^4-3-ketosteroids. J Chromatogr 168:526–529

Hornsby PJ (1982) Regulation of 21-hydroxylase activity by steroids in cultured bovine adrenocortical cells: possible significance for adrenocortical synthesis. Endocrinology 111:1092–1101

Hornsby PJ, Aldern KA (1984) Steroidogenic enzyme activities in cultured human definitive zone adrenocortical cells and resultant differences in adrenal androgen synthesis. J Clin Endocr Metab 58:121–127

Hunter IR, Walden MK, Heftmann E (1979) High-pressure liquid chromatography of androgens. J Chromatogr 176:485–487

Hutchens TW, Wiehle RD, Shahabi NA, Wittliff JL (1983) Rapid analysis of estrogen receptor heterogeneity by chromatofocussing with high-performance liquid chromatography. J Chromatogr 266:115–118

Hyder SM, Wiehle RD, Brandt DW, Wittliff JL (1985) High-performance hydrophilic-interaction chromatography of steroid hormone receptors. J Chromatogr 327:237–246

Imaizumi N, Morimoto S, Kigoshi T, Uchida K, Hosojima H, Yamamoto I (1984) Application of reversed-phase high-performance liquid chromatography for radioimmunoassay of plasma 18-hydroxycorticosterone. J Chromatogr 308:295–300

Janocko L, Hochberg RB (1983) Estradiol fatty acid esters occur naturally in human blood. Science 222:1334–1336

Johnson PR, Bowers LD (1982) Measurement of urinary estriol by immobilized enzymes and box car chromatography. Analyt Chem 54:2247–2250

Jones G (1983) Chromatographic separation of 24(R),25-dihydroxyvitamin D_3 and 25-hydroxyvitamin D_3-26,23-lactone using a cyano-bonded phase packing. J Chromatogr 276:69–75

Kabra PM, Tsai LL, Marton LJ (1979) Improved liquid-chromatographic method for determination of serum cortisol. Clin Chem 25:1293–1295

Kaplan LA, Hodnadel DC (1983) Unconjugated estriol in serum as measured by liquid chromatography with electrochemical detection, compared with radioimmunoassay. Clin Chem 29:1463–1466

Kautsky MP (ed) (1981) In: Steroid analysis by HPLC. Recent applications. Chromatographic Science Services, Vol 16, Marcel Dekker, New York Basel

Kawasaki T, Maeda M, Tsuji A (1979) Determination of plasma and urinary cortisol by high-performance liquid chromatography using fluorescence derivatization with dansyl hydrazine. J Chromatogr 163:143–150

Kawasaki T, Maeda M, Tsuji A (1981) Determination of 17-oxosteroids in serum and urine by fluorescence high-performance liquid chromatography using dansyl hydrazine as a pre-labeling reagent. J Chromatogr 226:1–12

Kawasaki T, Maeda M, Tsuji A (1982a) Determination of 17-hydroxycorticosteroids in urine by fluorescence high-performance liquid chromatography using Dns-hydrazine as a pre-column labeling reagent. J Chromatogr 232:1–11

Kawasaki T, Maeda M, Tsuji A (1982b) Determination of 17-oxosteroid glucuronides and sulphates in urine and serum by fluorescence high-performance liquid chromatography using dansyl hydrazine as a pre-labeling reagent. J Chromatogr 233:61–68

Kim CY, Sugiyama T, Okamoto M, Yamano T (1983) Regulation of 18-hydroxycorticosterone formation in bovine adrenocortical mitochondria. J Steroid Biochem 18:593–599

Kirdani RY, Pontes EJ, Murphy GP, Sandberg AA (1982) A micromethod for the measurement of estrogen receptors using high pressure liquid chromatography. J Steroid Biochem 17:311–317

Kirdani RY, Pontes EJ, Murphy GP, Sandberg AA (1984) Correlation of estrogen and androgen receptor status in prostatic disease measured by high pressure liquid chromatography. J Steroid Biochem 20:401–406

Kessler MJ (1982a) High performance liquid chromatography of steroid metabolites in the pregnenolone and progesterone pathways. Steroids 39:21–32

Kessler MJ (1982b) A rapid high performance liquid chromatography system for the separation of gonadal steroids. J Liquid Chromatogr 5:125–139

Kling OR, diZerega GS (1982) Quantitative separation of estrogens, androgens and progestogens using reversed-phase high-pressure liquid chromatography. Steroids 40:381–388

Kobayashi Y, Ogihara T, Yamamura Y, Watanabe F, Kiguchi T, Ninomiya I, Kumahara Y (1984) Identification of 19-nor-deoxycorticosterone in normal rat serum by enzyme immunoassay and gas chromatographymass spectrometry. Endocrinol Japon 31:49–53

Kraan GPB, Derks HJGM, Drayer NM (1980) Quantification of polar glucocorticoids in the urine of pregnant and non-pregnant women: a comparison with 6α-hydroxylated metabolites of cortisol in neonatal urine and amniotic fluid. J Clin Endocr Metab 51:754–758

Kramer RE, McCarthy JL, Simpson ER, Waterman MR (1983) Effects of ACTH on steroidogenesis in bovine adrenocortical cells in primary culture – increased secretion of 17-hydroxylated steroids associated with a refractoriness in total steroid output. J Steroid Biochem 18:715–723

Lambert WE, De Slypere J-PM, Jonckheere JA, Vermeulen A, DeLeenheer AP (1983) Improved liquid-chromatographic determination of serum cortisol with double internal standardization compared to radioimmunoassay and fluorimetry, and evaluated by isotope dilution/mass spectrometry. Analyt Biochem 134:216–223

Lansdorfer M, Clements NC, Wittliff JL (1983) Use of high-performance liquid chromatography in the evaluation of the synthesis and binding of fluorescein-linked steroids to estrogen receptors. J Chromatogr 266:129–139

Lantto O (1982) Radioimmunoassay and liquid-chromatographic analysis for free cortisol in urine compared with isotope dilution-mass spectrometry. Clin Chem 28:1129–1132

Levitz M, Young BK (1977) Estrogens in pregnancy.Vitamins & Hormones 35:109–147

Lewbart ML, Elverson RA (1982) Determination of urinary free cortisol and cortisone by sequential thin-layer and high-performance liquid chromatography. J Steroid Biochem 17:185–190
Lin JT, Heftmann E, Hunter IR (1980) High-performance liquid chromatography of the reaction products of progesterone. J Chromatogr 190:169–174
Lin JT, Heftmann E (1981) High-performance liquid chromatography of naturally occurring estrogens. J Chromatogr 212:239–244
Lin JT, Heftmann E (1982) Comparison of adsorption and reversed-phase partition high-performance liquid chromatography for the separation of androgens. J Chromatogr 237:215–224
Loche S, Porcelli F, Rosen M, Feffer M, Stoner E, New MI (1984) Clinical applications of the rapid high-performance liquid chromatographic determination of serum cortisol. J Chromatogr 317:377–382
Lundmo PI, Sunde A (1984) Rapid analysis of C_{19}-steroid metabolism by high-performance liquid chromatography and in-line monitoring of radioactivity. J Chromatogr 308:289–294
Madhok TC, Leung BS (1983) Characterization of uterine estrogen receptors by size-exclusion and ion-exchange high-performance liquid chromatography. Biochem Biophys Res Commun 115:988–994
Madhok TC, Leung BS (1986) Analysis of rat uterine estrogen receptors by high-performance liquid chromatography methods. Hormone Res 23:235–243
Majors RE (1980) Recent advances in HPLC packings and columns. J Chromatogr Sci 18:488–511
Makin HLJ (1984a) Methods of steroid analysis: group estimations and separation techniques. In: Makin HLJ (ed) Biochemistry of Steroid Hormones, 2nd edn. Blackwells, Oxford, pp 478–514
Makin HLJ (ed) (1984b) Biochemistry of Steroid Hormones, 2nd edn. Blackwells, Oxford
Mancilla J, Gil L (1984) A simple, reliable and rapid HPLC method to separate and quantify androstenedione, testosterone and hydroxytestosterones. Analyt Lett 17:873–886
Marston SA, Waterman MR, McCarthy JL (1985) 3β-hydroxysteroid dehydrogenase-isomerase activity in bovine adrenocortical cells in culture: lack of response to ACTH treatment. J Steroid Biochem 22:789–794
Matsuzawa T, Sugimoto N, Ishiguro I (1981) A simple micromethod for determining human serum cortisol by high-pressure liquid chromatography using 0.1 ml serum. Analyt Biochem 115:250–253
MacLusky NJ, Naftolin F, Krey LC, Franks S (1981) The catechol estrogens. J Steroid Biochem 15:111–124
McDermott M, Latif S, Morris DJ (1983) The metabolism of aldosterone in kidney. J Steroid Biochem 19:1205–1211
McDermott MJ, Freiberger M, Latif SA, Morris DJ (1985) The synthesis of reduced metabolites of aldosterone by subcellular fractions of rat kidney: effects of antimineralocorticoids. J Steroid Biochem 23:503–509
McGinley PA, Braughler JM, Hall ED (1982) Determination of methylprednisolone in central nervous tissue and plasma using normal-phase high-performance liquid chromatography. J Chromatogr 230:29–35
Mellon-Nussbaum SH, Ponticorvo L, Schatz F, Hochberg RB (1982) Estradiol fatty acid esters. The isolation and identification of the lipoidal derivative of estradiol synthesized in the bovine uterus. J Biol Chem 257:5678–5684
Messeri G, Cugnetto G, Moneti G, Serio M (1984) Helix pomatia-induced conversion of some 3β-hydroxysteroids. J Steroid Biochem 20:793–796
Milano G, Moll JL, Formento JL, Francoual M, Krebs BP, Namer M, Boublie JL, Lalanne CM (1983) Simultaneous micromeasurement of steroid receptors in breast cancer. Brit J Cancer 48:579–584
Momoi K, Okamoto M, Fujii S, Kim CY, Miyaki Y, Yamano T (1983) 19-hydroxylation of 18-hydroxy-11-deoxycorticosterone catalysed by cytochrome P-$450_{11\beta}$ of bovine adrenocortex. J Biol Chem 258:8855–8860
Monder C, Bradlow HL (1980) Cortoic acids – exploration at the frontier of corticosteroid metabolism. Rec Progr Hormone Res 36:345–391
Monder C, Purkaystha AR, Farhi RL, Han CA (1982) Oxidation of cortisol to hydroxy acid metabolites by liver cytosol. J Steroid Biochem 16:613–616

Morris DJ, Tsai R (1981) Chromatographic separation of aldosterone and its metabolites. Advances in Chromatography 19:261–285
Morris DJ, Kenyon CJ, Latif SA, McDermott M, Goodfriend TL (1983) The possible biological role of aldosterone metabolites. Hypertension 6 (suppl I):I35–I40
Mourey TH, Siggia S (1980) Microparticulate bonded pyrrolidone for high-performance liquid chromatographic separation of estrogenic steroids in urine. Analyt Chem 52:881–885
Nakamura J, Yakata M (1982) Two-cycle liquid-chromatographic quantitation of cortisol in urine. Clin Chem 28:1497–1500
Nakamura J, Yakata M (1983) Clinical evaluation of the liquid-chromatographic determination of urinary free cortisol. Clin Chem 29:847–851
Nice EC, O'Hare MJ (1978) Selective effects of reversed-phase column packings in high-performance liquid chromatography of steroids. J Chromatogr 166:263–267
Nilsson B (1983) Extraction and quantitation of cortisol by use of high-performance liquid affinity chromatography. J Chromatogr 276:413–417
Nishikaze O, Iwata J (1986) Direct determination of sulfate conjugates of 17-oxosteroids in urine by liquid chromatography. Clin Chem 32:835–839
Novotny M, Alasandro M, Konishi M (1983) Microcolumn liquid chromatography of benzoyl derivatives of steroid metabolites. Analyt Chem 55:2375–2377
Novotny M, Karlsson KE, Konishi M, Alasandro M (1984) New biochemical separations using precolumn derivatization and microcolumn liquid chromatography. J Chromatogr 292:159–167
Numazawa M, Osada R, Tsuji M, Osawa Y (1985) High-performance liquid chromatographic determination of aromatization of 16α-hydroxylated androgens with human placental microsomes. Analyt Biochem 146:75–81
O'Hare MJ, Nice EC, Magee-Brown R, Bullman H (1976) High-pressure liquid chromatography of steroids secreted by human adrenal and testis cells in monolayer culture. J Chromatogr 125:357–367
O'Hare MJ, Nice EC, Capp M (1980) Reversed- and normal-phase high-performance liquid chromatography of 18-hydroxylated steroids and their derivatives. Comparison of selectivity, efficiency and recovery from biological samples. J Chromatogr 198:23–29
O'Hare MJ, Nice EC (1981) Analysis of steroid hormones in adrenal and testicular cells and tissues. In: Kautsky MP (ed) Steroid analysis by HPLC. Recent applications (Chromatographic Science series, vol 16). Marcel Dekker, New York Basel, pp 277–322
O'Hare MJ, Nice EC, McIlhinney RAJ, Capp M (1981) Progesterone synthesis, secretion and metabolism by human teratoma-derived cell-lines. Steroids 38:719–737
Oka K, Minagawa K, Hara S, Noguchi M, Matsuoka Y, Kono M, Irimajiri S (1984) On-line extraction, evaporation, and injection for liquid chromatographic determination of serum corticosteroids. Analyt Chem 56:24–27
Okamoto M, Momoi K, Fujii S, Yamano T (1982) 18,19-Dihydroxydeoxycorticosterone; a novel product of cytochrome P-$450_{11\beta}$-catalysed reaction. Biochem Biophys Res Commun 109:236–241
Oliphant C, Bowers LD (1982) High performance liquid chromatographic analysis of urinary estriol conjugates. Clin Chem 28:1551–1552
Ost L, Falk O, Lantto O, Bjorkhem I (1982) Simultaneous determination of prednisolone and cortisol in serum by HPLC and by isotope dilution-mass spectrometry. Scand J Clin Lab Invest 42:181–187
Patterson DG, Patterson MB, Culbreth PH, Fast DM, Holler JS, Sampson EJ, Bayse DD (1984) Determination of steroid hormones in a humanserum reference material by isotope dilution-mass spectrometry: a candidate definitive method for cortisol. Clin Chem 30:619–626
Pavlik EJ, van Nagell JR, Donaldson E, Hanson M (1982a) Some practical aspects of the separation of estrogen and progesterone receptors by size-exclusion high-performance liquid chromatography which are relevant to quantitation. J Steroid Biochem 17:553–558
Pavlik EJ, van Nagell JR, Muncy M, Donaldson ES, Hanson M, Kenady D, Rees ED, Taliwalker VR (1982b) Rapid analysis of estrogen and progesterone receptors using gel-exclusion high-performance liquid chromatography. Biochemistry 21:139–145

Pavlik EJ, Nelson K, van Nagell JR, Robinson JC, Hanson MB, Donaldson ES, Flanigan RC, Kenady DE (1985) Steroid receptor analysis by size-exclusion liquid chromatography: considerations for the clinical laboratory. Clin Chem 31:537–545
Perry LA, Al-Dujaili EAS, Edwards CRW (1982) A direct radioimmunoassay for 11-deoxycortisol. Steroids 39:115–128
Petersen MC, Nation RL, Ashley JJ (1980) Simultaneous determination of betamethasone, betamethasone acetate and hydrocortisone in biological fluids using high-performance liquid chromatography. J Chromatogr 183:131–139
Pfeiffer DG, MacLusky NJ, Barnea E, Naftolin F, Krey LC, Loriaux DL, Merriam GR (1986) Are catechol oestrogens obligatory mediators of oestrogen action in the central nervous system? I. Characterization of pharmacological probes with different receptor binding affinities and catechol oestrogen formation rates. J Endocrinol 110:489–497
Pontes JE, Karr JP, Kirdani RY, Murphy GP, Sandberg AA (1982) Estrogen receptors and clinical correlations with human prostatic disease. Urology 19:399–403
Prescott WR, Boyd BK, Seaton JF (1982) High performance liquid chromatographic separation of the two estrogen isomers of estradiol with electrochemical detection. J Chromatogr 234:513–516
Purdy RH, Durocher CK, Moore PH, Rao PN (1980) Analysis of metabolites of progesterone in bovine liver, kidney, kidney fat, and milk by high-performance liquid chromatography. J Steroid Biochem 13:1307–1315
Purdy RH, Durocher CK, Moore PH, Rao PN (1981) Analysis of progestins by HPLC. In: Kautsky MP (ed) Steroid analysis by HPLC. Recent applications (Chromatographic Science series, vol 16). Marcel Dekker, New York Basel, pp 81–104
Purkaystha AR, Martin KO, Goldberg A, Monder C (1982) Oxidation of corticosteroids to steroidal 20-hydroxy-21-oic acids by mouse liver. J Steroid Biochem 17:51–59
Racz K, Varga I, Kiss R, Glaz E (1984) ACTH sensitivity of isolated pathological adrenocortical cells: variability of responses in aldosterone, corticosterone, deoxycorticosterone and cortisol production. J Steroid Biochem 20:1187–1194
Ramirez LC, Millot C, Maume BF (1982) Sample purification using aC_{18}-bonded reverse-phase cartridge for the quantitative analysis of corticosteroids in adrenal cells by high performance liquid chromatography or gas chromatography-mass spectrometry. J Chromatogr 229:267–281
Reardon GE, Caldarella AM, Canalis E (1979) Determination of serum cortisol and 11-deoxycortisol by liquid chromatography. Clin Chem 25:122–126
Redhwi AA, Anderson DC, Smith GN (1981) A simple method for the separation of vitamin D metabolites from serum extracts. Acta Endocrinol (Kbh): 97 (suppl 243), Abstract 21
Rhodes CJ, Claridge PA, Trafford DJH, Makin HLJ (1983) An evaluation of the use of Sep-Pak C_{18} cartridges for the extraction of vitamin D_3 and some of its metabolites from plasma and urine. J Steroid Biochem 19:1349–1354
Richards GE, Griffiths MD, Hoelzer DJ, Kessler MJ, Dan M, Marshall RN, Pottkotter LE (1983) Virilising adrenal adenoma in a 2-year old girl. Analysis of tumor steroids using high performance liquid chromatography. American J Dis Child 137:741–744
Roos RW (1980) High pressure liquid chromatographic separation and identification of estrogens. J Assn Off Analyt Chem 63:80–87
Roos RW, Mewick T (1980) Application of dansyl derivatization to the high-pressure liquid chromatograph identification of equine estrogens. J Chromatographic Sci 18:626–630
Rose JQ, Jusko WJ (1979) Corticosteroid analysis in biological fluids by high performance liquid chromatography. J Chromatogr 162:273–280
Saito Z, Mimoh N, Hifumi S, Takeda R (1983) Analysis of corticosteroids in human adrenal tissue by high pressure liquid chromatography. Clin Chim Acta 131:239–246
Schmidt GJ, Vandemark FL, Slavin W (1978) Estrogen determination using liquid chromatography with precolumn fluorescence labelling. Analyt Biochem 91:636–645
Schmidt GJ (1981) Analysis of estrogens by HPLC. In: Kautsky MP (ed) Steroid analysis by HPLC. Recent applications (Chromatographic Science series, vol 16). Marcel Dekker, New York Basel, pp 145–172
Schoneshofer M (1977) Simultaneous determination of eight adrenal steroids in human serum by radioimmunoassay. J Steroid Biochem 8:995–1009

Schoneshofer M, Dulce HJ (1979) Comparison of different high-performance liquid chromatographic systems for the purification of adrenal and gonadal steroids prior to immunoassay. J Chromatogr 164:17–28

Schoneshofer M, Fenner A, Altinok G, Dulce HJ (1980) Specific and practicable assessment of urinary free cortisol by combination of automatic high-pressure liquid chromatography and radioimmunoassay. Clin Chim Acta 106:63–73

Schoneshofer M, Fenner A, Dulce HJ (1980) Interferences in the radioimmunological determination of urinary free cortisol. Clin Chim Acta 101:125–134

Schoneshofer M, Fenner A, Dulce HJ (1981) Assessment of eleven adrenal steroids from a single serum sample by combination of automatic high-performance liquid chromatography and radioimmunoassay (HPLC-RIA). J Steroid Biochem 14:3377–3386

Schoneshofer M, Maxeiner J, Fenner A (1981) Quantitation of pregnenolone and 17-hydroxypregnenolone in human serum by automatic high-performance liquid chromatography and subsequent radioimmunoassay. J Chromatogr 224:229–237

Schoneshofer M, Skobolo R, Dulce HJ (1981) Reliability of the estimation of serum cortisol by high-performance liquid chromatography. J Chromatogr 222:478–481

Schoneshofer M, Weber B (1982) Determination of urinary free 20α-dihydrocortisol by high-performance liquid chromatography. Fresenius Z Analyt Chem 311:426

Schoneshofer M, Weber B, Dulce HJ (1982) Estimation of free urinary aldosterone and 18-hydroxycorticosterone by a combination of automatic high-performance liquid chromatography and radioimmunoassay. J Chromatogr 227:492–496

Schoneshofer M, Kage A, Weber B (1983) New "on-line" sample-pretreatment procedure for routine liquid chromatographic assay of low-concentration compounds in body fluids, illustrated by triamcinolone assay. Clin Chem 29:1367–1371

Schoneshofer M, Weber B, Nigam S (1983) Increased urinary excretion of free 20α- and 20β-dihydrocortisol in a hypercortisolemic but hypocortisoluric patient with Cushing's disease. Clin Chem 29:385–389

Schoneshofer M, Weber B (1983) Specific estimation of fifteen unconjugated, non-metabolized steroid hormones in human urine. J Steroid Biochem 18:65–73

Schoneshofer M, Kager A, Weber B, Lenz I, Kottgen E (1985) Determination of urinary free cortisol by "on-line" liquid chromatography. Clin Chem 31:564–568

Schoneshofer M, Dhar TK, Ioanides D (1986) Total urinary estriol determined by "on-line" liquid chromatography with ultra violet detection. Clin Chem 32:1948–1950

Schoneshofer M, Kage A, Eisenschmid P, Heilmann P, Dhar TK, Weber B (1986a) Automated liquid chromatographic determination of the 20-dihydro isomers of cortisol and cortisone in human urine. J Chromatogr 380:267–274

Schoneshofer M, Weber B, Oelkers W, Nahoul K, Mantero F (1986b) Measurement of urinary free 20α-dihydrocortisol in biochemical diagnosis of chronic hypercorticoidism. Clin Chem 32:808–810

Schreiber JR, Weinstein DB, Hsueh AJW (1982) Lipoproteins stimulate androgen production by cultured rat testis cells. J Steroid Biochem 16:39–43

Scott NR, Dixon PF (1979) Determination of cortisol in human plasma by reversed-phase high-performance liquid chromatography. J Chromatogr 164:29–34

Scott NR, Chakraborty J, Marks V (1980) Determination of prednisolone, prednisolone, and cortisol in human plasma by high-performance liquid chromatography. Analyt Biochem 108:266–268

Seki T, Yamaguchi Y (1983) New fluorimetric detection method of corticosteroids after high-performance liquid chromatography using post-column derivatization with glycinamide. J Liquid Chromatogr 6:1131–1138

Seki T, Yamaguchi Y (1983) New fluorimetric determination of 17-hydroxycorticosteroids after high-performance liquid chromatography using post-column derivatization with benzamidine. J Chromatogr 305:188–193

Sekihara H (1985) 19-hydroxyandrostenedione and 6β-hydroxyandrostenedione: new steroids regulated by the reninangiotensin system in man. J Steroid Biochem 20:383–385

Shackleton CHL, Taylor NR, Honour JW (1980) In: An atlas of gaschromatographic profiles of neutral urinary steroids in health and disease. Packard-Becker BV, Delft

Shackleton CHL, Whitney JO (1980) Use of Sep-Pak cartridges for urinary steroid extraction: evaluation of the method for use prior to gas chromatographic analysis. Clin Chim Acta 107:231–243

Shackleton CHL, Mattox VR, Honour JW (1983) Analysis of intact steroid conjugates by secondary ion mass spectrometry (including FABMS) and by gas-chromatography. J Steroid Biochem 19:209–217

Shackleton CHL (1984) Steroid synthesis and catabolism in the fetus and neonate. In: Makin HLJ (ed) Biochemistry of steroid hormones, 2nd edn. Blackwells, Oxford, pp 441–477

Shackleton CHL (1986) Profiling steroid hormones and urinary steroids. J Chromatogr 379:91–156

Shah H, Saranko A-M, Harkonen M, Adlercreutz H (1984) Direct solid-phase fluoroenzyme immunoassay of 3β-pregnene-3α,20-α-diol 3α-glucuronide in urine. Clin Chem 30:185–187

Shihabi ZK, Andrews RL, Scaro J (1982) Liquid chromatographic assay of urinary free cortisol. Clin Chim Acta 124:75–83

Shimada K, Kumai M, Shinkai H, Nambara T (1981) The simultaneous determination of the products of estrogen 2- and 4-hydroxylase action; the use of high-performance liquid chromatography with electrochemical detection. Analyt Biochem 116:287–291

Shimada K, Tanaka T, Nambara T (1981) Studies on steroids. CLXV. Determination of isomeric catechol oestrogens in pregnancy urine by high-performance liquid chromatography with electrochemical detection. J Chromatogr 223:33–39

Shimada K, Xie F, Nambara T (1982) Studies on steroids. CLXXIX. Determination of estriol 16- and 17-glucuronides in biological fluids by high-performance liquid chromatography with electrochemical detection. J Chromatogr 232:13–18

Shimada K, Tanaka T, Nambara T (1984) Studies on steroids. C. Determination of 17-ketosteroid sulphates in serum high-performance liquid chromatography with electrochemical detection using pre-column derivatization. J Chromatogr 307:23–28

Shimada K, Yumura Y, Terashima E, Nambara T (1984a) In vitro metabolic conjugation of catechol estrogens. J Steroid Biochem 20:1163–1167

Shimada K, Xie F, Terashima E, Nambara T (1984b) Separation of catechol estrogen monoglucuronides and monosulfates by high-performance liquid chromatography with electrochemical detection. J Liquid Chromatogr 7:925–934

Shimizu K, Amagaya S, Ogihara Y (1983) Analysis of corticosterone in the serum of mice and rats using high-performance liquid chromatography. J Chromatogr 272:170–175

Siggia S, Dishman R (1970) Analysis of steroid hormones using high-resolution liquid chromatography. Analyt Chem 42:1223–1229

Simonian MH, Capp M (1984) Reversed-phase high-performance liquid chromatography of steroid 3-sulfates and the corresponding unconjugated steroids. J Chromatogr 287:97–104

Slikker W, Lipe GW, Newport GD (1981) High-performance liquid chromatographic analysis of estradiol-17β and metabolites in bilogical media. J Chromatogr 224:205–219

Smith MD (1979) High-performance liquid chromatographic determination of hydrocortisone and methylprednisolone and their hemisuccinate esters in human serum. J Chromatogr 164:129–137

Smith MD (1981) Determination of synthetic adrenocorticosteroids in pharmaceutical preparations and biological fluids by HPLC. In: Kautsky MP (ed) Steroid analysis by HPLC. Recent applications (Chromatographic Science series, vol 16). Marcel Dekker, New York Basel, pp 105–144

Stoks PG, Benraad TJ (1983) Rapid separation of plasma steroids by reversed-phase high-performance liquid chromatography with timed collection of fractions. J Chromatogr 276:408–412

Stoner E, Loche S, Mirth A, New MI (1986) Clinical utility of adrenal steroid measurement by high-performance liquid chromatography in pediatric endocrinology. J Chromatogr 374:358–362

Street JM, Trafford DJH, Makin HLJ (1985) Extraction and fractionation of bile acids and their conjugates using pre-packed microparticulate silica cartridges (Sep-Pak SIL and Bond-Elut C18). J Chromatogr 343:259–270

Sunde A, Lundmo PI (1982) Separation of 5α-reduced androgens by high-performance liquid chromatography. J Chromatogr 242:381–384

Suzuki K, Kadowaki A, Tamaoki B (1980) A rapid measurement of Δ^5-3β-hydroxysteroid dehydrogenase activity coupled with Δ^4-Δ^5 isomerase by high-performance liquid chromatography. J Endocrinol Invest 4:441–444

Takada M, Terakawa N, Aono T, Kurachi K, Tsuji M, Matsumoto K (1984) Divergent effects of prolactin on 4-ene-5α-reductase activity and production of C_{19}-steroids from progesterone in immature rat ovary. J Steroid Biochem 20:657–661

Tan L, Hrycay EG, Matsumoto K (1983) Synthesis and properties of the epimeric 6-hydroperoxyandrostenediones, new substrates/inhibitors of human placental aromatase. J Steroid Biochem 19:1329–1338

Tan L, Navennec P, Gevry G (1984) De novo biosynthesis of 6 beta-hydroperoxyandrostenedione in human placental microsomes. Biochem Biophys Res Commun 118:805–813

Taylor JT, Knotts JG, Schmidt GJ (1980) Determination of urinary placental estriol by reversed-phase liquid chromatography with fluorescent detection. Clin Chem 26:130–132

Tetsuo M, Eriksson H, Sjovall J (1982) Gas chromatographic-mass spectrometric analysis of endogenous levels of estradiol in plasma and in cytosol from rat uterus. J Chromatogr 239:287–300

Thijssen JHH, Van den Berg JHM, Adlercreutz H, Gijzen AHJ, DeJong FH, Meijer JC, Moolenaar AJ (1980) The determination of cortisol in human plasma: evaluation and comparison of seven assays. Clin Chim Acta 100:39–46

Thompson DL, Ewing LL; Lasley BL (1983) Oestrogen excretion by in vitro perfused testis: species comparison and factors effecting short-term secretion: J Endocrinol 96:97–105

Trefz F, Byrd D, Kochen W (1975) Quantitative determination of cortisol in human plasma by high-pressure liquid chromatography. J Chromatogr 107:181–189

Tscherne RJ, Capitano G (1977) High-pressure liquid chromatographic separation of pharmaceutical compounds using a mobile phase containing silver nitrate. J Chromatogr 136:337–341

Tsuji K, Binns RB (1982) Micro-bore high-performance liquid chromatography for the analysis of pharmaceutical compounds. J Chromatogr 253:227–236

Ulick S, Chu MD, Land M (1983) Biosynthesis of 18-oxocortisol by aldosterone-producing adrenal tissue. J Biol Chem 258:5498–5502

Van den Berg JHM, Mol CR, Deelder RS, Thijssen JHH (1977) A quantitative assay of cortisol in human plasma by high performance liquid chromatography using a selective chemically bonded stationary phase. Clin Chim Acta 78:165–172

Van der Hoeven T (1980) Separation of metabolites of testosterone hydroxylation by high performance liquid chromatography. J Chromatogr 196:494–497

Van der Hoeven T (1981) Testosterone oxidation by rat liver microsomes: effects of phenobarbitone pretreatment and the detection of seven metabolites by HPLC. Biochem Biophys Res Commun 100:1285–1291

Van der Wal S, Huber JFK (1982) Separation of steroid conjugates by high performance liquid chromatography. J Chromatogr 251:289–310

Wada A, Okamoto M, Nonaka Y, Yamano T (1984) Aldosterone biosynthesis by a re-constituted cytochrome P-$450_{11\beta}$ system. Biochem Biophys Res Commun 119:3365–3371

Walters DG, Foster PMD, Cottrell RC (1981) High-performance liquid chromatography of progesterone and its metabolites. J Chromatogr 219:152–155

Watanabe K, Yoshizawa I (1983a) Clinical analysis on steroids. XXV. Assay of estradiol 17-sulfate 2-hydroxylase activity by high-performance liquid chromatography. J Chromatogr 277:71–77

Watanabe K, Yoshizawa I (1983b) Evidence of 2-hydroxylation of estradiol-17β 17-glucuronide by male rat liver microsomes. Steroids 42:163–170

Watanabe K, Yoshizawa I (1985) Clinical analysis of steroids. XXXI. Assay of oestradiol 17-sulphate 4-hydroxylase activity by high-performance liquid chromatography with electrochemical detection. J Chromatogr 337:114–120

Weber B, Schoneshofer M (1982) Analysis of unconjugated cortisone and 6β-OH-cortisol in human urine by high performance liquid chromatography. In: Molnar I (ed) Practical aspects of modern HPLC. Walter de Gruyter, Berlin, pp 63–75

Weinberger R, Koziol T, Millington G (1984) Improved dansylation procedure for the determination of ketosteroids by liquid chromatography. Chromatographia 19:452–456

Weisman Y, Bar A, Root A, Spirer Z, Golander A (1984) Rapid diagnosis of congenital adrenal hyperplasia by high performance liquid chromatography. Clin Chim Acta 138:1–8

Wittliff JL (1985) Separation and characterization of isoforms of steroid hormone receptors using high-performance liquid chromatography. In: Moudgil VK (ed) Molecular mechanism of steroid hormone action. Walter de Gruyter, Berlin New York, pp 791–813

Witzgall H, Hassan-Ali S (1980) Paper chromatographic systems for the separation of aldosterone, 18-hydroxycorticosterone, 18-hydroxydeoxycorticosterone and deoxycorticosterone. J Chromatogr 198:70–75

Wood AW, Ryan DE, Thomas PE, Levin W (1983) Regio- and stereoselective metabolism of two C_{19} steroids by five highly purified and reconstituted rat hepatic cytochrome P-450 isoenzymes. J Biol Chem 258:8839–8847

Wu MC, Takagi K, Okuyama S, Ohsawa M, Masahashi T, Narita O, Tomoda Y (1986) Determination of serum Δ^5-3β-hydroxysteroid sulpates by combined high-performance liquid chromatography and immobilised 3β,17β-hydroxysteroid dehydrogenase in column form. J Chromatogr 377:121–129

Yamada K, Aizawa Y (1983) A simple method for testosterone measurement by high-performance liquid chromatography-UV spectrophotometry. J Pharmacol Methods 10:167–173

The Mass Spectrometer as a Detector for High-Performance Liquid Chromatography: Problems of Interfacing

D. E. Games, and E. D. Ramsey

Introduction

The objectives of this chapter are to cover the current status of and future prospects for combined high-performance liquid chromatography/mass spectrometry (HPLC/MS) with particular emphasis on its ability to assist in the analysis of organic compounds relevant to endocrinological research. Initially, the currently most useful methods for HPLC/MS and techniques which hold promise for future development will be discussed. This will be followed by a description of areas relevant to endocrinology where HPLC/MS has been successfully applied. The off-line combination of HPLC with mass spectrometry, whereby peaks of interest are collected from the liquid chromatograph and are then transferred to the mass spectrometer, has been extensively used (Desiderio and Fridland), but will not be covered in this review.

Gas chromatography (GC) and its combination with mass spectrometry (GC/MS) has had widespread use in endocrinology. However the techniques suffer from the limitation that the compounds under investigation have to be vaporized prior to analysis, thus limiting its applicability to compounds which are volatile and thermally stable, or can be converted into such derivatives by the use of suitable chemical reactions. HPLC does not suffer from this restriction and hence has more universal analytical applicability. However it suffers from the lack of a universal detection system. Combination of a liquid chromatograph with a mass spectrometer can in theory provide such a detector and also enable the types of measurement currently perfomed by GC/MS to be undertaken, i.e. peak identification and improved sensitivity, specificity and selectivity for quantitative analysis.

The ideal system for HPLC/MS should place no restriction on the chromatography. It should be able to handle both normal and reversed mobile phases, with the multitude of modifiers (e.g. acids, bases, volatile and involatile buffers and ion-pair reagents) which are used to improve chromatographic performance. Also it should be remembered that when compounds of varying polarity are being studied gradient systems are necessary.

In the development of an interface for HPLC/MS there are three major problems to be addressed:

a) In conventional HPLC mobile phase flow rates are of the order of 1–2 ml min^{-1}. These flow rates can be much greater (5 ml min^{-1}) if fast HPLC is being performed or much lower (50 μl min^{-1}) if microbore HPLC is used. When conventional HPLC is used, since a mass spectrometer works under vacuum, the gas volumes generated from the mobile phase are much greater than can be handled directly by the mass spectrometer.

b) Compounds studied by HPLC cover a wide spectrum of structural types. They can be divided mass spectrometrically into three classes:
   - Thermally stable volatile compounds which are amenable to gas chromatographic study. However, HPLC is the method of choice for analysis because of factors such as, less sample work up, better sensitivity or selectivity. These compounds are readily amenable to study by electron impact (EI) and chemical ionization (CI) mass spectrometry.
   - Thermally labile and/or low volatility compounds which are not amenable to EI and/or CI mass spectral study using a direct insertion probe. The remainder will fail to provide molecular weight information by this approach and will require use of special ionization techniques such as desorption chemical ionization (DCI), field desorption (FD), thermospray ionization, fast atom bombardment (FAB) or ^{252}Californium plasma desorption. The technique used depending on the particular class of molecule under investigation.
   Compounds not currently amenable to study by mass spectral techniques. Usually these are very large molecules whose molecular weight exceeds the mass range of current mass spectrometers.

   Thus either some form of universal ionization technique is required or a variety of ionization methods will have to be used to cover a range of compounds. But at present not all compounds amenable to HP can possibly be handled because of mass spectral limitations.

c) Retention of chromatographic performance. This is the easiest of the problems to overcome and involves the use of minimal dead volumes in the construction of the interface.

It is not the intention of the authors to provide a fully comprehensive survey of all the approaches which have been studied for HPLC/MS. There are a number of excellent recent reviews (McFadden, 1980; Arpino and Guiochon, 1979; Curry, 1982; Willoughby and Browner, 1982; Aprino, 1983; Games, 1983; Majors, 1982; Desiderio and Fridland) which provide a detailed survey of the studies in the area. The next section will concentrate on assessing the relative merits of the currently more successful approaches and recent developments which show promise for future evelopment.

Interfaces for HPLC/MS

There are two basic philosophies in HPLC/MS. One approach introduces the mobile phase and solute (or a portion of it) into the ion source of the mass spectrometer, where in most cases the mobile phase is used to assist in the ionization of the solute. The other approach uses a transport system, where the mobile phase is removed, the solute is then vaporized and then ionized or is directly ionized.

Before moving to a detailed consideration of the various interfaces and their capabilities it is important to bear in mind the two main types of mass spectrometer in current use for organic analysis, since the choice of instrument can be important for the type of analysis contemplated. Quadrupole mass spectrometers operate at low source voltage and use a quadrupole mass analyzer. Until relatively recently their mass range was limited to 1000 u, but now many are commercially available with a mass range of up to 2000 u. Because of their lower source voltages they are easier to interface to a liquid chromatograph. Magnetic instruments operate at much higher source voltages and use magnetic and electric fields for mass analysis. They usually have a higher mass range than the quadrupole instruments and are capable of high resolution studies.

Interfaces for HPLC/MS where Solvent is not Removed

These approaches to HPLC/MS can be subdivided according to whether the interface attempts to effect enrichment of the solute relative to the solvent or the eluent from the liquid chromatograph is fed directly into the mass spectrometer ion source.

a) Enrichment of Solute

Early approaches were based on the use of the types of separator used for GC/MS i.e. use of a series of silicon membranes (Jones and Yang, 1975) or a jet separator (Takeuchi et al., 1978). Neither of these systems were effective for the handling of low volatility polar compounds. Modification of the latter approach by use of a nebulizing gas has resulted in the development of a system which can handle underivatized, amino acids, small peptides and mono- and di-saccharides (Yoshida et al., 1982). Solvent mediated CI spectra are obtained, but it appears that little enrichment is effected. Since flow rates of the order of 16 μl min^{-1} only can be handled the system is restricted to use of microbore HPLC if optimal sensitivity is to be obtained. The system appears to be virtually indistinguishable from some direct liquid introduction systems *(v. i.)*.

Studies using molecular beam techniques and aimed at developing an HPLC/MS interface capable of handling mobile phase flow rates of the order of 1 ml min^{-1} have resulted in the development of a more effective HPLC/MS interface and a new form of mass spectral ionization technique "thermospray ionization" (Blakley and Vestal, 1983). The eluent from the liquid chromatograph, which must contain an ionic modifier (0.1 M ammonium acetate appears to be most effective) passes through a stainless steel capillary which is brazed to an electrically heated copper block. A pumping line connected to a mechanical pump is sited opposite the exit of the capillary and the source housing is electrically heated. This results in the production of a supersonic jet of vapour, which contains both positive and negative ions. The jet stream is sampled into the mass spectrometer via a sampling cone with a small diameter. Mobile phase flow rates up to 2 ml min^{-1} can be handled and mass spectra have been obtained from a wide range of mass spectrometrically difficult molecules, e.g. peptides (Blakley et al., 1980a, b) and glucuronides (Liberato et al., 1983), using the technique as a stand alone ionization method and by HPLC/MS. Commercial systems are available from Vestec Corporation, VG Analytical, Finnigan MAT, Hewlett-Packard and Spectros. Most systems in use are on quadrupole mass spectrometers, but Spectros offer systems on magnetic instruments. Figure 1 shows a diagram of the system developed by Finnigan MAT. The eluent from the liquid chromatograph enters an electrically heated stainless steel capillary. Careful control of the temperature of the capillary results in the emergence of the solvent as a very fine spray, which enters the jet chamber. In the jet chamber the fine droplets evaporate and during this process charge exchange occurs between salt ions and the organic sample. Both positive and negative ions can be formed and the jet stream is sampled into the mass spectrometer by means of a sampling cone. The large volume of vapour formed can be pumped through a mechanical pump fitted with a liquid nitrogen cold trap, because the jet emerging from the tip of the vaporizer acts as its own ejector pump and provides a strong unidirectional motion towards the mechanical pump. It is usually necessary to empty the solvent from the cold trap at the completion of each day's work. Sensitivity is compound dependent and is related to proton/electron affinity. This is illustrated by recent studies of carbamate pesticides where detection limits for full scan spectra ranged from 1 to 100 ng (Voyksner et al., 1984). Solvent composition also affects compound sensitivity, the higher the percentage of water the better the sensitivity. An additional complication is that optimal vaporizer temperatures vary with solvent composition. A feature of the Finnigan system is the incorporation of a micro processor control module which enables the vaporizer temperature to be accurately set and monitored, in addition the temperatures of the aerosol and jet can be monitored. The module also enables temperature programming of the vaporizer to be effected, which is necessary for gradient elution

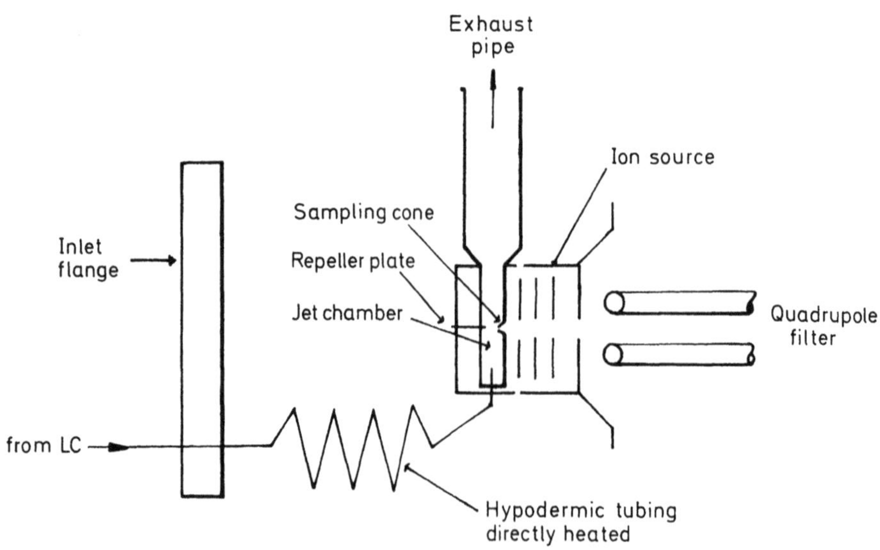

Fig. 1. Diagram of Finnigan MAT thermospray source

studies. The necessity of having ammonium acetate buffer present to effect thermospray ionization may cause problems for some liquid chromatographic studies. However the problem can be readily overcome by postcolumn addition of buffer through a coaxial tee (Voyksner et al., 1984).

Thermospray is an excellent method for HPLC/MS studies of compounds which give good sensitivity with the technique. It appears to have particular promise for the quantification of biologically important compounds and as an aid to their identification. Because of the simplicity of the spectra obtained with many classes of compound its utility for identification purposes particularly of unknowns may be limited. However if used with tandem mass spectrometric techniques e.g. triple quadrupole instruments this problem should be overcome to some extent.

Direct Introduction of the Eluent from the Liquid Chromatograph

One of the most widely used methods for HPLC/MS is the "direct liquid introduction" approach (Arpino et al., 1974a, b). Conventional mass spectrometers configured for chemical ionization (CI) can only directly handle mobile phase flow rates of the order of 10 μl min^{-1}. In this approach a fraction of the eluent from the liquid chromatograph is fed into the mass spectrometer ion source. Ionization of the solvent is effected by use of a filament and the ions formed then ionize the solute molecules by ion-molecule reactions. The result is production of positive or negative CI spectra from the solutes. Because of the presence of solvent ions the mass window has to be set above 100 u or higher to obtain satisfactory total ion current traces. This approach works best with reverse-phase HPLC using acetonitrile, methanol or water, or combinations of these as mobile phase. Acetonitrile is preferred to methanol since the latter sometimes causes problems with filament lifetimes. The technique was initially developed on magnetic instruments (Arpino et al., 1974a, b).

Early systems consisted of a capillary glass tube with a restriction at the end and there were problems in their use, because the tip of the tubing became rapidly plugged, particularly if low volatility compounds were being studied. Replacement of the capillary tube by a nonviscous type restriction such as a pinhole of 5 μ diameter in a diaphragm overcomes this problem and this principle is incorporated into the system manufactured by Hewlett-Packard and Nermag (Merela, 1980; Arpino et a., 1981). Incorporation of a heated desolvation chamber also assists in the handling of more difficult molecules (Melera, 1980; Arpino et al., 1983; Sugnaux et al. 1983).

HPLC/MS systems of the direct liquid introduction type have been used for studies of a wide range of compound types (Edmonds et al., 1983) and show no problems in handling a wide range of volatile modifiers for HPLC. A major advantage is that systems are relatively cheap and easy to construct (Evans and Williamson, 1981; Bruins and Drenth, 1983; Schäfer and Levsen, 1981; Henion and Wachs, 1981). Commercial systems are capable of routine use and show minimal thermal decomposition of sample. Optimization of the conditions is important, since spectra vary with changes in source pressure, solvent composition and temperature (Voyksner et al., 1982b). These variations may cause problems when gradient elution is being used. Because CI spectra are produced, many classes of compound will give spectra dominated by $(M+1)^+$ ions with few fragment ions. Whilst this can be advantageous for quantitative studies, it can cause problems when analysis is being undertaken where the chemical nature of the compounds is unknown or compounds with the same molecular weights are present. One method of improving the data in this regard, is the use of tandem mass spectrometry, whereby collisional activation is performed on the $(M+1)^+$ ion and the spectrum thus produced contains more structural information. This approach has been demonstrated with a tandem quadrupole instrument (Voyksner et al., 1982a).

An alternative method for effecting nebulization with interfaces of the direct liquid introduction type is to have a coaxial flow of gas. An approach of this type was discussed in the section on enrichment of solute and other systems have been reported (Evans and Williamson, 1981; Apffel et al., 1983). This approach does not appear to have any particular advantage over effecting nebulization through a hole in a diaphragm, and unless microbore HPLC is used only a portion of the HPLC eluent can be handled.

In order to overcome the problem of having to discard a large portion of the eluent from the liquid chromatograph for direct liquid introduction, a preconcentration direct liquid introduction HPLC/MS system has been developed (Christensen et al., 1983). The liquid stream is concentrated as it flows down a resistance-heated stationary wire. For normal-phase HPLC the concentrated eluent is drawn into the mass spectromenter ion source through a metal capillary tube. This approach did not work for reversed-phase systems and ultrasonic vibration was used to effect nebulization of the eluent. The system is marketed by Extranuclear and has been used for quantification of valproic acid in serum and heptanoic acid in shale oil process water.

If an atmospheric pressure ion (API) source is used on the mass spectrometer all of the eluent from a conventional HPLC column can be handled. Ionization is effected by use of a corona discharge or a ^{63}Ni source (Carroll et al., 1975). The ionized solvent undergoes ion molecule reactions with the solute and the ions produced are sampled through a pin-hole into a quadrupole mass analyzer. Because solute has to be vaporized at atmospheric pressure low volatility compounds could not be handled with this early approach. There are also problems in obtaining satisfactory total ion current traces because of the presence of solvent ion clusters in the spectra. Use of a nebulizer to introduce the HPLC eluent enables more difficult compounds to be handled (Henion et al., 1982). The spectra produced are usually very simple being dominated by $(M+1)^+$ ions. Use of a triple quadrupole system to effect collisional activation produces more informative spectra (Henion et al., 1982). An alternative approach which appears very effective for ionic compounds is use of liquid ion evaporation (Thomson et al., 1982). The solution is nebulized into air through a fine stainless steel hypodermic needle, and a charge is induced on the sprayed droplets by means of a small electrode placed close to the sprayer. Spectra have been obtained from many mass spectrometrically difficult molecules and amino acids and biogenic amines have been analysed using this approach by HPLC/MS

(Shushan et al., 1983). HPLC/MS systems of these types are commercially available from Sciex.

Other studies have improved sample introduction methods for atmospheric pressure ionization and may find utility for HPLC/MS. Fragmentation can be induced and the number of cluster ions produced reduced by variation of a potential at the sampling aperture (Tsuchiya and Taira, 1980). Use of a nebulizing gas for sample introduction enables more difficult molecules to be handled (Kambara, 1982). An alternative approach is the use of metastable argon ions to effect ionization, together with application of a high voltage (Tsuchiya and Kuwabara, 1984).

There are a number of interesting developments in this type of approach to HPLC/MS which could result in further improvements. A direct liquid introduction/thermospray interface has been described (Covey and Henion, 1983) and shown to work in both modes although the thermospray operation was not routine. Incorporation of a vacuum line attached directly to the ion source block and use of a heated capillary has resulted in considerable improvements to the flow rates which can be accommodated for direct liquid introduction HPLC/MS (Arpino and Beaugrand, 1985). Spectra were obtained from mass spectrometrically difficult molecules with the filament on and with the filament off a spectrum was obtained from tetrabutyl ammonium bromide, indicating that data similar to that obtained by thermospray ionization can be obtained by what is termed as "hot direct liquid introduction".

Electrospray ionization (Mack et al., 1970; Gienic et al., 1984) provides mass spectra from difficult molecules. It is closely related to the liquid ion evaporation technique discussed earlier. The technique has recently been used for HPLC/MS with a quadrupole instrument (Whitehouse et al., 1984). The eluent from the HPLC column is electrosprayed from ground potential into a bath gas to produce a dispersion of ions that passes into a vacuum system by way of a small supersonic jet. A core portion of the free jet flow carries ions into a high vacuum chamber containing the mass spectrometer. LC/MS data has been obtained from nucleotides, peptides, nucleosides and quinazolines. The technique has the advantage over thermospray ionization that an ionic modifier is not required. However it currently works best with flow rates of 10 μl min^{-1} which limits its most effective use to studies using microbore columns.

Interfaces for HPLC/MS where Solvent is Removed

This approach has evolved from a system where the eluent from the liquid chromatograph was fed onto a moving stainless steel wire (Scott et al., 1974). Solvent was removed in two vacuum locks and the solute was then flash vaporized by electrical heating of the wire into a combined electron impact (EI) chemical ionization (CI) source. The problem was that less than 1% of the effluent from the liquid chromatograph was taken up by the wire resulting in poor sensitivity. Use of a continuously moving belt overcomes the problem (McFadden, 1976). Initially belts were made of stainless steel but this causes problems of mechanical wear on the interface and thermal decomposition of the solute. Current systems use Kapton, which is a polymide material and is capable of withstanding high temperatures. Systems of this type are capable of handling most normal mobile phases at flow rates of 1 ml min^{-1}. Problems are encountered in handling reversed-phase systems at normal flow rates, incorporation of an infrared heater (McFadden, 1980) assists but may cause problems such as thermal decomposition and distillation of solute and in most cases incorporation of a splitting device is necessary. One approach to improving the handling of aqueous mobile phases is incorporation of a continuous liquid extractor, whereby a more volatile solvent is fed to the interface (Karger et al., 1979). Incorporation of this device does however result in degradation of chromatographic performance and it is probably not universally applicable. The most effective approach is to use some form of spray deposition where the bulk of the solvent has been removed prior to the eluent going on to the belt. Both nebulizing (Smith, Johnson, 1981; Hayes et al., 1983) and thermospray deposition (Hardin et al., 1984) devices have been described. An alternative approach is to reduce the mobile phase flow rate by use of microbore columns (Games et al., 1982; Alcock et al., 1982; Foster et al., 1984).

Systems of this type show excellent retention of chromatographic performance (Games et al., 1982; Hayes et al., 1983) and can handle a wide range of volatile modifiers without problems. Involatile modifiers can be handled with the addition of a scrubber system, or incorporation of an on-line continuous extraction interface (Kirby et al., 1981). A major advantage of this approach is the provision of both EI and CI spectra, the value of the former cannot be overemphasized if structural studies are being pursued. The major disadvantage of the approach is that thermal decomposition of labile compounds occurs at low levels (Cairns et al., 1983). Deactivation of the belt by silanizing it (Beattie et al.) or treating with Carbowax 20 M (Van der Greef) can assist with this problem. Systems of this type are commercially available. The earliest (McFadden, 1976), is available from Finnigan MAT on their quadrupole range of instruments. It does not enter into the mass spectrometer ion source, with a resultant loss of sensitivity and difficulty in handling compounds of the type that are not amenable to direct insertion probe study. VG Analytical have developed a system which enters into the mass spectrometer ion source (Millington et al., 1980) and is available on both their magnetic and quadrupole instruments. Their approach results in improved EI sensitivity but as yet there is no evidence that it extends the range of compounds amenable to study. The most recent system, available from Finnigan MAT on their magnetic instruments, also enters the ion source and as a result of careful design shows improved sensitivity and an ability to handle compounds which require desorption chemical ionization (DCI) for their analysis (Dobberstein et al., 1983). The performance of these systems has recently been compared (Games et al., 1984).

Moving belt interfaces have been used for the solution of a wide range of problems (Edmonds, 1983). Sensitivity for thermally stable compounds is comparable with that obtained from interfaces of the direct liquid introduction type. For thermally labile molecules lower detection limits can be reached with interfaces of the direct liquid introduction type. Advantages of moving belt interfaces over direct liquid introduction interfaces include the provision of both EI and CI spectra, stability of spectral quality when gradient elution is used and the ability to use normal phase eluents.

A transport interface which uses a continuous stainless steel belt, which is perforated with small holes has been described (Privett and Erdahl, 1981). Solvent is removed in an evaporator and the remaining solute is transported to a reactor where it is converted into hydrocarbon by catalytic reduction with hydrogen. CI spectra are obtained of the hydrocarbons produced and volatile compounds can be handled without use of the catalytic reduction. Impressive analyses of complex lipid mixtures have been reported. However the system is really more of a liquid chromatographic detector than a true HPLC/MS system.

Earlier we referred to limitations in the range of compounds which can be handled by moving belt systems designed for provision of EI and CI data. Use of surface ionization methods is an attractive approach to extend the range of compounds amenable to study. The most promising investigations in this area are the recent studies of laser desorption (LD) and secondary ion mass spectrometry (SIMS) (Hardin et al., 1984; Fan et al., 1984), and those of other groups with fast atom bombardment (FAB) mass spectrometry (Lewis and Brooks, 1983; Stroh et al., 1984). The viability of using FAB for on-line HPLC/MS has been shown using peptide mixtures (Stroh et al., 1984). Sensitivity and quality of total ion current traces are both areas which require improvement. In the case of LD and SIMS the former gives more consistent molecular weight information but the latter technique provides better total ion current traces and more structurally useful fragment ions. Currently one cannot describe any of these approaches as being full developed, however each show promise and one awaits further development with interest.

Supercritical Fluid Chromatography/Mass Spectrometry (SFC/MS)

Although not strictly with the remit of this chapter one would be remiss not to bring the reader's attention to the future potential of this combination. SFC has better efficiency than HPLC and although it is not as efficient as capillary GC, samples do not require vaporization for analysis, thus, a much wider range of compounds can be analyzed than by GC. The advantage of SFC over HPLC for combination with a mass spectrometer is that much lower gas volumes are generated making interfacing easier. Studies with columns of the conventional HPLC type were not very successful (Randall and Wahrhaftig, 1981), although good interfacing using a direct liquid introduction system has recently been achieved (Henion). To date the best data in the literature is with capillary SFC/MS (Smith et al., 1982). Both EI and conventional CI spectra are obtainable. As studies in the area increase, undoubtedly a wide range of compounds will prove to be amenable to study by this approach.

Applications of HPLC/MS

A comprehensive bibliography of publications on HPLC/MS covering the literature to late 1982 has appeared (Edmonds et al., 1983) and the abstracting journal Liquid Chromatography Mass Spectrometry Abstracts (Aylott) provides an on-going source of references and abstracts of papers published on HPLC/MS. Although the perfect universal system for HPLC/MS has not yet been developed, selection of an appropriate interface for the problem to be addressed usually results in the provision of useful data which can assist in solution of the problem. Currently in most laboratories the choice rests with using either a moving belt, direct liquid introduction or thermospray interface. This section will review areas relevant to endocrinology where interfaces of these types have been used or have, on the basis of running model compounds, the potential for use in the solution of problems.

Steroids

Steroid have been used by a number of groups to demonstrate interfaces for HPLC/MS. A mixture of 5-α-3-androstanone, estrone methyl ether and androstanolone was used to demonstrate the viability of the first direct liquid introduction system (Arpino et al., 1974b) and a mixture of oestrone, oestradiol and oestriol in a similar exercise with a moving belt system (Games et al., 1978). A number of groups have demonstrated an ability to analyse corticosteroids (Melera, 1980; Yamauchi et al., 1980; Yoshida et al., 1980; Henion, Wachs, 1981; Sugnaux et al., 1983). A moving belt interface has also been used to confirm the presence of corticosteroids in horse urine samples (Houghton et al., 1981). In this case negative ammonia CI was found to be the ionization method of choice. However there are problems with analysis of compounds of this type with moving belt systems, since thermal decomposition occurs at low levels (Cairns et al., 1983) and modern interfaces of the direct liquid introduction or thermospray types would appear to be preferable if low level studies are contemplated.

Moving belt interfaces do have some advantages in analysis of steroids in that if the interface is on a magnetic instrument medium resolution accurate mass data can be obtained. A mixture of cholest-4-en-3-one testosterone and pregnenolone gave accuracies of mass measurement on their molecular ions and some other ions in their spectra of between 10–15ppm at a resolution of 7000 with 250 ng in each component injected onto the column of the liquid chromatograph (Baty and Willis, 1982). Also since EI spectra can be obtained computerized library searches can be used to assist in identification.

A most impressive demonstration of the utility of HPLC/MS in this area has been the use of a direct liquid introduction system in the identification of marine sterols (Sugnaux and Djerassi, 1982; Arpino, 1983). A series of epidioxy steroids gave abundant protonated molecular ions and the technique was used to identify compounds of this type in extracts of marine sponges.

A more difficult proposition is the analysis of conjugated steroids. A series of estrogen glucuronides have been studied using ammonia CI with a moving belt interface (Cairns and Siegmund, 1982). Only two of the compounds gave molecular weight information however characteristic spectra were obtained which enabled the aglycone to be identified and a characteristic ion at *m/z* 194 corresponding to the ammonia adduct of the glucuronic acid portion should enable characterization to be made. These studies were performed on the earliest moving belt interface and the most recent interface of this type will undoubtedly provide better data. As yet no other studies have been reported on steroid glucuronides, however glucuronides of other classes of compound have been successfully analysed with moving belt (Games, Lewis, 1980), direct liquid introduction (Kenyon et al., 1981) and thermospray (Liberato et al., 1983) interfaces. Bile acids and a glycine conjugate give protonated molecular ions with moving belt interfaces (Games et al., 1980), however the taurine conjugates failed to provide molecular weight information. It will be of considerable interest to see the range of steroid conjugates which are amenable to study by thermospray HPLC/MS, perhaps sulphates will prove to be amenable to study with this approach.

Biogenic Amines

Dopamine (DA), noradrenaline (NA) and one of their metabolites 3,4-dihydroxyphenylacetic acid (DOPAC) have been studied using a direct liquid introduction system (Milon, Bur, 1983). Direct injection provided $(M+1)^+$ ions for all three compounds, however when introduced with the *n*-octylsulphonate ion pair used for HPLC only DOPAC produced ions. Initial studies were performed without a desolvation chamber, but incorporation of this device overcame the problem providing spectra characteristic of the ion pair and amine. Unfortunately no online studies were reported of a mixture of the amines. Ion evaporation with an API source has been used to analyze a mixture of norepinephrine, epinephrine and dopamine (Sushan et al., 1983).

Thermospray HPLC/MS has been investigated as a method for the analysis of L-carnitine ($Me_3\overset{+}{N}$ $CH_2CH(OH)CH_2$-CO_2^-), acetylcarnitine and propionylcarnitine (Yergey et al., 1984). On-line studies yielded $(M+H)^+$ ions for all the compounds with detection limits in the selected ion monitoring mode below 0.1 nmol. The feasibility of performing quantitative studies in biological fluids was demonstrated using isotopically labelled compounds as internal standards.

Amino Acids and Peptides

t-Butyl derivatives of tryptophan have been analyzed by direct liquid introduction HPLC/MS (Dixon, 1982) and the potential of a vacuum nebulizing interface for studying underivatized amino acids has been shown (Yoshida et al., 1982). Ion evaporation with an API source has been used to analyze a mixture of creatine, valine, leucine and isoleucine (Shushan et al., 1983). We have studied mixtures of aromatic amino acids using a moving belt interface (Games et al., 1984). Figure 2 shows the computer reconstructed total ion current trace obtained from a mixture of five amino acids by ammonia CI HPLC/MS. All the compounds gave $[M+1]^+$

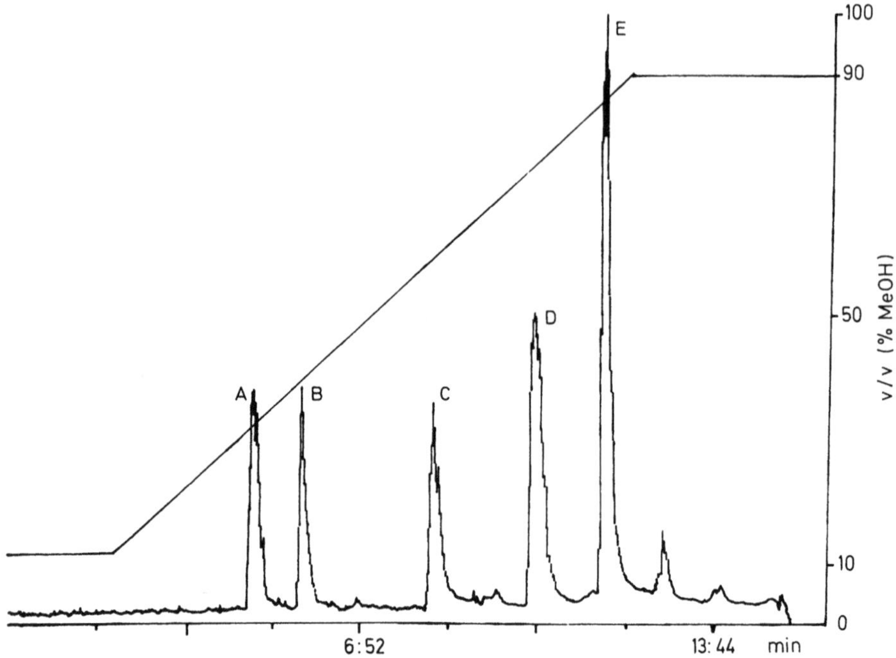

Fig. 2. Computer reconstructed total ion current trace obtained by ammonia CI HPLC/MS of a mixture of *p*-hydroxyphenylglycine, B phenylglycine, C tyrosine, D β-phenylalanine, E tryptophan. A Spherisorb ODS 5 250 × 4.7 mm column was used with a gradient of methanol + water. The mobile phase was maintained at pH 2.3 using trifluoroacetic acid and ammonia and the flow rate was 1 ml min^{-1}

ions and the spectrum obtained from tryptophan is shown in Figure 3. In order to obtain spectra of the quality shown of the order of 5 μg of each, amino acid had to be injected on column. At lower levels although good total ion current traces were obtained $[M+1]^+$ ions were weak or absent. The reasons for this poor sensitivity are twofold: At the start of the gradient a very high percentage aqueous mobile phase was being used and thus less than 5% of the eluent could be handled by the interface. Secondly the compounds are thermally labile and at low concentrations on the belt thermal decomposition occurs. Use of 1 mm i. d. liquid chromatographic columns improved the situation since all the eluent could be handled by the interface. Similar quality spectra were obtained under these conditions with 200 ng of each amino acid injected onto the column. PTH derivatives of amino acids proved to be an easier proposition, providing characteristic EI spectra at the pmol level when microbore HPLC/MS was used. Thermospray ionization is an excellent method for providing mass spectra of amino acids (Blakley et al., 1980). Using a Finnigan MAT thermospray interface we have studied the mixture of aminoacids mentioned earlier with the addition of p-aminophenylalanine. Because of the availability of temperature programming for the capillary heater we were able to use a gradient system in these studies. The chromatographic conditions had to be modified to incorporate ammonium

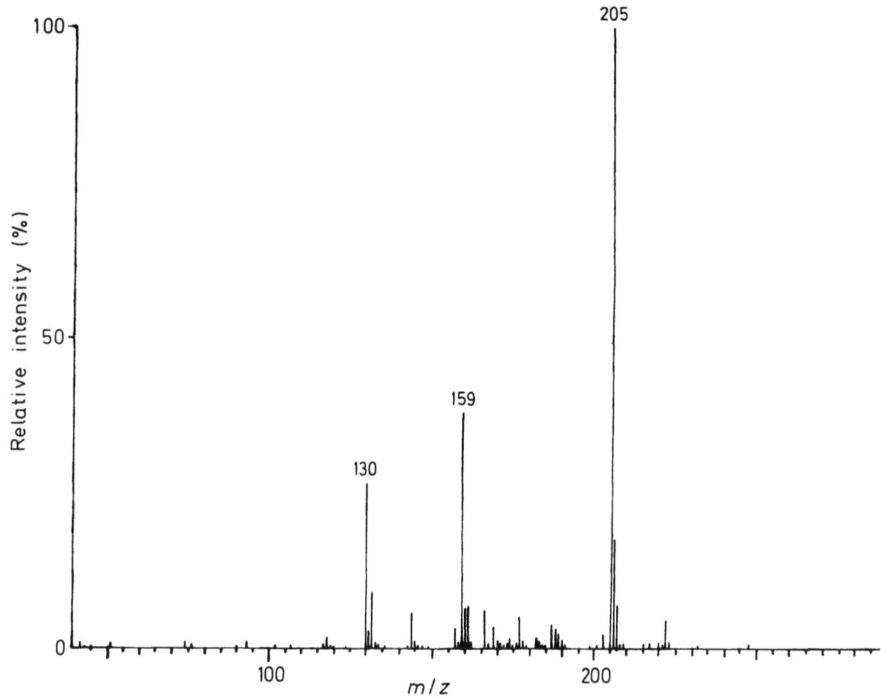

Fig. 3. Ammonia CI mass spectrum of tryptophan obtained from Figure 2

acetate buffer, an excellent total ion current trace was obtained, however only five amino acids were detected, p-hydroxyphenylglycine being absent. When injected in solution directly into the thermospray source this compound gave a spectrum but sensitivity was very poor. We are presently unable to explain the behaviour of this compound, which is particularly baffling in that with an earlier version of the source it gave excellent spectra. Another notable feature of this study was the variation in sensitivity between early and late eluting peaks. The former giving sensitivities vastly superior to the moving belt system, whilst with the latter compounds the moving belt gave better sensitivities. This is due to variation in the amount of water present, high percentage aqueous systems producing better sensitivity. In order to maintain sensitivity throughout the run, post column addition of water using the approach described earlier (Voyksner 1984) is necessary.

The ability to obtain mass spectra of amino acids at the pmol level using thermospray ionization has recently been exploited as a method for obtaining sequence information from peptides (Pilosof et al., 1984). Underivatized peptide solutions are injected through a column containing immobilized carboxypeptidase Y and the amino acids released starting from the C-terminus of the peptide chain are directly transported by a continuously flowing aqueous buffer into the thermospray source of a mass spectrometer, where they are detected and their $[M+1]^+$ ions are quantified.

Underivatized peptides because of their low volability are difficult to handle by conventional EI or CI mass spectrometry. Although some simple underivatized peptides have been handled by a moving belt system (Games et al., 1978), it is not the method of choice. The combination of fast atom bombardment (FAB) and other surface ionization techniques (see earlier) show some promise for future development of interfaces of this type. FAB HPLC/MS has been performed on mixtures of underivatized peptides formed from the partial hydrolysis of antiamoebin I (Ac-Phe-Aib-Aib-Aib-Iva-Gly-Leu-Aib-Aib-Hyp-Gln-Iva-Hyp-Aib-Pro-Phol). The peptides yielded an abundance of sequence information in their FAB spectra. At present the total ion current traces obtained with this approach are poor and μg amounts of peptide have to be injected onto the HPLC column.

Direct liquid introduction HPLC/MS has been used for the detection and sequencing of leu- and met-enkephalin (Kenyon, 1983). The positive and negative ion mass spectra obtained enabled the peptides to be readily identified and sequenced. This study was performed at the high picomole level and a conventional HPLC column was used. Use of a microbore system would considerably improve the detection limits. Figure 4 shows the computer reconstructed total ion current trace obtained in this study from a mixture of met- and leu-enkephalin and Figure 5 shows the positive ion mass spectra obtained from the two compounds. Thermospray ionization has also been shown to be effective for

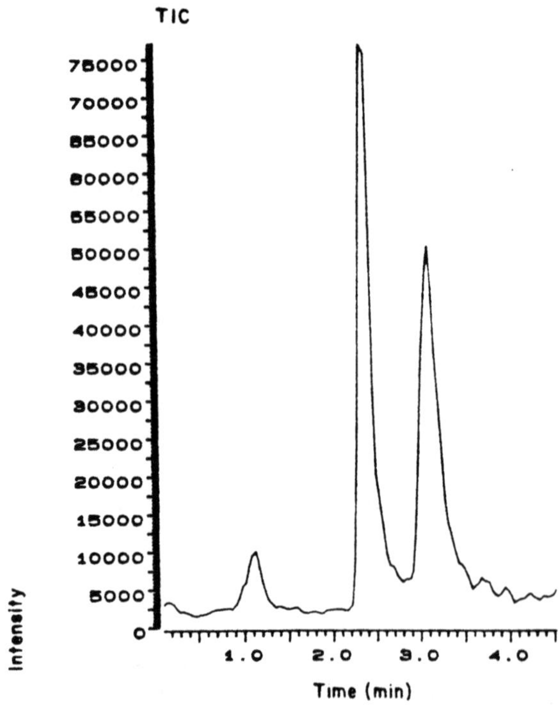

Fig. 4. Total ionization chromatogram for a mass range of 150–650 u, negative ion direct liquid introduction HPLC/MS analysis of met- and leuenkephalin (5 μg of each injected). Reproduced from Kenyon, 1983, with permission of John Wiley and Sons. Copyright (1983) John Wiley & Sons Ltd.

handling leu-enkephalin (Blakley, 1980a; Covey, Henion, 1983). The spectra reported in these papers is much simpler than that obtained by direct liquid introduction and there is an abundant $[M+Na]^+$ ion as well as the $[M+1]^+$ ion, but little sequence information. Larger peptides have been studied by this technique, renin substrate showed a $[M+1]^+$ ion at *m/z* 1758 (Blakley, Vestal, 1983) but no sequence information. In another study (Pilosof et al., 1984) even though the mass range of the mass spectrometer was not large enough to show $[M+H]^+$ ions, a $[M+2H]^{2+}$ ion at *m/z 833* was observed for α-melanocyte stimulating hormone (α-MSH) and a $[M+3H]^{3+}$ ion at *m/z* 1162 for glucogen. On-line HPLC/MS was performed on a mixture containing 2nmol each of renin substrate, α-MSH and [Sar^1, Ala^8] angiotensin II using mass chromatography of the doubly charged protonated molecular ions for the first two compounds and the $[M+1]^+$ ion for the angiotensin all three compounds were detected. Thermospray HPLC/MS clearly has considerable potential analysis of underivatized peptides.

One of the problems with studying underivatized peptides with any mass spectral ionization technique is that ions which can provide sequence information are not always present. FAB mass spectrometry is probably the best technique for provision of this information but even here complete sequence information is not always present. The best approach, to be sure of obtaining sequence information is to study derivatized peptides. This approach has been widely used for sequencing peptides and proteins by studying the derivatized hydrolysis products either with GC/MS or by introduction of sample on a direct insertion probe (Biemann, 1980). In the former case the size of peptide which can be studied is limited by volatility, whereas in the latter case only relatively simple mixtures can be handled. HPLC/MS has the potential to handle complex mixtures with less restriction of peptide size. Studies of using mixtures of N-acetyl peptide methyl esters with a direct liquid introduction system showed the viability of this approach although molecular weight and sequence peaks were not obtained in all cases (Dawkins et al., 1978). N-acetyl-N,O-permethylated peptides provide excellent sequence information. Two groups have investigated mixtures of peptides derivatized in this way on a moving belt system (Yu et al., 1981; Roepstorff et al., 1983). In one case isobutane CI was used (Yu et al., 1981) and in the other ammonia CI (Roepstorff et al., 1983), in both cases molecular weight and N- and C-terminal sequence ions are present enabling unequivocal sequence of the peptides. One problem with the approach is the production of under and over methylation artefacts (Yu, 1983; Roepstorff, 1983), however use of a combination of deuterated and non-deuterated reagents would overcome problems in interpretation of spectra. We have used the technique to identify enzymic hydrolysis products of peptides. Presently of the order of 40 nmol of peptide is required, however with improvements in the permethylation procedure and use of spray deposition or microbore HPLC with the belt interface we anticipate requiring much smaller amounts of peptide.

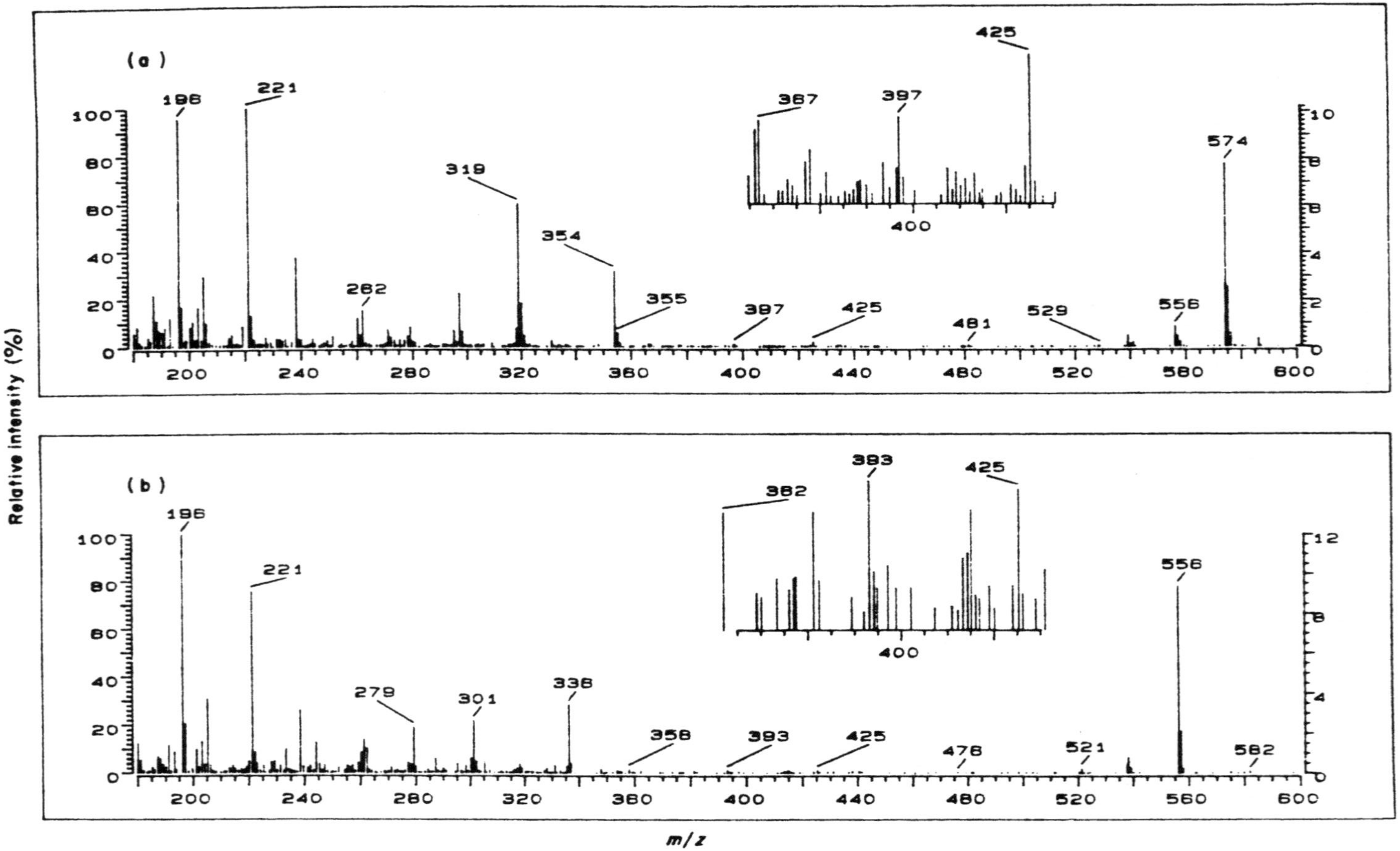

Fig. 5. Positive ion mass spectra for meta- and leu-enkephalin separated as in Figure 4 a) met-enkephalin, b) leu-enkephalin. The peak at m/z 196 arises from the TEAA buffer used. Reproduction form Kenyon, 1983, with permission of John Wiley and Sons. Copyright (1983) John Wiley & Sons

Conclusions

As yet there is no one HPLC/MS system which is capable of handling all the compounds which are amenable to liquid chromatographic study, nor does such a system appear to be forthcoming in the near future. Currently one is faced with a choice between interfaces of the moving belt, direct liquid introduction or thermospray types, or a combination of such systems. All three types of system are now in use in many laboratories on a relatively routine basis and are solving problems. Often it is necessary to modify chromatographic conditions to obtain optimal HPLC/MS data. Choice of system depends on the types of problem to be addressed and the compounds which require analysis. Readers contemplating purchase of such systems would be well advised to ensure that the system really can do HPLC/MS on their compounds, they should not be satisfied with a demonstration which just injects sample into a liquid stream or spots a solution of it onto a moving belt interface. There can be marked differences between on-line and off-line data.

With regard to selection of an appropriate interface the following points are now well established. Moving belt systems have the advantage of providing both EI and CI spectra. The former being a universal detection system, whilst the latter can provide a measure of selectivity, molecular weight information and improved sensitivity in the negative mode for some classes of compound. By the very nature of the interface it is unlikely to achieve the sensitivities obtainable by the direct coupling of a capillary gas chromatographic column to the mass spectrometer. Use of microbore HPLC or spray deposition device do, however, enable acceptable sensitivities to be achieved. Both normal and reversed mobile phases and a wide range of modifiers can be handled and spectra can be obtained from a lower mass range than other interfaces, although the presence of some higher mass chromatographic modifiers can restrict the mass range. The upper limit of compounds which can be handled by the best system is the same as those which provide molecular weight information by DCI. Hence it is pointless attempting to use such a system for the analysis of e. g. large underivatized peptides or tetra- or larger saccharides. The range of compounds which can be handled can be extended by use of surface ionization methods, but this approach is not yet fully developed. The major problem is thermal decomposition of thermally labile compounds particularly at low levels. Hence quantification of such compounds is unlikely to be very successful and detection limits will be high.

Interfaces of the direct liquid introduction type have restricted detection limits if conventional HPLC columns are used. However, sensitivity obtainable with microbore HPLC approaches that of capillary GC/MS. Data is restricted to solvent mediated CI spectra, although a reagent gas can also be introduced into the source. Much more care has to be exercised with this system in quality of solvents and filters must be incorporated into the system to prevent blocking of the diaphragm. Because of the presence of solvent reagent ions the mass range of spectra is limited at the lower mass end. The approach works best with reversed phase systems, although data has been obtained with normal phase systems. A major advantage over moving belt systems is that there is less evidence of thermal decomposition of thermally labile compounds. The range of compounds amenable to study on current evidence appears to be similar to that on moving belt interfaces.

Thermospray ionization HPLC/MS has the advantages over the other approaches in that it works best with high percentage aqueous mobile phases and can provide molecular weight information from mass spectrometrically difficult compounds. It is easy to use, but the ionization process is selective which may cause problems when mixtures of unknowns are being analyzed and the lower end of the mass range is restricted because of the presence of solvent ions. Sensitivity for compounds which work well with the technique is excellent. If normal phase HPLC is being undertaken it may be possible to add buffer and water post-column to effect ionization, although this may cause problems with sample coming out of solution. So far there is no evidence for improved sensitivity in the negative mode for compounds which have high electron affinities. A further problem is the simplicity of the spectra produced which will mean that for most effective use the technique is used with a tandem mass spectrometer.

Of the other approaches to HPLC/MS. Use of API systems show promise but there are still problems with obtaining good total ion current traces and the approach works best when known compounds are being analyzed. Electrospray/ion evaporation show promise for the handling of more difficult molecules and may prove to be more universal than thermospray.

Combined HPLC/MS has established itself as an effective technique which can assist in the solution of a wide range of problems. It is hoped that researchers in the endocrinology area will be stimulated to explore the possibilities of HPLC/MS in their investigations.

References

Alcock NJ, Corbelli L, Games DE, Lant MS, Westwood SA (1982) Liquid chromatography/mass spectrometry using glass lined stainless steel microbore columns. Biomed Mass Spectrom 9:499–504

Apffel JA, Brinkman UAT, Frei RW, Evers EAI (1983) Gas-nebulized direct liquid introduction interface for liquid chromatography/mass spectrometry. Anal Chem 55:2280–2284

Arpino P, Baldwin MA, McLafferty FW (1974a) Liquid chromatography-Mass spectrometry II – Continuous monitoring. Biomed Mass Spectrom 1:80–82

Arpino PJ, Dawkins BG, McLafferty FW (1974b) A liquid chromatography/mass spectrometry system providing continuous monitoring with nanogram sensitivity. J Chromatogr Sc 12:574–578

Arpino PJ, Guiochon G (1979) HPLC/MS Coupling. Anal Chem 51:682A–701A

Arpino PJ, Guiochon G, Krien P, Devant G (1979) Optimization of the instrumental parameters of a combined liquid chromatrograph-mass spectrometer, coupled by an interface for direct liquid introduction I: Performance of the vacuum equipment. J Chromatogr 185:529–547

Arpino PJ, Krien P, Vajta S, Devant G (1981) Optimization of the instrumental parameters of a combined liquid chromatograph-mass spectrometer, coupled by an interface for direct liquid introduction II: Nebulization of liquids by diaphragms. J Chromatogr 203:117–130

Arpino PJ (1983) Mass spectrometric detectors . In: Vickrey TM (ed) Liquid Chromatography Detectors. Marcel Dekker, New York, p 243–322

Arpino PJ, Bounine JP, Dedieu M, Guiochon G (1983) Optimization of the instrumental parameters of a combined liquid chromatograph-mass spectrometer coupled by an interface for direct liquid introduction. IV: A new desolvation chamber for droplet focusing or Townsend discharge ionization. J Chromatogr 271:43–50

Arpino PJ, Beaugrand C (1985) Design and construction of HPLC/MS interfaces utilizing fused-silica capillary tubes as vacuum nebulizers. Int J Mass Spectrom Ion Phys, 61:275–298
Aylott IC, Editor Liquid Chromatography Mass Spectrometry Abstracts, PRM Science and Technology Agency Ltd, 261A Finchley Road, London NW3 6LU
Baty JD, Willis RG (1982) HPLC-MS detection of steroids and drugs at low and medium resolution. Anal Proc 19:251–253
Beattie IG, Games DE, Gilbert J, Startin J, unpublished work
Biemann K (1980) Amino acid sequence in oligopeptides and proteins. In: Waller GR, Dermer OC (eds) Biochemical applications of mass spectrometry. First supplementary volume, John Wiley & Sons, New York, p 469–525
Blakley CR, Carmody JJ, Vestal ML (1980a) A new soft ionization technique for mass spectrometry of complex molecules. J Amer Chem Soc 102:5931–5933
Blakley CR, Carmody JJ, Vestal ML (1980b) Liquid chromatograph-mass spectrometer for analysis of nonvolatile samples. Anal Chem 52:1636–1641
Blakley CR, Carmody JC, Vestal ML (1980c) Combined liquid chromatograph/mass spectrometer for involatile biological samples. Clin Chem 26:1467–1473
Blakley CR, Vestal ML (1983) Thermospray interface for liquid chromatography/mass spectrometry. Anal Chem 55:750–754
Bruins AP, Drenth BFH (1983) Experiments with the combination of a micro liquid chromatograph and a chemical ionization quadrupole mass spectrometer, using a capillary interface for direct liquid introduction. Some theoretical considerations concerning the evaporation of liquids from capillaries into vacuum. J Chromatogr 271:71–82
Cairns T, Siegmund EG (1982) Characterization of glucuronides by chemical ionization mass spectrometry with ammonia as reagent gas. Anal Chem 54:2456–2461
Cairns T, Siegmund EG, Stamp JJ, Skelly JP (1983) Liquid chromatography mass spectrometry of dexcamethasone and betamethasone, Biomed Mass Spectrom 10:203–208
Carroll DI, Dzidic I, Stillwell RN, Haegele KD, Horning EC (1975) Atmospheric pressure ionization mass spectrometry: Corona discharge ion source for use in liquid chromatograph-mass spectrometer – computer analytical system. Anal Chem 47:2369–2373
Christensen RG, White E, Meiselmann VS, Hertz HS (1983) Quantitative trace analysis by reversed-phase liquid chromatography-mass spectrometry. J Chromatogr 271:61–70
Covey T, Henion J (1983) Direct liquid introduction/thermospray interface for liquid chromatography/mass spectrometry. Anal Chem 55:2275–2280
Curry ZF (1982) Recent developments in HPLC/MS interconnection. J Liquid Chromatogr 5:(Suppl 2) 257–272
Dawkins BG, Arpino PJ, McLafferty FW (1978) Polypeptide sequencing by liquid chromatography mass spectrometry. Biomed Mass Spectrom 5:1–6
Desiderio DM, Fridland GH (to be published) A review of combined liquid chromatography and mass spectrometry. J Liquid Chromatogr
Dixon DJ (1982) Applications of the direct coupling of liquid chromatography with mass spectrometry to biochemical problems. Analysis 10:343–348
Dobberstein P, Korte E, Meyerhoff G, Pesch R (1983) Investigation of an HPLC/MS interface for EI-, CI- and FAB-ionization. Int J Mass Spectrom Ion Phys 46:185–188
Edmonds CG, McCloskey JA, Edmonds VA (1983) A bibliography of combined liquid chromatography mass spectrometry. Biomed Mass Spectrom 10:237–252
Evans N, Williamson JE (1981) The construction and use of simple interfaces for combined liquid chromatography mass spectrometry. Biomed Mass Spectrom 8:316–321
Fan TP, Hardin ED, Vestal ML 1984 Direct comparison of secondary ion and laser desorption mass spectrometry on bio-organic molecules in a moving belt HPLC-MS system. Anal Chem 56:1870–1876
Foster MG, Meresz O, Games DE, Land MS, Westwood SA (1983) Capillary gas chromatographic mass spectrometric and microbore liquid chromatographic mass spectrometric studies of a test well sample from a landfill site. Biomed Mass Spectrom 110:338–342
Games DE, Gower JL, Lee MG, Lewis IAS, Pugh ME, Rossiter M (1978) Scope of HPLC-MS of "soft" ionization MS in quanitation. In: Reid E (ed) Blood drugs and other analytical challenges. Vol 7 of Methodological surveys of Biochemistry. Ellis Horwood, Chichester, p 185–194

Games DE, Eckers C, Gower JLK, Hirter P, Knight ME, Lewis E, Rao KRN, Weerasinghe NC (1980) Combined HPLC-MS and its potential in clinical studies. In: Lawson AM, Lim CK, Richmond W (eds) Current developments in the clinical applications of HPLC, GC and MS. Academic Press, London p 97–118
Games DE, Lewis E (1980) Combined liquid chromatography mass spectrometry of glucosides, glucuronides, sugars and nucleosides. Biomed Mass Spectrom 7:433–436
Games DE, Hewlins MJE, Westwood SA, Morgan DJ (1982a) Extracolumn band spreading in high-performance liquid chromatography-mass spectrometry using a moving belt interface. J Chromatogr 250:62–67
Games DE, Lant MS, Westwood SA, Cocksedge MJ, Evans N, Williamson J, Woodhall BJ (1982b) Microbore high-performance liquid chromatography-mass spectrometry. Biomed Mass Spectrom 9:215–224
Games DE (1983) High-performance liquid chromatography/mass spectrometry (HPLC/MS) In: Giddings JC, Grushka E, Cazes J, Brown PR (eds) Advances in chromatography, Vol 21, Marcel Dekker, New York, p 1–39
Games DE, McDowall MA, Levsen K, Schafer KH, Dobberstein P, Gower JL (1984a) A comparison of moving belt interfaces for liquid chromatography mass spectrometry. Biomed Mass Spectrom II:87–95
Games DE, Ramsey ED, Catlow DA, Woodhall BJ, Roepstorff P (1984b) Combined HPLC/MS of amino acids and peptides. 32nd Annual Meeting on Mass Spectrometry and Allied Topics, San Antonio, Texas, paper MOC 9
Gienic J, Mack LL, Nakamae K, Gupta C, Kumar V, Dole M (1984) Electrospray mass spectroscopy of macro molecules: Application of an ion-drift spectrometer. Biomed Mass Spectrom 11:259–268
Hardin ED, Fan TP, Blakley CR, Vestal ML (1984) Laser desorption mass spectrometry with thermospray sample deposition for determination of nonvolatile biomelocules. Anal Chem 56:2–7
Hayes MJ, Lankmayer EP, Vouros P, Karger BL, McGuire JM (1983) Moving belt interface with spray deposition for liquid chromatography/mass spectrometry. Anal Chem 55:1745–1752
Hayer MJ, Schwartz HE, Vouros P, Karger BL, Thurston Jr AD, McGuire J/M (1984) Gradient liquid chromatography/mass spectrometry using microbore columns and a moving belt interface. Anal Chem 56:1229–1236
Henion JD, personal communication
Henion JD (1978) Drug analysis by continuously monitored liquid chromatography/mass spectrometry with a quadrupole mass spectrometer. Anal Chem 50:1687–1693
Henion JD, Maylin GA (1980) Drug analysis by direct liquid introduction micro liquid chromatography mass spectrometry. Biomed Mass Spectrom 7:115–121
Henion JD, Wachs T (1981) Micro liquid chromatography/mass spectrometry diaphragm probe interface. Anal Chem 53:1963–1965
Henion JD (1981) Continuous monitoring of total micro HPLC eluant by direct liquid introduction HPLC/MS. J Chromatogr Sci 19:57–64
Henion JD, Thomson BA, Dawson PH (1982) Determination of sulfa drugs in biological fluids by liquid chromatography/mass spectrometry/mass spectrometry. Anal Chem 54:451–456
Houghton E, Dumasia MC, Wellby JK (1981) The use of combined high performance liquid chromatography negative ion chemical ionization mass spectrometry to confirm the administration of synthetic corticosteroids to horses. Biomed Mass Spectrom 8:558–564
Jones PR, Yang SK (1975) A liquid chromatograph/mass spectrometer interface. Anal Chem 476:1000–1003
Kambara H (1982) Sample introduction system for atmospheric pressure ionization mass spectrometry of nonvolatile compounds. Anal Chem 54:143–146
Karger BL, Kirby DP, Vouros P, Foltz RL, Hidy B (1979) On-line reversed phase liquid chromatography-mass spectrometry. Anal Chem 51:2324–2328
Kenyon CN (1983) Sequencing of underivatized peptides by direct liquid inlet liquid chromatography mass spectrometry. Biomed Mass Spectrom 10:535–543
Kenyon CA, Melera A, Erin F (1981) Utilization of direct liquid inlet HPLC/MS in studies of pharmacological and toxicological importance. J Anal Toxicol 5:216–230

Kirby DP, Vouros P, Karger BL, Hidy B, Petersen B (1981) On-line liquid chromatography-mass spectrometry of ion pairs. J Chromatogr 203:139–152

Lewis IAS, Brooks PW (1983) Soft ionisation HPLC-MS 31st Annual Meeting on Mass Spectrometry and Allied Topics, Boston, Mass, 850–D851

Liberato DJ, Fenselau CF, Vestal ML, Yergey AL (1983) Characterization of glucuronides with a thermospray liquid chromatography/mass spectrometry interface. Anal Chem 55:1741–1744

Mack LL, Kralik P, Rheude A, Dole M (1970) Molecular beams of macroions II. J Chem Phys 52:40977–40986

Majors RE (1983) Special report on HPLC/MS liquid chromatography and HPLC magazine 1:488–495

McFadden WH (1980) Liquid chromatography/mass spectrometry systems and applications. J Chromatogr Sci 18:97–115

McFadden WH, Schwartz HL, Evans S (1976) Direct analysis of liquid chromatographic effluents. J Chromatogr 122:389–396

Melera A (1980) Design, operation and applications of a novel HPLCMS CI interface. Adv Mass Spectrom 8B:1597–1616

Millington DS, Yorke DA, Burns P (1980) A new liquid chromatography-mass spectrometry interface. Adv Mass Spectrom 88:1819–1825

Milon H, Bur H (1983) Mass spectrometry of biogenic catecholamnine ion pairs by direct liquid introduction. J Chromatogr 271:83–92

Pilosof D, Kim HY, Dyckes DF, Vestal ML (1984) Determination of nonderivatized peptides by thermospray liquid chromatography/mass spectrometry. Anal Chem 56:1236–1240

Pilosof D, Kim H-Y, Vestal ML, Dyckes DF (to be published) Direct monitoring of sequential enzymatic hydrolysis of peptides by thermospray mass spectrometry. Biomed Mass Spectrom

Privett OS, Erdahl WL (1981) Lipid analysis by liquid chromatography-mass spectrometry. Methods in Enzymology 72:56–108

Randall LG, Wahrhaftig AL (1981) Direct coupling of a dense (supercritical) gas chromatograph to a mass spectrometer using a supersonic molecular beam interface. Rev Sci Instrum 52:1283–1295

Roepstorff P, McDowall MA, Games MPL, Games DE (1983) Peptide sequencing by combined liquid chromatography/mass spectrometry. Int J Mass Spectrom Ion Phys 48:197–200

Schafer KH, Levsen K (1981) Direct coupling of a micro high-performance liquid chromatograph and a mass spectrometer. J Chromatogr 206:245–252

Scott RPW, Scott CG, Muroe M, Hess Jr J (1974) Interface for on-line liquid chromatography-mass spectrometry analysis. J Chromatogr 99:395–405

Shushan B, Fulford JE, Thomson BA, Davidson WR, Danylewych LM, Ngo A, Nacson S, Tanner SD (1983) Recent applications of triple quadrupole mass spectroetry to trace chemical analysis. Int J Mass Spectrom Ion Phys 46:225–228

Smith RD, Johnson AL (1981) Deposition method for moving ribbon liquid chromatograph-mass spectrometer interfaces. Anal Chem 53:739–740

Smith RD, Fjeldsted JC, Lee ML (1982) Direct fluid injection interface for capillary supercritical fluid chromatography-mass spectrometry. J Chromatogr 247:231–243

Stroh JG, Cook JC, Rinehart Jr KL, Lewis IAS (1984) Liquid chromatography/fast atom bombardment mass spectrometry (HPLC/FABMS) 32nd Annual Meeting on Mass Spectrometry and Allied Topics. San Antonio, Texas. Paper MPA 20

Sugnaux FR, Djerassi (1982) Liquid chromatography-mass spectrometry of polar lipids: Comparison of on-line method using direct liquid introduction interface with off-line method using desorption/chemical ionization. J Chromatogr 251:189–201

Sugnaux FR, Skrabalak DS, Henion JD (1983) Direct liquid introduction micro-liquid chromatography-mass spectrometry coupling. Optimization of droplet desolvation and instrumental parameters for high sensitivity. J Chromatogr 264:357–376

Takeuchi T, Hirata Y, Okumura Y (1978) On-line coupling of a micro liquid chromatograph and mass spectrometer through a jet separator. Anal Chem 50:659–660

Thomson BA, Iribane JV, Dziedzic PJ (1982) Liquid ion evaporation/mass spectrometry/mass spectrometry for the detection of polar and labile compounds. Anal Chem 54:2219–2224

Tsuchiya M, Taira T (1980) A new method of ionization for organic compounds in the liquid phase under atmospheric pressure. Int J Mass Spectrom Ion Phys 34:351–359

Tsuchiya M, Kuwabara H (1984) Liquid ionization mass spectrometry of nonvolatile organic compounds. Anal Chem 56:14–19
Van der Greef J, personal communication
Voyksner RD, Hass JR, Bursey MM (1982a) An on-line liquid chromatography/mass spectrometry/mass spectrometry experiment. Anal Lett 15:1–12
Voyksner RD, Parker CE, Hass JR, Bursey MM (1982b) Effects of pressure temperature and solvent composition in analysis by direct liquid introduction liquid chromatography/mass spectrometry. Anal Chem 54:2583–2586
Voyksner RD, Bursey JT, Pellizzari ED (1984) Post-columns addition of buffer for thermospray liquid chromatography/mass spectrometry identification of pesticides. Anal Chem 56:1507–1514
Whitehouse CM, Fenn JB, Dreyer RN (1984) The electrospray ion source (ESPIS) as an HPLC/MS interface. 32nd Annual Meeting on Mass Spectrometry and Allied Topics. San Antonio, Texas, paper MPA 6
Whitehouse CM, Fenn JB, Yamashita M (1984) An electrospray ion source (ESPIS) for mass spectrometry of fragile organic specis. 32nd Annual Meeting on Mass Spectrometry and Allied Topics. San Antonio, Texas, paper TOB 8
Willoughby RC, Browner RF (1982) Combining liquid chromatography with mass spectrometry. In: Lawrence JF (ed) Trace Analysis, vol 2, Academic Press, New York, p 69
Yamauchi E, Mizuno T, Azuma K (1980) Direct coupling interface for high sensitivity HPLC/MS. Shibak Shitsuryo Bunseki 28:227–234
Yorgey AL, Liberator DJ, Millington DS (1984) Thermospray liquid chromatography/mass spectromentry for the analysis of L-carnitine and its shortchain acyl derivatives. Anal Biochem 139:278–283
Yoshida Y, Yoshida H, Tsuge S, Takeuchi T, Mochizuki K (1980) Direct measurement of mass fragmentograms for eluents from a micro-liquid chromatogram using an imporoved nebulizing interface. J High Res Chromatogr & CC 3:16–20
Yoshida H, Matsumoto K, Itoh K, Tsuge S, Hirata Y, Mochizuki K, Kokubun N, Yoshida Y (1982) Improvement of vacuum nebulizing interface for direct coupling micro-liquid chromatograph with mass spectrometer and some applications to polar natural organic compounds. Fresenius Z Anal Chem 311:674–680
Yoshida H, Matsumoto K, Itoh K, Tsuge S, Hirata Y, Mochizuki K, Kokubun N, Yoshida Y (1982) Improvement of vacuum nebulizing interface for direct coupling micro-liquid chromatograph with mass spectrometer and some applications to polar natural organic compounds. Fresenius Z Anal/Chem 311:674–680
Yu TJ, Schwartz H, Giese RW, Karger BL, Vouros P (1981) Analysis of N-acetyl-N.O.S-permethylated peptides by combined liquid chromatography-mass spectrometry. J Chromatogr 218:519–533
Yu TJ, Karger BL, Vouros P (1983) High-performance liquid chromatographic mass spectrometric examination of C-methylation artifacts from the permethylation reaction of peptides. Biomed Mass Spectrom 10:633–640

Subject Index

MIX
Papier aus verantwortungsvollen Quellen
Paper from responsible sources
FSC® C105338

If you have any concerns about our products,
you can contact us on
ProductSafety@springernature.com

In case Publisher is established outside the EU,
the EU authorized representative is:
Springer Nature Customer Service Center GmbH
Europaplatz 3, 69115 Heidelberg, Germany

Printed by Libri Plureos GmbH
in Hamburg, Germany